T0330041

Construction Graphics

Construction Graphics

A Practical Guide to Interpreting Working Drawings

THIRD EDITION

Keith Bisharat

WILEY

Library of Congress Cataloging-in-Publication Data Applied for:

Hardback ISBN: 9781394222261

Cover Design: Wiley
Cover Image: Concrete foundation © Jordan Lye/Getty Images,
Architectural drawing used with permission of John Wiley & Sons, Inc.

Contents

Preface to the Third Edition

In this edition of *Construction Graphics*, I added a chapter that seeks to illustrate how the ideas offered throughout the first fourteen chapters might be applied to a specific portion of a construction project, to put those ideas into a recognizable context. I also added another appendix section, a commentary on trends in construction, including the increasing use of modular construction. As always, I relied on friends in the design professions and the construction industry, including (again), Kevin Wilcox and Ron Migliori, senior principals at, respectively, Comstock Johnson Architects and Buehler Engineering. Both were very helpful in describing how they use digital modeling to enhance the services they provide owners. Isaac Dye, a former student and now the Operations Manager at Alston Construction, and Alston Sacramento's very talented digital modeling team headed by Ricky Molina, showed me the remarkable array of digital tools they use, including unmanned aerial vehicles, 3D laser scanning, LIDAR, and robotic layout; their work is truly impressive. My friend Ron Nurss, owner of Blueline Construction (the subcontractor who did the concrete work on the project shown mostly in the latter half of Chapter 8), is a discerning and very helpful industry observer. I have also benefited from the very thoughtful input of my friend Dave Kimball, from whom I learned production framing years ago and whose own career spanned many successful years as a carpenter, superintendent, and construction company owner. Dr. Tarek Salama, a Construction Management professor, and Peter Ouchida, an adjunct professor of Civil Engineering, both at California State University, Sacramento, provided valuable input on how the book is used in their classes. One of my great pleasures, after over thirty years at the university, is seeing the numerous achievements of our hard-working students who have graduated from the Sac State Construction Management Program; their accomplishments are extensive and noteworthy and are a source of great pride for me.

KEITH BISHARAT,
Sacramento, California,
December 2024

Preface to the Second Edition

The changes to the second edition of this textbook were made in response to the input that a number of users of the book were good enough to provide, as well as the compulsive modifications that seem to be inherent in the creative process. The book retains its essential character and focus (what do drawings mean to builders?); however, the original work has been supplemented in several ways. Commentary regarding Building Information Modeling (BIM), project delivery systems, interpreting working drawings, and the similarities between residential and commercial building construction drawings has been added.

The appendixes have gone from two sections to seven, and a number of the drawings have been improved or created anew. Appendices A and B include, respectively, thoughts on the take-off process and the answers to the questions that follow each chapter. Appendix C is a three-part section that compares residential and commercial building drawings and includes some representative drawings from a residential drawing set and comments on the nature of the information in the set. Appendix D is devoted to interpreting commercial building drawings and suggestions for extracting information from a drawing set and putting it into a useful format. Appendix E addresses welding in structural steel frames and welding symbology. The common weld types have been listed in a short commentary and shown in examples taken from actual structural drawings. Appendix F contains a copy of the new Construction Specifications Institute (CSI) 49 division MasterFormat, and while the number of design professional's offices using the new MasterFormat is currently limited, CSI's success in convincing the industry to standardize bodes well for its use in the future. Appendix G describes how to execute a site section, an essential skill in earthwork and grading, that was not part of the chapter on sitework. It should be mentioned that certain of the figures, tables, and captions in the text contain a considerable amount of useful information and bear some study. Those that incorporate a series of captions and images such as 8.1a–g, 8.4a–q, 8.5a–h, 8.6a–i, 8.7a–f, 10.4a–u, 11.5a–gg, 11.8a–h, 12.9a–u, 12.10a–l, 14.3a–k, 14.4a–e and Table 11.2 were designed to show the progression of drawings in a set or elucidate some other point. As with the previous edition, I hope people will continue to submit their suggestions as to how the book can be improved.

A wise old Sacramento contractor once said to me, "There's no such thing as a self-made millionaire; somebody had to help."[1] That humble statement certainly applies to my work on this book, too. Among the people who were instrumental in supporting the project are Bob Christenson, Chief Operating Officer of Panattoni Construction Company; Dr. Braja Das, Dean of the College of Engineering and Computer Science at California State University, Sacramento; and Dr. Joan

---

[1] Eugene Carson, Sr., founder of Kit Contractors, Inc.

Construction Graphics: A Practical Guide to Interpreting Working Drawings, Third Edition. Keith Bisharat.
© 2025 John Wiley & Sons Inc. Published 2025 by John Wiley & Sons Inc.

Al-Kazily, Chair of the Civil Engineering Department, who felt strongly enough about the original project to commit resources toward it. Bruce Yoakum and Mike Borzage (colleagues of mine at California State University, Chico) and Hal Johnston at California Polytechnic State University, San Luis Obispo, were helpful in their reviews of the initial proposal and generous in their insights as to the content and structure of the book. Donald Nostrant continues to be a reliable sounding board on topics inside and outside of construction while he pursues a fascinating career as an emeritus faculty member. I join a long line of construction industry representatives and educators who have benefitted from friendships with Professor Rovane Younger of Chico State, and Emeritus Professor Richard A. Nickles of California State University, Sacramento, both thoughtful people and keen observers of the industry. Our many interesting and engaging discussions about education and the construction industry over the last sixteen years have, I am sure, found their way into this work.

My friends Ron Nurss, Chuck Smith, Rob Cacioppo, Jerry Cova, and Sean Riley of Blueline Construction, and Irv and Brian Ballance, of Ballance Construction, were generous with their time and insights into the construction process and construction graphics. Kevin Wilcox and Don Comstock of Comstock Johnson Architects,[2] Ron Migliori, Todd Thorp, and Tom Krejci of Buehler and Buehler Structural Engineers, Jim Amundson of Hope Engineering Structural Engineers, Stan Gibbons of Entelechy, Gary Tamano and Marcus Chaw of Tamano and Chaw Residential Design, Terry Frizzell of PFS Consulting, and the folks at the HLA Group, Landscape Architects, were enthusiastic contributors of their graphic work. A long line of construction management students—a hardy bunch all in all—have wrestled with my classes for over fifteen years and have helped me understand and appreciate the challenges they face in assimilating the information related to construction graphics—there is a lot to know on this subject.

I sincerely appreciate the editing work that Ruth Younger and Janice Borzendowski made to the first edition, and Amy Handy's very keen eye in the second. I also appreciate the confidence that Rob Garber, Joel Stein, and now Amanda Miller demonstrated by supporting this project through two iterations. Mike Olivo and Daniel Magers were impressive in their diligence and Jim Harper's wit and creativity make him a terrific partner in this project. Bob Hilbert was a calm and helpful production editor.

I owe special thanks to my parents, who encouraged my siblings and me to wonder, and I am again indebted to my wife, Patti, who bore the lion's share of the domestic workload while managing her own busy career, and to Adrienne and Ian, my wonderful children, who tolerated all-too-frequent and periodically lengthy periods of absence from our family activities.

KEITH BISHARAT
Sacramento, California

---

[2]The project that is the subject of the drawings reproduced in this text under Comstock Johnson Architects' name was a design collaboration between Comstock Johnson Architects serving as the architect of record and LPA Sacramento, Inc., which acted as the design architect.

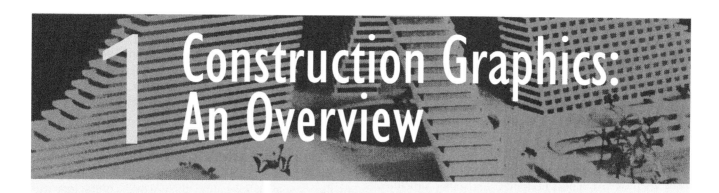

1 Construction Graphics: An Overview

Key Terms

American Institute of Architects (AIA)

Architect/engineer (AE)

As-built drawings

Building codes

Computer-aided design and drafting (CADD)

Construction Specifications Institute (CSI)

Dead load

Delivery systems

Design-build

Design contract

Dynamic load

Engineering contractor

General building contractor

Geotechnical report

Hardscape

Hydrostatic load

Integrated services delivery

Live load

National Cad Standard (NCS)

Planning codes

Prime contractor

Prime design professional

Shop drawings

Specialty contractors

Static load

Substructure

Superstructure

Temporary structures

Value engineering

Work breakdown structure

Zoning codes

Key Concepts

- The probability of differences arising between people engaged in an endeavor rises exponentially with the number involved, as demonstrated by the following formula: 2^{n-1}, where n is the number of participants and 1 is the single circumstance in which all participants agree.

Objectives

- Describe many of the challenges that design professionals face in their work.
- List the principal participants in a construction project, and identify their roles.
- List and describe the principal categories of construction project.
- Describe the typical educational path of architects, engineers, and contractors.
- Differentiate the design professional's and constructor's responsibilities.

Construction Graphics: A Practical Guide to Interpreting Working Drawings, Third Edition. Keith Bisharat.
© 2025 John Wiley & Sons Inc. Published 2025 by John Wiley & Sons Inc.

Introduction

Graphics are indispensable to anyone trying to communicate a vision or describe an object to an audience. Ideas commonly originate as pictures, which their creators then translate into words or words and pictures when they want to share them with others (see Figures 1.1a and 1.1b).

Indeed, combining pictures with words is the only reliable way to describe some objects, particularly complex ones such as machines, buildings, and other structures. Because professionals involved in designing construction projects must communicate particularly complex ideas, they typically do so through graphics. Therefore builders—whose responsibilities include interpreting the drawings and other documents that design professionals create, determining the means and methods of construction, and executing construction work in a safe and profitable manner—must have a thorough understanding of the graphics commonly used in construction projects.

Figure 1.1a Graphics are used to give life to a design concept, which may grow out of simple sketches such as this. (Sketch courtesy of the estate of Victor H. Bisharat.)

Figure 1.1b The sketch takes model form after considerable design work has been executed. (Photograph courtesy of the estate of Victor H. Bisharat.)

Graphics and Builders

In the course of executing their work, contractors use drawings, specifications, and other documents extensively, and are responsible for producing shop drawings, as-built drawings, and drawings for temporary structures. They are responsible for effectively communicating work requirements to their employees and subcontractors. Constructors who fully understand the project they are planning to build and can effectively communicate the processes required to build it, orally, graphically, and in other ways, dramatically reduce the risk in the undertaking. Those who are alert to what the design professional is trying to achieve in a project and can offer less costly solutions to the design without compromising its integrity—a process known as *value engineering*—give themselves a significant edge over their competitors. Additionally, the increasing popularity of the design-build project delivery system (*integrated services delivery*), which requires the design professional and constructor to collaborate on the design from the inception of a project, suggests that the constructor who is attuned to the interests and concerns of the design professional will be a sought-after partner.

Challenges Facing the Design Professional: Telling a Complicated Story

Architects and engineers have the very challenging task of conceiving solutions to the complex design problems their clients have, then communicating them to the people who bring the project to reality. They face a staggering number of issues in the course of their work. Every client has a unique set of requirements to identify, evaluate, and accommodate. Although they often borrow ideas from other projects, many architects develop solutions from "out of the blue," that is, from their imaginations. Their concepts are usually a melding of the owner's needs and their own artistic proclivities. Architects are well aware of the publicity associated with their work—after all, buildings are the most visible art—and this awareness frequently affects their decision making. Indeed, the principal criterion for selecting a particular architect is often the architect's design "signature."

During the design development stage, in addition to the owner, myriad permitting authorities and community groups have the opportunity to influence a project's design. For example, a project located on a major waterway within a large city might involve 10 or more public organizations—federal, state, and local agencies—that have jurisdiction over various aspects of it. In some jurisdictions, this means that the architect must develop a photograph of a prospective building design and insert it in a digitized photo of the neighborhood, so that planners can evaluate its visual impact prior to approving it. The context of a project requires the design professional to consider, for example, the effects that weather and surrounding buildings will have on a building project—or, in the case of a bridge, such considerations as traffic flow, kinesthetic effect, and safety. The size and configuration of a site, access and egress to it, and even the nature of its soil play a significant role in the project design. The single greatest influences on projects are the time and money that owners allocate to them. And construction as an industry is unique—tens of thousands of parts must be transported to and assembled on the site by armies of skilled workers, directed and supervised by management teams that change with every project. Consequently, no two projects are exactly alike; all are prototypes.

Recent developments in computer technology enable architects to develop highly unusual designs—projects so unique as to be impossible to draw using conventional board drafting methods (see Figure 1.2).

Although computer-aided design and drafting (CADD), when employed to take advantage of its strengths, has contributed significantly to document production, it is commonly used to produce the conventional two-dimensional depictions of the project (the drawings). The full capabilities of existing software, including its potential to produce dimensionally sound pictorial representations of the project, have yet to be widely used.

Computer software companies are currently developing and refining programs that potentially will give designers and builders much more effective control over the design and construction processes. Formerly called "intelligent building models" (variously dubbed "parametric modeling" or "object-oriented" programs) and "object linking and embedding technology" (OLE) and now more commonly referred to as BIM (Building Information Modeling), this software enables the user to define and store detailed parameters of building assemblies (such as foundations, walls, ceilings, and equipment) within the drawings. This makes retrieving many kinds of information quick and easy. BIM software creates a digital building model, a substantial database that correlates myriad building features to graphic images, and it holds significant advantages for design professional and builder alike. Chief among them is the linking of related elements, components, or systems in the digital model. In conventional board drafting or CADD-produced drawings, the potential for discrepancies within drawings is high: changes made in one drawing must also be made in all the other drawings in which the subject of the drawings is depicted, and even in a modest project that could mean making changes in 5, 10, even 15 drawings or more depending on the nature of the change. Imagine the effect of changing the location of a key structural column on the ground floor of a 30-story building to please a major lessee! All the related architectural drawings and especially the structural drawings have to reflect the change.

In drawings produced with BIM software, changes on one drawing are automatically made in the related drawings. That same linking enables the user of the digital model to take sections wherever it is desired. The "anatomy" of the project at any place within the model is visible—the builder is no longer dependent on the judgment of the design professional to select the location of major and minor building sections. One of the key challenges presented by conventional drawings—seeing a three-dimensional object (in the mind's eye) using a series of contiguous two dimensional depictions of it—is more easily overcome by a model (digital or otherwise). Another feature of BIM, justifiably touted by its proponents, is what is known as "clash detection"—identifying the locations where two elements (say an HVAC duct and a structural member) erroneously occupy the same space. Such clashes are not unusual in the interstitial space (between suspended ceilings and the floor above) of commercial buildings. When integrated with scheduling software, a 3D digital model is a useful analytical and training tool (dubbed 4D CADD). Mechanical

Figure 1.2 The Experience Music Project in Seattle, Washington, represents a growing genre of projects that will test the ingenuity and skill of the construction community. (Frank Gehry, Architect; Hoffman Construction, Contractor; photo by the author.)

specialty contractors, who seem to be leading the application of 3D modeling to their work, translate standard two-dimensional drawings into a 3D model for use in component manufacturing (Computer Integrated Manufacturing) and for billing and inventory purposes.

As with other technology though, BIM is not without drawbacks, including the need for electrical power and expensive computing equipment, the costs of user education, a fairly steep learning curve required for proficiency, and a lack of realism. The conceptualization value of a model may help builders understand the design professional's instructions more readily (assuming the instructions are well done), but understanding what one is being directed to build is overcoming only half the problem. Contractors will still need to develop the construction plan, collect and organize all the resources required to build the project, and lead the complex and diverse team of participants to a timely and profitable conclusion. It is worth noting that this technology appears to work best when the *prime design professional*, design consultants, the builder, and key specialty contractors collaborate early in the design development stage. (Some owners mandate the use of BIM and a delivery system that involves contractors early in the design process—refer ahead to page 13 for an explanation of delivery systems.) Thus BIM may have more of an impact on delivery systems and interdisciplinary collaboration than on the construction process itself. It is worth mentioning that the involvement of more parties early in the design phase introduces design liability issues that may take some time to resolve.

The form that construction drawings take in the future—and the medium used during construction—remains to be seen. For the time being, however, architects and engineers continue to produce, on paper, two-dimensional depictions of the projects they design even when BIM is used in the project design. These influences make it easy to understand why the design and construction professions are error-prone and why they will remain lively fields of practice in the future.

■ Dialects and the Standardization of Drawings

It is useful to think of construction graphics as a language that participants in the construction process must "speak" fluently, even if with varying "accents." As oral communication varies with the individual using it, so does graphic communication. Consequently, we see considerable variation in how drawing sets are organized, how each sheet is formatted and annotated, and how graphic symbols are used. This lack of consistency adds difficulty to an already complicated process and increases the potential for misunderstandings between builder, owner, and design professional.

The publication of the first *Architectural Graphic Standards (AGS)*, written by Charles George Ramsey and Harold Reeve Sleeper in 1932,[1] was an early effort to standardize graphics for construction projects. Today, *AGS* is a widely used compendium of planning, design, construction, graphic, and other reference information that focuses primarily on design data. *ConDoc*, developed by Onkal Guzey and James Freehof of the American Institute of Architects (AIA), the principal professional organization of the architectural community, is a design presentation standard that offers a uniform drawing set organization, sheet format and numbering methodology, and a keynote system that facilitates the prompt communication of graphic and specification information. Though *ConDoc* has contributed significantly to better quality control and drawing productivity, it is not universally employed. Many design offices insist on using their own formats and drawing set organization.

As recently as 1990, the Construction Specifications Institute (CSI) discovered that few standards related to construction drawings actually exist. Having had considerable success with its MasterFormat indexing system (a widely used list of construction products and activities oriented toward how parts are specified, purchased, and installed), CSI turned its attention in 1994 to developing a single construction drawing standard, which it named the *Uniform Drawing System (UDS)*. The *UDS* has become a fundamental part of the *National Cad Standard (NCS)*, an evolving collaboration of CSI, AIA, the U.S. Department of Defense Tri-Service CADD/GIS Technology Center and the United States Coast Guard, the Facility Information Council of the National Institute of Building Sciences (NIBS), the U.S. General Services Administration (GSA), and the Sheet Metal and Air Conditioning Contractors' National Association (SMACNA). These organizations signed a memorandum of understanding in 1997 to develop and promulgate the use of the NCS. The Facility Information Council of NIBS provides the forum for the effort; CSI is contributing its UDS, the AIA its CAD layering guidelines, and the Tri-Service CADD/GIS Technology Center its plotting standards. The NCS is a much-needed single standard for the organization and presentation of drawing sets; a format for individual sheets, schedules, and diagrams; and standards for color, keynotes, material symbols, and CAD layering.[2] Though not yet widely used, the NCS is a work in progress that is gaining acceptance.

[1]Ramsey and Sleeper (2000).
[2]National Institute of Building Sciences/American Institute of Architects/ Construction Specifications Institute/Tri-Service and the U.S. Coast Guard (2001).

■ Participants in the Design Process: Owners, Design Professionals, and Contractors

Designing and constructing projects frequently involves hundreds of participants; however, there are three principal players in every project: the owner, the design professional, and the contractor.

Owners

Although the list of potential construction project owners is nearly infinite, the short list includes governments (federal, state, and local), districts (school, irrigation, and reclamation), for-profit and nonprofit corporations, partnerships, and individuals. Construction projects occur when a representative of one of these groups seeks to mitigate a need or realize an idea. The owner's role generally is to provide the site; finance both the design and construction of the improvements; give timely, accurate feedback to the design professional; and operate the facility. Owners commonly engage the services of a design professional to conceive the design and produce the construction documents. Owners place contractors under contract to execute the work described in the construction documents.

The Design Professional and Design Consultants

Design professionals—primarily architects and engineers—offer a wide variety of services to project owners. Traditionally, they have created projects and produced construction documents and contract administration on behalf of owners, under service agreements called *design contracts*. The tumult in the design profession in the last couple of decades has prompted architects and engineers to diversify the services they offer for a fee. Many now include facility life-cycle analysis, recycling and management, as well as practice management in the range of services they provide.

The number of different design professionals involved in producing construction drawings varies according to the type and complexity of the project. Individuals or very small organizations generate most drawings for homes. In some states, laypersons may design homes and duplexes without a design professional's license. Developing the design for a hospital, performance center, or a manufacturing facility, in contrast, may require many highly specialized design professionals who, after rigorous examination, have been licensed by the states in which they do business.

The core participants commonly responsible for the design of building construction projects include architects and landscape architects, and geotechnical, civil, structural, mechanical, and electrical engineers.

Architects, whose authority to design projects derives from state licensing boards, have the daunting task of identifying their clients' problems during the predesign phase and describing their solutions to them, using pictures and words, during the design phase. Architects conceive the physical attributes of a project and incorporate local land-use ordinances and applicable building code requirements into their designs. Their interests and professional responsibilities are focused primarily on how a project looks (esthetics) and how it works as a product (that is, will it protect its users from the elements and from injury during catastrophic events such as earthquake and fire? Does it fit effectively into its environment? Does it fulfill the owner's needs?).

The number of specialists and the variety of services that design consultants offer is substantial; however, architects commonly hire structural, mechanical, and electrical engineers for significant portions of building design work—areas of specialty for which they frequently do not have the training, license, or personnel. Large design firms, however, frequently have in-house engineering capability, which gives them more market share, greater efficiency, and more control over the design process. Such organizations are commonly referred to as *architect/engineer* (AE) firms. Regardless of the size and organizational structure of the office, the overall responsibility and liability for the design of a project reside with the architect, who becomes known as the *prime design professional* (the "prime" designer or contractor is the term given to the entity that signs a contract with the project owner).

Although many civil engineers are qualified to prescribe the treatment required to prepare soil for a project, geotechnical engineers are registered professional engineers who are required to devote several more years to practice and/or additional education after becoming licensed civil engineers before they can legally call themselves Geotechnical Engineers. They are hired by owners to investigate a project site and produce a comprehensive evaluation of its soil conditions, which are recorded in a geotechnical report. Geotechnical engineers commonly investigate the past uses of a site and its hydrology, identify its soil types, determine whether and to what extent a site is contaminated, and delineate any procedures that the contractor must follow to prepare the soil for its intended role. For example, soils must be made stable and competent to bear the weight of structures and vehicular traffic for years, and soils may be used to encapsulate solid waste and to line excavations and earthen structures that will contain water.

A host of participants in the design and building process use the geotechnical report. The structural engineer uses the report to design the foundation of a structure; the landscape architect uses the report to develop the specifications for the planting and irrigation of landscaped areas; and the contractor and subcontractors use the report to determine the costs of earthwork (such as excavation, soil preparation, pile-driving, and foundation work) and evaluate the risk associated with it.

The principal concern of geotechnical engineers is how the soil will perform over time with the planned activities imposed on it. Their contracts with the owner normally require them to prepare the geotechnical report, and monitor, inspect, and approve earthwork while it is being performed. Additionally, the geotechnical engineer resolves issues that arise in the course of construction, such as the mitigation of contaminated soil that might not have been apparent during the site investigation. Beyond these functions, they do not typically get involved in design.

Civil engineers typically produce most of the construction documents related to engineering construction (streets and highways, sewer and water treatment plants, harbors, dams, bridges, and utilities). They must be licensed by the state in which they perform design work. On commercial building projects, the civil engineer plays a relatively limited design role, normally taking responsibility for on-site grading, drainage, and paving plans and specifications; for off-site improvements (driveways, gutters, curbs, and sidewalks along a public thoroughfare); and for the design of certain on-site underground utilities (sewer lines, fire system supply, storm drainage, domestic water supply). Civil engineers often cite the standard specifications of the city, county, or state in which the project is located, particularly in the design of off-site improvements. These specifications are frequently tried-and-true specifications that are developed by state departments of transportation (which invest considerable funding in research) and are often wholly adopted by public works departments at the local level.

Structural engineers specialize in the design of foundations (piles, caissons), substructures (habitable portions of a structure that are below ground, such as basements), and superstructures (the portion of the project above grade, or above water in the case of bridges built across bays, lakes, and rivers). Like civil engineers, structural engineers are licensed by the states in which they do business, but they are frequently required to have specialized education and training beyond that of the civil engineer. Structural engineers—frequently hired by

"It is generally recognized that an architect is a combination of artist and engineer, with the 'art' aspect emphasized . . . most architects have an initial and abiding concern about a project's aesthetic appeal, impact, and propriety.

—RALPH LIEBING, *Architectural Working Drawings*, 4th edition (Wiley, 1999)

"The qualities that most clearly set architecture apart from other established professions are its close ties to the arts and its similarities to artistic endeavors. Creativity is crucial to all professions, but for the architect it is of the highest priority. Moreover, architects produce objects that are fixed in space, highly public, and generally long lasting.

—DANA CUFF, PH.D., *Architect's Handbook of Professional Practice*, 12th edition (American Institute of Architects, 1994)

architectural firms for their expertise—are focused on the performance of the structural system under various loading conditions that fall under two classifications—static and dynamic loading. Static loading comprises dead loads (gravitationally imposed loads resulting from the weight of the structure and its permanent equipment) and live loads (mobile loads that are not necessarily present at all times). Furniture, snow, hydrostatic pressure (the pressure at any point exerted on a surface by a liquid at rest), and a building's occupants are examples of live loads. Dynamic loads, such as seismic activity and wind can occur suddenly, and they vary in intensity, duration, and location.

Structural engineers are responsible for protecting the lives and property of project users in a cost-effective way. Although their focus is on the performance of a structure under the loading conditions just mentioned, they should also be aware of the esthetics of the project.

Mechanical engineers involved in building project design are responsible for plumbing, sewerage and piping systems, and for heating, ventilating, and air conditioning systems (HVAC). Mechanical engineers commonly form consultant agreements with the A/E to develop and describe the plumbing and HVAC systems in buildings, which are designed to ensure the comfort and health of building occupants. Plumbing and sewerage systems provide an adequate source of water for human consumption and sanitation, and effectively dispose of wastes generated in the building. The heating, cooling, ventilating, and air conditioning equipment is used to control environmental comfort factors such as the temperature of the ambient air in a building, the mean radiant temperature of the surrounding surfaces, the relative humidity of the air, pureness of the air, and air motion. HVAC and plumbing systems in building projects present a significant design challenge, particularly in the distribution of conditioned air and piping through the structure. The involvement of mechanical engineers in the design process increases dramatically when they are involved in industrial construction projects, such as refineries, manufacturing facilities, chemical plants, waste and water treatment plants. Indeed, they may hold the prime design professional role on these projects. Mechanical engineers concentrate on the performance of the systems they design.

Electrical engineers are involved in the design of a variety of construction projects, including massive power generation and distribution systems for state and federal governments, cogeneration power plants, and building construction projects, to name a few. As with the other engineers, electrical engineers must be licensed by the state in which they conduct business. In building construction projects, these engineers design the electrical service and communications systems on the site, as well as the site lighting, usually at the request of the A/E. They also design the service and distribution systems inside the structures. In addition, electrical engineers must design and clearly spell out the type and location of electrical equipment and cabinetry and the means of distributing and controlling the power. Those engineers who work for the local utility company frequently control the design of the off-site system (the portion found in public utility easements). Electrical engineers focus on the proper sizing of the system, the location of the equipment, the distribution of the power, and the safety of the end user.

Landscape architects, also licensed by the state, specialize in developing ornamental landscaping plans, which includes selecting trees, shrubs, ground cover, and grasses, and designing the irrigation system required to support them. The landscape architect's work may also include some site improvements (such as walkways, garden structures, screens, fencing, and water features, all of which are referred to generally as *hardscape*). Landscaping plays an important role—not only for the visual beauty it brings to a project, but for the beneficial effects that a well-conceived and -executed design can have on the energy consumption of a building, as well as on air and water pollution.

Contractors

Although the term "contractor" is loosely applied to anyone who earns income from constructing things, sole proprietorships, partnerships, corporations, and joint ventures are the common legal entities that assume responsibility and liability for constructing projects under contract with the owner. Many states regulate contractors through licensing boards, which assure the health, welfare, and safety of the public through education, testing, and, where applicable, the enforcement of state license laws.

There are distinct categories of contractor:

- *Engineering contractors* construct engineering projects such as highways, bridges, and industrial construction projects.
- *General building contractors* produce residences, multiple-family projects, commercial and civic buildings, and/or retail spaces.

- *Specialty contractors* focus on one portion of a project, such as plumbing, sheet metal and air conditioning, roofing, insulation, tile, floor coverings, or elevators.

The contractor who signs a construction contract with an owner is called the *prime contractor*. The prime contractor, for a variety of reasons, frequently hires specialty contractors for portions of the work, who become subcontractors under the construction contract. Plumbing, mechanical, and electrical specialty contractors are commonly hired in this fashion.

▪ Varying Professional Viewpoints: Legal Responsibilities, Education, Training, and the Consequences of Diversity

The term "professional" has numerous definitions, among them "engaged in a specific activity as a source of live-lihood," "performed by persons receiving pay," "and having great skill or experience in a particular field or activity."[3] The term is commonly used simply to describe anyone being paid to do something. However, as it pertains to architects, engineers, lawyers, doctors, and now many construction people, "professional" means "one who has an assured competence in a particular field or occupation" as determined by education, training, and rigorous examination.

Education of the Design Professional

An architect or engineer can follow any of several paths to professional registration; however, the common one is formal postsecondary education in an accredited architectural or engineering program. Long and often arduous courses of study, architecture and engineering programs prepare students for internships after graduation from college, which qualifies them for registration board examinations. In the case of architecture program graduates, the internship must be acceptable to the governing jurisdiction: Acceptable usually means three years in an environment that requires "diverse experience" (practice in a variety of areas), after which the professional exam may be taken.

Individuals desiring an engineer's license must acquire six to eight years' work experience, depending on the state, and must pass the Fundamentals of Engineering (FE) exam to become an Engineer in Training and to qualify for the Professional Engineer (PE) exam. Credit against the work experience requirement may be granted for academic study in an accredited engineering program, but some work experience is required.

---

[3]The definitions in this paragraph come from the *American Heritage Dictionary of the English Language*, Houghton Mifflin Company, 2000.

Geotechnical engineers must have four years of "responsible charge" in geotechnical engineering after licensure as a civil engineer to qualify for the geotechnical engineering exam. Structural engineers must have three years of work experience in structural engineering after licensure as a civil engineer in order to sit for the Structural Engineers' (SE) exam. Passing these exams gives examinees the authority to use the title Geotechnical Engineer or Structural Engineer in their practices. Additional benefits are conferred on holders of SE licenses; in California, structural design work on schools and hospitals is restricted to this class of engineer.

Engineering and architectural programs approach education from considerably different directions. In general, however, architectural programs are distinguishable from engineering programs in that they incorporate liberal studies, design, and architectural theory and history with the physical sciences, mathematics, and technology. Engineering students are normally required to go into greater depth in physics, chemistry, mathematics, and engineering classes than do their counterparts in architecture. They consequently have less time for liberal studies. While generalizing has its drawbacks, architectural education is a "generalist" education that encourages a holistic view of problem solving, with design and human behavior at its center. Engineering education, on the other hand, focuses on physical phenomena and the behavior of systems.

Education of the Contractor

Construction is one of few industries in which people with little education, minimal money, and limited experience can still find good compensation either as employees, vendors, or contractors. In some states, it is possible to build without a license. States that regulate the industry frequently require license applicants to demonstrate the requisite experience, knowledge, and financial wherewithal to contract for construction services.

Owners, partners, and managing employees on commercial building, heavy and highway, and industrial construction projects tend to hold baccalaureate degrees in construction management, civil engineering, architecture, or allied fields. The past several decades have seen an increase in the number of educational institutions offering construction management as an academic pursuit, replete with accreditation boards that assure the academic quality of the programs. Still in their infancy, construction management programs vary widely in their curricula, but are similar to architecture and engineering programs in their duration and level of difficulty.

Professional Licensing

As noted, many states regulate contractors through license boards, but until 1996, no professional licensing body

> "Contractors do not know the thinking, reasoning, and rationale behind the design and documentation of the project."
>
> —Ralph Liebing, *Architectural Working Drawings*, 4th edition (Wiley, 1999).

existed for contractors. In 1994, the American Institute of Constructors (AIC), the professional organization representing the construction management discipline, undertook to institute a professional licensing program similar to that of civil engineers. By 1996, AIC had developed its exams and began testing in numerous sites across the nation. AIC's Certified Professional Constructor (CPC) program includes verification of an appropriate education or equivalent experience and practice at an advanced level, prior to applicants' sitting for two rigorous examinations, either the basic or the advanced CPC exam.

Consequences of Diversity

A professional education is as much a socialization process as it is an educational process. Consequently, students of various disciplines learn the values of the group they choose to join. These values manifest themselves in how each participant in the construction process views the others, and are frequently the source of conflict and misunderstanding. It is well worth the effort for each of these professionals to understand the values of the others; all are codependant and play equally vital roles in the construction process. Accepting that the differing responsibilities and proclivities of each participant are likely to manifest themselves in their work is a critical step to cooperation and productivity on a construction site.

Design Professionals and Contractors: A Comparison

The management responsibilities of the prime design professional and prime contractor are similar in many ways. The differences between them tend to occur in their professional responsibilities.

The prime design professional assumes responsibility and legal liability for managing the design process and its

> "Architects tend to have a more global or overall view and concept of a project; many consultants become totally engaged in their narrow realm of work, without thinking of others."
>
> —Ralph Liebing, *Architectural Working Drawings*, 4th edition (Wiley, 1999).

"The CAD system has been wonderful for architects. But do they know how to build what they're drawing? Constructability is one of the biggest problems I find."

—George M. Grant, V.P, Halmar Builders, in the article "Listening to Contractors," *Architectural Record*, February 1998.

outcome (the contract documents), whereas the constructor assumes responsibility and liability for the construction process and its product (the project as described by the architect). The prime design professional transcribes clients' needs and desires into a form that is commonly used by the builder (drawings and specifications). Builders translate the designer's work into discrete processes to which a value in time and money can be assigned, then set about executing the work safely, to the prescribed quality, and within the allotted time and budget. Each hires specialists for portions of the work. For building projects, architects are commonly the prime design professional, and may hire design consultants such as structural, mechanical, and electrical engineers. Builders commonly are prime contractors, and they also hire specialists for parts of the work, who are commonly known as *subcontractors*. The designer evaluates materials in terms of their performance characteristics and esthetic value; contractors evaluate materials in the context of their effects on costs and the installation process, as well as their performance characteristics. The prime design professional determines *what the project will be*; the prime contractor determines *how it will be constructed*. Both make necessary changes to the processes they control so that the resulting project will be as close to what was planned as possible.

Summary

Both graphics and text are required to successfully describe and construct even the simplest construction project. Graphics are particularly effective for size and shape description and the correlation of elements, components, and assemblies, and they represent a significant portion of the design professional's work. Text is critical to qualitative aspects of a project; in fact, it is virtually impossible to describe the quality of a component graphically. Due to the technical complexity of most projects and the numerous and varied participants in design and construction processes, ambiguity, errors, and omissions exist in the documents and in the correspondence related to them, even under the very best of circumstances. Misunderstandings are, therefore, a common occurrence. Without good graphic, written, oral, and listening skills, as well as respect for project participants and a results-oriented philosophy, owners, design professionals, and constructors alike will doom a construction project to unnecessary conflict.

CHAPTER I EXERCISES

1. Explain what the roles of the owner, design professional, and contractor entail in construction projects.
2. How do the perspectives of the contractor and design professional differ?
3. By what means do contractors and design professionals perform specialized aspects of their work?
4. List the common design professionals involved in construction projects.
5. Prepare a list of questions or discussion items pertaining to design and construction and attend a monthly meeting of the local chapter of the Construction Specifications Institute, the American Institute of Architects, the American Society of Civil Engineers, or builders' organizations. These organizations are very interested in students; in fact, many have student chapters on university campuses, and schedules of events are frequently listed on Web pages. For students with time constraints, CSI meetings may be the best use of time, as CSI's membership runs the gamut of participants in the design and construction processes: construction product vendors, architects, consulting engineers and builders. Members are generally very willing to share their views on the design and construction professions.

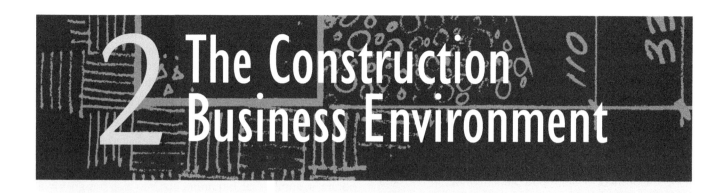

2 The Construction Business Environment

Key Terms

Addenda
Agency CM
Agreement
American Institute of
 Architects (AIA)
Approved set
As-builts (record set)
Associated General
 Contractors (AGC)
Best value
Bid set
Building codes
CM at Risk
Commercial building
 construction
Construction Specifications
 Institute (CSI)
Contract set
Delivery system
Design-bid-build
Design-build
Design development
Engineering construction
Engineers' Joint Contract
 Documents Committee
 (EJCDC)
General conditions
Industrial construction
Integrated services delivery
MasterFormat
Owner-builder
Planning codes
Preconstruction services
 agreement
Project delivery system

Project manual
Qualifications-based selection (QBS)
Residential construction
Schematic design
Specifications
Supplementary conditions

Key Concepts
- "Say it once in the right place."— Construction Specifications Institute (CSI) tenet.
- The Achilles' heel of many in the construction industry is the failure to adequately conceive, develop, and implement all elements of the management cycle: planning, organizing, executing, monitoring, and adapting.
- "If you can't measure it, you can't manage it."— Words of wisdom from the management industry.

Objectives
- Explain the extent to which the construction industry affects the national economy.
- Describe how construction projects originate and evolve.
- List the fundamental project delivery systems and explain their differences.
- Identify the principal components of a contract for construction and the role each plays.
- Discuss some of the challenges a builder faces when using construction drawings.

Construction Graphics: A Practical Guide to Interpreting Working Drawings,
Third Edition. Keith Bisharat.
© 2025 John Wiley & Sons Inc. Published 2025 by John Wiley & Sons Inc.

■ The Construction Industry

The construction industry accounts for up to 5% of the gross domestic product (GDP), or close to half a trillion dollars[1] (2001), the vast majority of which is privately financed. Considered a bellwether industry, it employs a significant percentage of the industrial workforce directly and indirectly. Construction is a fragmented industry made up primarily of small businesses, is easily penetrated by competition, and no significant advantage derives from being established. A disproportionate number of business failures overall occur in construction, due in part to:

- Mismanagement of resources
- Insufficient planning
- Poor accounting and collection practices
- Inadequate insurance for catastrophic losses
- Lack of experience in a particular type of construction
- Operating in unfamiliar geographic regions
- Too-rapid growth
- Loss of key personnel

Types of Construction

Residential Construction
Residential construction projects—single-family homes, duplexes, condominiums, low-and high-rise apartments—represent a significant percentage of the market. Although some states permit individuals without design professional licenses to design single-family and duplex structures, architects commonly design these projects.

Commercial Building Construction
Office buildings, retail centers, schools, universities, light industrial, and institutional projects are examples of the commercial building category, the documents for which are produced by architects.

Engineering Construction
Engineering construction consists of several types of projects, including highway and airfield construction (excavation, paving, drainage structures, bridges); heavy projects (water and wastewater treatment plants, tunnels, large bridges, dams); and utility construction (infrastructure improvements, underground systems). Engineers are normally the prime design professional on such projects.

Industrial Construction
Industrial construction projects are complicated structures constructed to manufacture or produce commercial products, and they are among the largest conceived and constructed. Refineries, chemical processing plants,

electrical generation facilities, and cement production facilities are included in this construction type. Design-build entities (organizations or joint ventures that take responsibility for both design and construction) with specialized expertise produce these projects.

Governing Influences: Project Design

Regulatory Law and the Design Process: Codes
As noted previously, design professionals must incorporate a number of requirements into the design for a project. One of the critical areas of concern—an area that justifies full-time work for some architects—is code research. Codes result from federal and state legislative mandates establishing regulatory agencies for a variety of industries, among them the insurance, real estate, and construction industries. Among the organizations creating building codes are the:

- International Conference of Building Officials (ICBO)
- Building Officials and Code Administrators International, Inc. (BOCA)
- Southern Building Code Congress International, Inc. (SBCCI)
- International Code Council (ICC)
- National Fire Protection Association (NFPA)
- International Association of Plumbing and Mechanical Officials (IAPMO)
- American National Standards Institute (ANSI)

The ICBO, BOCA, and SBCCI each produce national *model codes*, codes that are adopted into law by local governments. The ICBO publishes the Uniform Building Code (UBC), popular in the western United States and much of the Midwest. The National Building Code is a BOCA product that is widely used in other areas of the Midwest and in the East. The SBCCI produces the Standard Building Code, which is used in the South and Southeast. While the codes themselves vary somewhat, their purpose—to establish minimum standards for health and safety in building construction—is the same. Fortunately, the first unified model code in U.S. history, called the International Building Code (IBC), was published in 2000 and will no doubt be adopted nationally in the years to come. The IBC was a cooperative effort of BOCA, SBCCI, and ICBO representatives under the direction of the International Code Council.

Accessibility
The effort to bring people with various disabilities into the mainstream of life goes back over four decades, originating in such efforts as the President's Committee on Employment of the Handicapped in the late 1950s, the Architectural Barriers Act of 1968, the Civil Rights Act of 1968, the Rehabilitation Act of 1973, the Uniform

---

[1]Bureau of Economic Analysis, U.S. Department of Commerce.

Federal Accessibility Standard (1984), the Federal Fair Housing Act of 1988, and the Americans with Disabilities Act of 1990 (ADA). These efforts affect not only building construction, but even the features of Web pages on the Internet. Companion legislation at the state and local levels frequently parallels the federal government's efforts.

Unfortunately, a single standard has yet to emerge— vestiges of various laws are in effect somewhere, and regulatory efforts continue to evolve. Compliance, therefore, requires thoroughly researching the regulations that are applicable to a given project.

Energy Use

Although interest in energy efficiency in building construction has existed for years, concern for the amount of energy that the United States consumes increased dramatically in the early 1970s. At that time, the United States was prompted to focus domestic policy on conserving natural resources, particularly oil, and the government launched a concerted effort to preserve resources. Federal, state, and local governments offered incentives to reduce energy consumption, resulting in numerous energy-efficiency programs in the manufacturing, transportation, and construction sectors, among others. Many states now mandate energy consumption levels for different building types, and require design professionals to demonstrate that their plans comply with the mandates. The increasing interest in "green" architecture (environmentally sound, sustainable design), life-cycle costing, and building recycling are creating numerous opportunities for both the construction and design professions.

Application of Regulatory Law

Whatever the project, the list of codes and regulations to which a given set of drawings complies is normally found in the first few pages of the drawings. The goals of most regulatory law—a better quality of life—are worthwhile for a modern society, but they are not without their drawbacks, including greater complexity in daily business functions, varying interpretations of their applicability, and additional cost.

Delivery Systems

The term *delivery system* refers to the legal structure that is employed to "deliver" a project to its owner from inception to completion. The various systems typically address both the design and the construction processes, and hybrids developed from the basic systems may include site acquisition, financing, occupancy, and facility management and operation as well. Most of the systems involve three main participants: the owner, the design professional, and the contractor. In *Project Delivery Systems for Construction* (2004), the Associated General Contractors

of America defines project delivery systems using a simple identification matrix. Two basic questions are asked and answered:

1. Are the Design and the Construction under separate contracts directly with the owner, or are they combined under one contract?
2. Is the final selection of the constructor based on criteria other than merely the lowest total construction costs?

Answering these questions results in the "defining characteristics" of delivery systems, which in turn results in three seminal types: the traditional *Design-Bid-Build* system, *CM at Risk*, and *Design-Build*. It should be pointed out here that "final selection based on criteria other than merely the lowest total construction costs" includes "**best value**" (criteria identifying the best value in light of the owner's needs, as established by the owner or owner's representative) and **Qualifications-Based Selection (QBS)**,[2] a selection method demanding that respondents identify the projects that are similar to the one the owner is contemplating, the key people involved in the project, and the places in which the respondent has been involved in similar projects. The price of services is specifically omitted. QBS has long been used in the selection of design professionals by project owners. The definition of these basic delivery systems follows:

Design-bid-build is the long-standing delivery system in which complete contract documents (the agreement, general and supplementary conditions, drawings, specifications, and miscellaneous documents such as addenda[3]) as well as bidding instructions are produced for the owner by a design professional or by staff that is familiar with the design and procurement processes (highway departments, for example). Upon completion of the documents, contractors are allowed to submit (usually) stipulated-sum bids for the construction work in a competitive bidding process. The contract for the work is awarded to the lowest responsive, responsible bidder. Many state public works laws still mandate this system.

CM at Risk is an arrangement wherein the design professional signs a design contract with the owner, the contractor signs a preconstruction services agreement with the owner, and the two entities collaborate on the design of the project. When the construction documents are complete, or mostly complete, the construction contract is let and a contractor (often the one that offered its preconstruction services) is selected according to the owner's

---

[2]This description of the focus of QBS is derived from *The Architect's Handbook of Professional Practice*, 13th ed., John Wiley & Sons (2001).
[3]Refer ahead to "The Construction Contract: Five Components" for a more detailed description.

procurement criteria (Best Value or QBS). The prevailing contractor then assumes the risk for construction.

Design-build (integrated services delivery) is the delivery system that imposes the responsibility for the design and construction of a project on a single entity (often a joint venture involving a design firm and a contractor), which becomes the single point of contact for the owner. Long a popular method of constructing very large projects for foreign governments and for industrial construction projects, design-build is making forays into public works contracting. Selection of the design-build entity may be based on the lowest bid, the Best Value, or QBS. Though it may not fit neatly into AGC's definition of delivery systems, a significant number of projects are produced on an ***Owner-Builder*** basis. An Owner-Builder acts as the builder and takes responsibility for coordinating the work. The design work is performed either by the owner or a design professional. Many single-family and duplex projects are constructed under this system, as are a number of commercial projects as well. Developer/owners of apartment projects, for example, frequently hire a design professional to produce drawings and specifications, then construct projects with their own forces and trade contractors, which they own and operate or sell to investors. These developers are called *merchant builders*. AGC defines this class of builder as Program Managers (see below).

AGC's defining criteria are useful in that they establish the three basic delivery systems from which the myriad hybrid delivery systems—which AGC refers to as "management systems"—are born. Among the popular hybrids are:

Agency CM involves a fourth party (the construction manager) that is charged with the task of administering the construction contract on behalf of the owner. The architect performs the design work and the contractor builds the project. The CM system has become very popular with public organizations lacking staff with experience in the construction process.

Program Management As defined by AGC,[4] program managers are involved with all five phases of a project, including locating and purchasing real estate, entitlement research and other development issues, financing the project, project design, project construction, and occupancy. Program managers are involved with and responsible for large and involved projects.

In the United States, considerable flexibility to negotiate binding agreements is given to players in the free-market economy, to the extent that the delivery systems themselves can be (and frequently are) created for a particular project: Aspects of the basic systems just listed can be combined to suit parties' needs. Each

system has its advantages, disadvantages, and proper application. See Appendix H for more information.

▪ Evolution of a Building Construction Project

Design Development

As noted earlier, a project comes into being when an owner or owner representative seeks to resolve a need or realize an idea. Figure 2.1 graphically displays the evolution of a construction contract.[5]

In the design-bid-build delivery system, the contractual arrangement under which the majority of public building projects in the United States are created, the architect contracts with the owner to develop the construction documents for a project. This long and complicated effort normally begins with a "predesign" effort known as *programming*, a process through which the owner's ideas, needs, functions, and limitations are "discovered." During this phase, the designer collects and analyzes data, and develops a design concept. The principal focus of programming is to state problems and identify issues that the design process should address. Design—comprising the schematic design, design development, and construction drawings stages—begins with the schematic design effort, during which the designer develops the conceptual design, the scale, and the relationship of building components.

In contrast, the documents for public infrastructure projects—water and sewage treatment plants, for example—may be prepared by private civil engineering firms in cooperation with public utilities departments. In the case of state highway projects, transportation department staffs frequently develop the drawings and specifications. These projects follow markedly different paths to fruition than do public building projects.

The result of the schematic design phase is a clearly designed concept, presented to the owner in a preliminary site plan, floor plans, exterior elevations, and critical sections (all two-dimensional depictions of the project). Since a principal goal of the presentation is to help the owner visualize the project, other tools such as perspective drawings and models (electronic and scale models) are frequently used. Additionally, the designer develops outline specifications, a statistical summary of the design area, and an assemblies cost estimate (for which CSI's UniFormat indexing system is especially helpful). The approval of the owner at this stage is critical, because the work produced during the schematic design phase will be further refined in the design development stage that follows. The documents that derive from the design development phase include fully developed floor

---

[4]*Project Delivery Systems for Construction*, 2nd ed., AGC of America (2004).

[5]This information is derived in part from the *Architects' Handbook of Professional Practice*, 12th ed., American Institute of Arichtects (1994).

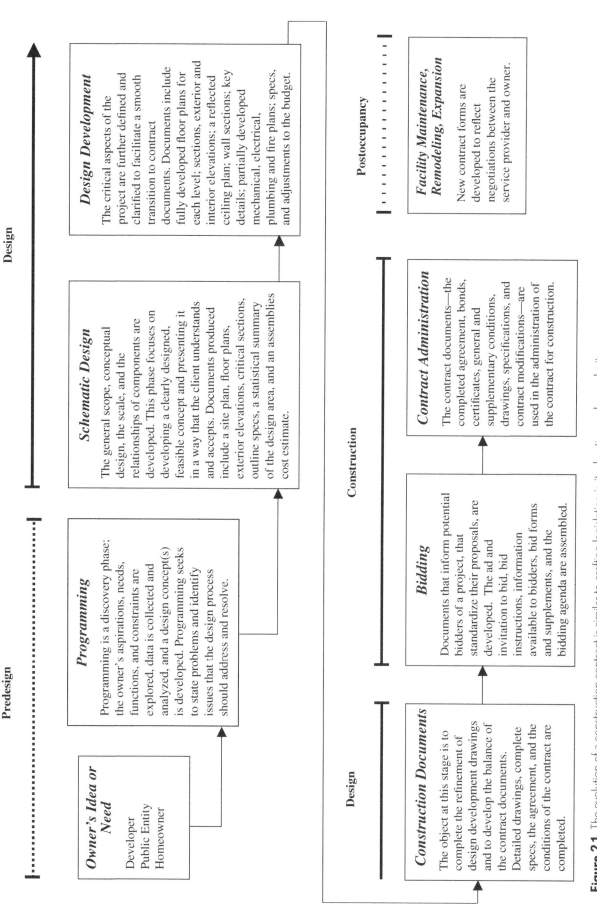

Figure 2.1 The evolution of a construction contract is similar to crafting legislation in its duration and complexity.

Predesign

Design

Owner's Idea or Need

Developer
Public Entity
Homeowner

Programming

Programming is a discovery phase; the owner's aspirations, needs, functions, and constraints are explored, data is collected and analyzed, and a design concept(s) is developed. Programming seeks to state problems and identify issues that the design process should address and resolve.

Schematic Design

The general scope, conceptual design, the scale, and the relationships of components are developed. This phase focuses on developing a clearly designed, feasible concept and presenting it in a way that the client understands and accepts. Documents produced include a site plan, floor plans, exterior elevations, critical sections, outline specs, a statistical summary of the design area, and an assemblies cost estimate.

Design Development

The critical aspects of the project are further defined and clarified to facilitate a smooth transition to contract documents. Documents include fully developed floor plans for each level; sections, exterior and interior elevations; a reflected ceiling plan; wall sections; key details; partially developed mechanical, electrical, plumbing and fire plans; specs, and adjustments to the budget.

Postoccupancy

Facility Maintenance, Remodeling, Expansion

New contract forms are developed to reflect negotiations between the service provider and owner.

Construction

Design

Construction Documents

The object at this stage is to complete the refinement of design development drawings and to develop the balance of the contract documents. Detailed drawings, complete specs, the agreement, and the conditions of the contract are completed.

Bidding

Documents that inform potential bidders of a project, that standardize their proposals, are developed. The ad and invitation to bid, bid instructions, information available to bidders, bid forms and supplements, and the bidding agenda are assembled.

Contract Administration

The contract documents—the completed agreement, bonds, certificates, general and supplementary conditions, drawings, specifications, and contract modifications—are used in the administration of the contract for construction.

plans, detailed exterior and interior elevations, wall sections, reflected ceiling plans, critical details, specifications, and partially developed mechanical, electrical, plumbing, and fire system drawings. The budget developed during the schematic phase is refined as well.

The drawings and specifications that are actually used to construct the project, as well as the documents used in bidding and administering it, are fully developed in the construction drawings phase. Many design firms produce a project manual, which consists of bidding documents (for competitively bid projects), the agreement, general and supplementary conditions, specifications, and miscellaneous documents such as bonds, certificates of insurance and compliance, and addenda. The project manual is an effective way to organize all the pertinent project information that is produced in an 8½" × 11" format. The drawings are commonly referred to as the "plans," "working drawings," and "blueprints," as well as other names. Figure 2.2 is a graphic display of the documents required in most building construction projects.

The Construction Contract: Five Components

A contract for construction is a complicated collection of documents that can number in the thousands of pages. It contains five fundamental components: the agreement, the general and supplementary conditions, the drawings, the specifications, and miscellaneous documents.

The Agreement

The agreement, which sets forth the legal relationship of the parties to the contract, is the component most often thought of as "the contract." A list of the parties to the contract, a brief project description, the contract type (such as stipulated or lump sum, cost plus a guaranteed maximum price, unit price), start and substantial completion dates, method of payment and consideration, and a list of the documents that make up the contract are all recorded (or should be) in the agreement.

General and Supplementary Conditions

Conditions are the guidelines to the administration of the contract. General conditions refer to boilerplate guidelines that exist in standard forms created by various professional organizations such as the American Institute of Architects (AIA), the Associated General Contractors (AGC), and the Engineers' Joint Contract Documents Committee (EJCDC). The AIA's Form A-201 and EJCDC's form 1910-8 are popular, "tried-and true" general conditions documents; that is, they have stood the tests of the courts and time. Supplementary conditions, also referred to as "special conditions," are written modifications or additions to the standard form conditions that reflect the contracting parties' special needs; they enable the parties to tailor the general conditions to suit their purposes.

Figure 2.2 CSI developed this appropriate graphic (a tower of paper) to illustrate the components of a project manual. Note that not all of the documents in a project manual are contract documents. (This document is published by the Construction Specifications Institute (CSI) and is used with permission from CSI, 2003. For those interested in a more in-depth explanation of this document and its use in the construction industry, contact the Construction Specifications Institute, 99 Canal Center Plaza, Suite 300, Alexandria, VA; 22314, 800-689-2900; 703-684-0300; CSINet URL: www.csinet.org.). While still relevant, this depiction of CSI's graphic has been superseded by the new MasterFormat, shown in Figure F-1 in Appendix F.

Drawings

The drawings are the graphic depiction of the proposed work. The value of pictures is evident when one attempts to verbally describe how to build even a simple object, which can take hundreds of words. It is possible to build a complex object without using any words at all, if one uses carefully presented images such as "exploded" diagrams (see Figure 2.3). Drawings identify the project, its location on the site, its exterior and interior characteristics, the relationship of the parts and the manner in which they are connected, and many of the materials used in its construction.

Specifications

Specifications are the guidelines to the quality of work and the materials used in a project. They are essential to a successful description of a project, and are inseparable from the drawings—at least from the standpoint of fully understanding the project. Quality is certainly impractical, if not impossible, to draw. Imagine graphically depicting a 2" maximum deviation from plumb (vertical) on a 500' (152 m)-tall building, using the few scales (1:200; $1/16" = 1'-0"$ that would result in a drawing small enough to fit on a standard 28" × 42" (841 mm × 1189 mm) sheet. Two inches at $1/16$th scale becomes $1/96$th of an inch on a drawing—thinner than most pencil leads, and, for all practical purposes, indiscernible to the naked eye. How would one distinguish walnut paneling from ebony or cherry wood on drawings—by having a symbol for every different species of finish wood? How would an architect describe the quality of an exterior plaster finish that s/he desires not to vary more than $1/8"$ in a 10' × 10' (0.003 m in a 3 m) square plane? These details must be described in words.

It is worth mentioning that Division 1 of the Construction Specifications Institute's MasterFormat indexing system focuses on administrative issues pertaining to all specifications sections. Part 1 of the individual specification section is focused on administrative issue as well (refer to the "Written Information" section of Chapter 6 for more information).

Miscellaneous Documents

The fifth fundamental component of a construction contract—the miscellaneous category—consists of all the documents not mentioned in the first four categories, including addenda (amendments to the contract that add or deduct work; change the scope of work; clarify, substitute, or otherwise change the terms of a contract before it is signed by the parties), certificates of compliance and insurance, bonds, and modifications to the contract.

Chronic Confusion: Errors and Omissions in Drawings and Specifications

Errors and omissions in drawings and specifications, as well as discrepancies in those documents, result from a variety of causes, including:

- Project complexity
- The design professional's project workload

- Inadequate office staffing
- Inadequate coordination efforts on the part of the designer
- Severe time constraints imposed by owners
- The design professional's lack of experience with the type of project being designed
- Inadequate compensation of the design professional by the owner
- Owner-induced changes
- Failure of the design team to fully explore the effects that changes have on the project overall and to properly account for them
- Poor quality-control systems

The footings being designed by the structural engineer may encroach on space occupied by the underground utilities in proximity to the footings.

The size of supply and return air ducts sometimes exceeds the space available in the structural members through which they are supposed to pass. The size and number of electrical conduits, especially where the direction of the conduit changes, may exceed the space relegated to it. It is not unusual to be unable to install plumbing and piping systems within the spaces allotted on the drawings.

Causes

Taking only the drawings and specifications under consideration for a moment, bear in mind that a large project may require hundreds of drawing sheets and a thousand pages of specifications to adequately describe it. As noted previously, the design and review process involves numerous participants, and producing reliable construction drawings requires countless iterations of drawings. A constructor might base the cost estimate on one set of drawings (the bid set), may become liable for the construction of the project based on another set (the contract set, which is the plan set identified in the legal agreement), and might actually build based on a third set (the approved set, the plans that have been approved by the permitting authorities). The subcontractors might base their estimates on any of the three, depending on when they became involved in the project. Almost without exception, time is too limited, both for the design professional and the constructor. It is not unusual for contractors to be given

"Architects tend to have a more global or overall view and concept of a project; many consultants become totally engaged in their narrow realm of work, without thinking of others."

—RALPH LIEBING, *Architectural Working Drawings*, 4th edition).

Figure 2.3 This diagram of a tankless water heater is a fine example of the effectiveness of the exploded diagram. (Drawing courtesy of Controlled Energy Corporation.)

three to six weeks to develop a stipulated-sum price for multimillion-dollar building projects.

Although computer technology has done a great deal to aid designers and constructors in their work, building projects are still largely depicted in two dimensions. Fully understanding what the project looks like requires a certain amount of "cognitive construction." One must have a vivid imagination to build a three-dimensional image in the mind's eye from numerous two-dimensional projections of it. Even experienced design professionals and constructors occasionally struggle with this, which is one reason that scale models (and increasingly, computer models) are used in design and construction. There are downsides to computer use in the production of standard (two-dimensional) drawings. Ralph Liebing, in the fourth edition of *Architectural Working Drawings*, expressed the following concerns related to current circumstances in the design field:

- Lack of construction knowledge on the part of those now producing the drawings, as well as lack of field experience
- Lack of knowledge of and regard for traditional (past) standards, techniques, values, and criteria for drawing production and for drawing intent and content
- Education, starting in high school and vocational-education, that overemphasizes the use of computer-aided drafting (CAD) by teaching only the manipulations of the machines
- Utter disregard, and almost disdain, on the part of students and far too many faculty, for producing quality documents with valid information correctly shown
- The lack of a will or impetus to correct these concerns by incorporating the capabilities of CAD with refined production methods and proper construction methods, materials, and details

Liebing adds: "Only a drafter/operator trained in architecture, engineering, and construction methods, materials, and allied technology can achieve suitable documentation." The Construction Specifications Institute *Manual of Practice* (1996 edition) makes these points regarding project documentation:

Incomplete coordination among design team members is common within the construction industry. Following are some of the problems that can result from incomplete coordination:

- *Duplications* (part or all of the same element of work is specified in more than one section of the specifications). Duplications violate the Construction Specifications Institute (CSI) principle of "Say it once and in the right place." Each duplication is an opportunity for discrepancies that alter the uniform basis for bidding and cause confusion. In turn, this can result in higher construction costs, faulty construction, disputes,

decreased document credibility, and increased exposure to liability.

- *Omissions* (information concerning an element of work that is inadvertently excluded from the construction documents). Omissions are often caused by lack of understanding of the responsibilities of the various design disciplines and the absence of a joint checklist.
- *Discrepancies* (conflicting information in the contract documents). Lack of knowledge by the team leader or team members of the bidding requirements, contract document requirements, agreements, insurance, bonds or the conditions of the contract increase the possibility of discrepancies and conflicts.

Solutions

One can do little about some causes of errors, omissions, and discrepancies in drawings. It is not likely, for example, that projects will become simpler; in fact, the trend is toward even more complicated design and construction, thereby perpetuating the need for numerous specialists and regulatory oversight in both the design and construction processes. The capability of the computer to present realistic pictures of a project—thereby simplifying the conceptualization challenges—is a quantum leap in communication, especially with the owner. That said, CAD is not a panacea for the problems inherent in the document production process.

Design professionals may uniformly adopt the National CAD Standard, which would simplify document analysis for builders, and they could improve office procedures and personnel training to reduce errors and omissions in drawings. However, these adjustments require a significant commitment in time and money. Design professionals might also consider developing strategic alliances with contractors, and consult with them during the design process (one perceived benefit of the design-build and CM at Risk delivery systems), or add summaries of their design program to the documents that contactors use in estimating project costs. Sharing the basis for design decision making would help the constructor analyze and resolve the inevitable discrepancies in construction drawings.

> "The use of CAD technology does not necessarily translate into better documents. . . . CAD [use] still needs the construction knowledge and the understanding of document content and intent to be successful. This, in turn, is a direct reflection of the operator's background, education, perspective, and contribution; these all must have a strong construction orientation."
>
> —RALPH LIEBING, *Architectural Working Drawings*, 4th edition).

The single greatest challenge facing designers and builders alike—and the best resolution to the endemic problems with construction drawings—is to educate owners as to the value of well-planned, clearly presented, well-coordinated construction documents and to convince them that higher up-front costs (mostly for design services and contract administration by the design professional, and lengthened bidding periods for contractors) are worth trading for a dispute-ridden construction process and lower-quality products.

What Contractors Do

Management, fundamentally, consists of analyzing opportunities or problems, establishing goals and developing a plan to achieve them, organizing, implementing the plan, monitoring outcomes, and making adjustments to achieve the goals originally set forth. Contractors exercise these functions on the construction process when they procure a project, estimate its costs, develop a construction plan, implement the plan, monitor the progress of the project against the original objectives, and make adjustments to keep the project on track.

Once a construction company has elected to compete for or has been invited to pursue a project—in itself a process requiring numerous decisions—it sets about determining the likely cost of the project. Organizational functions, which are generally divided into three basic categories—marketing, operations, and administration—are the means by which the purposes of an organization are carried out. Depending on the size of the company, these functions may be carried out by one or more people or by divisions within a company. Determining project costs is typically the purview of estimating departments, which—inasmuch as estimating is an integral part of project procurement—frequently operate under the auspices of the marketing division. It is the estimating department that gives construction drawings and specifications their initial close scrutiny.

The Estimating Process

This is not an estimating book—numerous references that describe the estimating process for various construction types are available. Nevertheless, several critical points regarding the estimating process and construction documents, particularly drawings and specifications, are worth mentioning.

How estimating is done varies from company to company; however, generally, it consists of methodically "deconstructing" a project according to a variety of "work breakdown" structures (systems for categorizing work according to specifications indexes such as CSI's MasterFormat, or by the location of work items in the project), analyzing the constituent parts, formulat-

ing plans for producing the product described by the designer, organizing the information in a way that is useful to the builder, and determining values in time and money for all aspects of the project. Doing this effectively requires knowledge of the materials and processes of construction and the ability to imagine the project under construction.

Construction costs can be effectively divided into two principal cost categories: direct and indirect costs. Direct costs (costs that correlate directly to furnishing or installing the physical aspects of a project) include labor, equipment, materials, subcontracts, and other costs (those costs that do not fit into the other four categories). Indirect costs, known variously as "general conditions," "preliminary and temporary," and "project overhead costs" consist of project supervision; project office costs (trailers, telephones, faxes, supplies, etc.); temporary utilities; site security; hoisting equipment; small tools; general office overhead and profit; and permits, taxes, insurance, and bonds. The information is compiled and formatted in a new order called the *contractor's estimate*, which in essence is a collection of goals.

Much of what the contractor is required to do is spelled out in the drawings and specifications; however, critical aspects of the required work are not a part of the drawings. Temporary structures such as access roads, falsework and scaffolding, and the excavation requirements of a project are good examples of such aspects.

Operations and Administration

The operations division of a company, which may assist in the development of the estimate, takes the completed estimate and assigns cost codes to specific line items, develops the detailed construction schedule, assigns the staff (which frequently writes the purchase orders, material contracts, and subcontracts), and mobilizes to construct and monitor the project until it is completed.

The administrative department or division of the company supports the field operations of the construction company as well as its home office by attending to legal matters, financing, investment and accounting, bonds and insurance, and the like. Project cost reports, for instance—critical tools required to manage the construction effort effectively—are generated by the accounting department, based on input from managers at the site, who exercise their best judgment as to the extent of completion of tens or hundreds of items.

Interpreting Construction Drawings and Specifications: Words of Caution

The drawings and specifications created by the architect or engineer are at the heart of the estimating process. However, as noted earlier, much of what has to be considered is not shown on the drawings at all. It is the contractor's responsibility to determine the means and methods of construction—the "how" of the project—whereas the designer is responsible for the "what" of the project. Contractors must therefore imagine the project under construction: the holes in the ground, the machinery required to accomplish much of the work, and the measures necessary to protect the skilled workers on site and a curious public, to name just a few considerations.

Scaled Drawings and the Magnitude of Work

Of particular significance to the constructor, as far as scaled drawings are concerned, is the tendency to overlook the magnitude of work represented by the drawings. Design professionals develop or use existing office standards that govern a host of tasks relevant to drawings, including drawing sheet size. The designs for a 100,000- (9290 m²-) and a 300,000- (27,870 m²-) square-foot concrete tilt-up warehouse, for example, would probably be recorded on sheets of the same size. The floor plan views (horizontal cuts through the building) for the larger of the two warehouses would have to be recorded on a minimum of two pages, since even at $1/128$ scale ($3/32" = 1'- 0"$), the footprint of a building 400' wide and 750' long (122 m × 229 m) would be $37\frac{1}{2}" \times 70\frac{1}{2}"$ (953 mm × 1791 mm) larger than the largest drawing sheet in use, and this figure does not take into account the margins that are commonly used in design offices. Conceiving the magnitude of such a project is very difficult when looking at a depiction of it that fits on a large desk. Consider that it takes about two and a half minutes, walking at four miles per hour (six and a half km/hr), just to walk from corner to corner in this building! The roof alone constitutes nearly seven acres (2.79 ha; tilt-ups well over 1 million square feet [92,900 m²] have been constructed)! More remarkable than these considerations is that even when the complete scope of this project—site work, underground utilities, paving, irrigation, landscape, and off-site improvements—is taken into consideration, it is a relatively small, simple project when compared to a manufacturing facility, prison, highway, or dam project. This same reality applies to vertical construction projects such as high-rises, the tallest of which is now over 1,600 feet (500 m)!

Half-Scale Reproductions

In an often dusty and chaotic environment such as a construction site, drawings and specifications that are printed on paper survive well. Unlike using the computer, reading and understanding printed graphics requires no artificial light, special tools, or electrical power, nor is much maintenance required, especially with modern reproduction technologies and paper. However, drawings for a large project can be cumbersome—a large set can weigh over 25 pounds (11.3 kg)—and sets are frequently bound separately by discipline (architectural, structural, mechanical, electrical) to make them manageable. Manageability as

well as cost savings on reproductions often motivate an owner to make half-scale reproductions available to the constructor. Although more difficult to read, half-scale drawings offer two benefits: they are easier to handle and less expensive for the contractor. What does not change in the drawing reduction is the designated scale, so when reviewing floor plans that were originally drawn to a $1/4" = 1'-0"$ (1:50) scale, for example, it is necessary to double the measurement indicated on the scale (the instrument) or use a $1/8" = 1'-0"$ (1:100) scale to determine dimensions (when they are not written). Drawing scales are often recorded in a legend on the drawings, too, and are an alternative to using the instrument known as a scale. One simply transfers the scale in the legend to some medium—a 3" × 5" (75 mm × 125 mm) card, for example—to decipher the sizes of objects and systems. As the drawing is reduced, so is the scale legend. It goes without saying that underestimating the size of something by 50 percent could have very costly ramifications.

Metric Drawings

Constructors transitioning from U.S. Customary Units to the International System of units (SI), or metric, face a significant mind-set adjustment. Since measurements for smaller objects, say a cast-in-place curb, will be given in millimeters on a drawing set using SI, the numbers resulting from quantity calculations can be substantial. The area in section of the curb shown in Figure 2.4, for instance, is 68 400 mm².

One hundred lineal feet of this curb (about 31 meters) amounts to 2,098,000,000 cubic millimeters (2.098 ×

Figure 2.4 Curb in SI. (Sketch by the author.)

10^9), or just over 2 cubic meters of concrete (1 cubic meter, the "buy" quantity for concrete when using SI units, is 1 billion cubic millimeters!). One could always convert from millimeters to meters, but that just transfers the complications to the other side of the decimal.

■ Summary

The construction industry represents a significant segment of the economy in the United States, and it is affected by a variety of influences, not the least of which are extensive involvement by a wide range of participants and regulatory laws that run the gamut from planning and building codes to licensing, civil rights legislation, and jobsite safety. The delivery systems are numerous and varied as well, reflecting considerable flexibility in how projects can be legally packaged for an owner.

Once a delivery system and a design professional have been selected for a project, the design professional sets about investigating the owner's needs and developing a written and graphic description of the work to be done. The drawings and specifications represent the majority of work produced by the architect/engineer; however, the construction contract is itself incomplete without supporting documents—the agreement, general and supplementary conditions, and course-of-construction documents such as change orders, material and subcontracts.

The technical complexity of buildings themselves, the complexity of the documents drawn up in support of them, project uniqueness, the numerous and varied participants, and their differing responsibilities and styles of communication conspire to make construction projects challenging and sometimes frustrating undertakings. Mutual respect among the participants, recognition and acceptance of the difficulties involved in construction, and a results-oriented focus go a long way toward reducing the challenge and frustration.

CHAPTER 2 EXERCISES

1. Identify some of the causes of errors, omissions, and discrepancies in construction documents.
2. What is meant by the term *delivery system*? Describe two common types.
3. What are the different construction types?
4. What caveats should a contractor bear in mind when reviewing construction drawings?
5. Describe the design sequence for a building construction project.
6. List the five components of a contract for construction and the contents in each.
7. Describe what contractors do once the decision to pursue a particular project has been made.

3 The Design Professional's Work Product

Key Terms

Addenda
Addendum
Agreements
Bidding documents
  Advertisements and
    invitations to bid
  Bidders' information
  Bid forms and
    supplements
  Instructions to bidders
Conditions
  General
  Supplementary or special
Hard and soft conversions
Hardscape
International System
    of Units (SI)
Modular construction
Project manual
SWPPS

Key Concepts

- Ambiguity is defined as "susceptible to more than one interpretation." Every effort should be made to avoid ambiguity in graphics, writing, and speech.
- Active listening and assuring that one has taken the guesswork out of all one's communication are helpful commitments in the design and constructions professions.

Objectives

- Identify the proper location of various kinds of graphic information within a construction drawing set.
- Identify the contents of the project manual.

Construction Graphics: A Practical Guide to Interpreting Working Drawings,
Third Edition. Keith Bisharat.
© 2025 John Wiley & Sons Inc. Published 2025 by John Wiley & Sons Inc.

The Project Manual

The design profession has experienced significant changes in the last couple of decades, in no small part due to the globalization of the economy, the increased sophistication and heightened demands of owners, a protracted recession in the 1990s, and the meteoric development of the personal computer. These developments have forced architects in particular to evaluate and redefine their practices and to search for different ways of making themselves useful to their clients. Forming strategic alliances with other design professionals, builders, and other service providers is one means by which architects are re-establishing their value to clients. How the profession will define itself in the coming years remains to be seen; however, it is likely that the traditional services—solving design problems and producing the contract documents used to construct clients' facilities—will continue to be in demand.

Although there are many formats, architects' design work frequently culminates in a project manual and the drawings. The project manual commonly contains bidding requirements,[1] contracting requirements,[2] and specifications.[3] The drawings are, of course, the graphic depiction of the work. Between these two packages, the builder has, in theory at least, a complete, relevant document set from which to construct a project. Addenda (clarifications, additions, deletions and modifications to a project prior to contract signing) are often produced in response to questions generated during the bid process and are communicated to bidders after the project manual has been issued.

Within the project manual, it is the specifications that are the most critical to actually constructing the project—the other documents establish the legal and administrative structure for bidding and construction. Specifications play a minor role in this textbook, but only because construction graphics is the focus. It is worth emphasizing that *fully understanding the project is impossible without both drawings and specifications*; they are co-dependents in the construction process.

Construction Drawings: Organization and Content

For hundreds of years, scaled drawings have been used to describe the myriad construction projects undertaken by humanity to provide shelter from the elements and from other people; to communicate; to produce, transport, and store things; to conduct business and the affairs of government; to control the effects of nature; and to memorialize people or events. By selecting representative units and developing some drawing standards over time, designers of these projects have been able to depict very large objects (or very small ones as well) on media that were small, portable, durable, easy to use at the construction (or manufacturing) site, and that—since the latter half of the nineteenth century—were easily reproduced.

The National CAD Standard (NCS) is intended to create a widely accepted single drawing standard for construction projects, an ambitious but worthwhile effort involving hundreds of design professionals and contractors. And although its future is by no means assured, the NCS is gaining popularity, and there appears to be enough impetus to develop a single standard that it is likely to survive. The following paragraphs reflect that likelihood.

There are many approaches to packaging construction drawings. Variations are a function of the project type, owner requirements, and design professional preferences. Commonly, however, design professionals present building project information in an order that reflects the construction sequence, more or less. There are minor differences between drawing sets. For instance, some offices insert landscape drawings and owner- or client-produced documents such as interior architecture toward the rear of the set. The NCS and drawings for many military buildings, in contrast, place landscape drawings adjacent to or within civil engineering drawings; interior architectural drawings (tenant improvement drawings) follow the architectural drawings in the NCS.

Some offices use a title sheet with general notes and a vicinity map, then add architectural, structural, mechanical, plumbing, electrical, and landscape drawings, in that order. In the NCS, structural drawings precede architectural drawings. Telecommunication, an ever-growing aspect of commercial buildings, enjoys its own division within the NCS, as do hazardous materials, survey and mapping, geotechnical information, process information (piping-intensive industrial work), resource drawings, drawings created by other disciplines, contractor drawings, and drawings created by operations personnel. There are offices that produce a project manual and drawings, then supplement the architectural sheets with a detail book consisting of $8\frac{1}{2}" \times 11"$ (210 mm $\times$ 297 mm) drawings. The National CAD Standard is a comprehensive organizational effort that is intended to be useful to a variety of users of facility documents. Consequently, it includes documents that have not traditionally been part of construction drawing sets, for example geotechnical information, resource, and contractor shop drawings (B, R, and Z discipline drawings in the NCS).

---

[1] The advertisement or invitation to bid, instructions to bidders, bidders' information including preliminary schedules, geotechnical data, an existing conditions description (property survey, description of existing buildings, hazardous materials report), bid forms and supplements (bid form, bid form supplements, representations, and certifications).

[2] The agreement, bonds (payment performance, and warranty), and certificates (payment and insurance certificates), general and supplementary conditions.

[3] The written description of products and installation requirements (qualitative guidelines).

Civil Engineering Drawings

There are other variations on the preceding order; for example, civil projects such as highway and bridge work may commence with a title sheet, followed by typical roadway sections, general layout plans, highway profiles, project-specific construction details, storm water pollution prevention plans (SWPPPs, or "sweepies" in the vernacular), erosion control drawings, drainage drawings (plans, profiles, and details, followed by quantities), utility layouts (existing and designed), staged construction (traffic handling, etc.), retaining and sound wall drawings with quantities, landscape drawings, electrical drawings, revisions to standard details, then bridge structures, the retaining walls correlated to them, and quantity summaries. This drawing set organization is partly a result of the relationship between some transportation department district offices, which may produce all but the bridge structure drawings, and the central transportation department office, which may take responsibility for bridge design.

Order within the Disciplines

Perhaps more critical than the drawing set organization is the order in which the individual drawings are presented within the common disciplines mentioned. Within each division (civil, architectural, structural, etc.), the information commonly goes from general to specific. In the NCS, general information (legends, notes) are followed by plans (horizontal sections or cuts), elevations (vertical views that show surfaces), sections (vertical cuts through buildings or major systems such as the walls), large-scale views (plans, elevations, or sections that show smaller portions of the project, but are not details), details (large-scale plans, elevations, and sections of very small portions of a project), schedules and diagrams (door, window, and finish schedules or matrices), two user-defined categories that serve miscellaneous needs, and, finally, pictorial drawings such as perspectives and isometric projections. Table 3.1 summarizes the sheet identification methodology promulgated by the NCS. Not surprisingly, there are variations on this organizational theme as well.

TABLE 3.1 NCS Drawing Sheet Identification Example

NCS Drawing Set Order	Discipline[1]	Discipline Subdivision[2]	Sheet Type[3]	Page Number[4]		Supplemental Drawings[5]	
	A*	A	N**	0	N	A/N	A/N
Simple Project[6]	A	—	1	0	2		
Complex Project[7]	A	D	1	0	2		
Example[8]	A	—	1	0	2	R	1

[1]*Level 1 discipline designators* by themselves are appropriate for simpler projects such as a midrise office building. The discipline designators, in sequence, are: G—General, H—Hazardous Materials, V—Survey Mapping, B—Geotechnical, W—Civil Works, C—Civil, L—Landscape, S—Structural, A—Architectural, I—Interiors, Q—Equipment, F—Fire Protection, P—Plumbing, D—Process, M—Mechanical, E—Electrical, T—Telecommunications, R—Resource, X—Other Disciplines, Z—Contractor/ Shop Drawings, and O—Operations.

[2]*Level 2 discipline designators* are reserved for more complicated projects such as hospitals. This space is filled in with a dash when the Level 1 designator is used by itself.

[3]There are 10 sheet types: 0—**General** (symbols legend, notes, etc.), 1—**Plans** (horizontal sections), 2—**Elevations** (vertical views of surfaces), 3—**Sections** (large vertical cuts such as wall sections), 4—**Large Scale Views** (plans, elevations, or sections that are not details, such as wall and stair sections), 5—**Details**, 6—**Schedules and Diagrams**, 7—**User-Defined** (miscellaneous category, e.g., generic details), 8—**User Defined** (yet another miscellaneous category), and 9—**Pictorial representations**, such as isometric and perspective drawings.

[4]These two columns are reserved for sheet sequencing within the discipline for numbers 1–99.

[5]The NCS provides the flexibility for users to designate these two columns to various uses, for example, if partial or complete revisions to a drawing are made, or drawings are produced after a sheet order has been established.

[6]A102 is the second sheet of architectural floor plans on a relatively simple project.

[7]AD102 describes the demolition of the assembly described on the second sheet of the floor plans. Designers of complex projects will use level 2.

[8]This designation identifies page 1 of revisions to the architectural floor plans, and is an example of how the user defined supplemental drawing columns could be used.

*The letter A stands for an alphabetical designation of some sort.

**The letter N stands for a numeric designation.

Content in Drawing Set Divisions

While the following order reflects the National CAD Standard, much of what is described pertains to drawings that follow other organizations.

Cover Sheets

Cover sheets are, simply, title pages, with the project title, the owner's name, the names of the design professionals involved, a pictorial of the project (frequently a perspective drawing), and similar content.

G: General Information

Site data, location map, energy compliance calculations, building code summary, project square-foot calculations, key plans, general notes, abbreviations, and an index of sheets all belong in the general information category.

H: Hazardous Materials

Hazardous materials occur in a variety of goods and materials, primarily in older structures. Asbestos, for example, occurs in a variety of materials (predating the 1970s), including floor coverings; plaster; piping insulation; ceiling tiles; plaster, floor, wall, and ceiling insulation; and myriad other materials, thus making sheet organization somewhat problematic in this discipline. The way the NCS is currently established, demolition of an HVAC or piping system in an older building would be described in mechanical drawings using the Level 2 discipline designator (see sheet identification paragraphs upcoming). The likelihood that asbestos would be encountered in the duct or piping insulation and sealed joints in older buildings is high. Sitework involving hazardous materials—hydrocarbons, for example—is likely to be described in the civil drawings. Just how this discipline will develop remains to be seen; however, the idea is to identify the location of hazardous materials so that their potential for harm is mitigated.

V: Survey/Mapping

These pages contain relevant survey and map information. It is the normally the owner's responsibility to provide the design professional with accurate information related to the real estate being developed. Vicinity maps, common on drawings sets, and general layout information are recorded in this sheet set.

G: Geotechnical Information

Providing information in a drawing on site soil conditions makes sense from several viewpoints. As mentioned in Chapter 1, a number of people make use of the geotechnical report in their analysis of work requirements, including the owner, the prime contractor, and subcontractors. Though including this information in a

drawing set is cumbersome, it is perhaps justified by the ready availability of the information.

W: Civil Works

The civil works category is, for all practical purposes, the same as civil drawings (the next one); however, it was added at the behest of government agencies that are responsible for civil work that encroaches upon the property of multiple landowners. A municipal pipeline project that impinges upon numerous landowners, for example, would be the appropriate project type to describe in this category.

C: Civil Drawings

In a drawing set that describes civil construction projects such as highway and street improvement projects, it is not necessary to distinguish the discipline from others—the entire project is "civil drawings." When the discipline is distinguished, however, it is generally when civil drawings form a part of a commercial building set. As mentioned in Chapter 1, civil engineers design and describe the off-site improvements (curbs, gutters, and sidewalks along public thoroughfares), on-site grading and paving requirements and underground utilities for building construction projects. Their work is recorded in the "C" sheets—civil sheets.

L: Landscape Drawings

Landscape drawings generally include planting plans and schedules (lists of plants, shrubs, and grasses), the irrigation system required to support the plants, and *hardscape*, (trellises, fences, site benches, patios, walkways, etc.) described in plan, elevation, section views, and details.

S: Structural Drawings

These drawings describe the elements, components, and assemblies of structural systems, and the manner in which they are connected, for a variety of projects. They are the construction equivalent of the skeleton, tendons, and muscle matter of the human body; in fact, "tendons" is a term used to describe the steel cables that are inserted in components, such as precast prestressed concrete piles and girders, and in cast-in-place post-tensioned concrete beams, bridge decks, and floor slabs. Architects frequently determine the basic structural system, since the functional arrangement of a project and the required esthetic treatment may dictate column spacing and therefore spans; however, the calculations and detailed structural design parameters are the bailiwick of the structural engineer.

A: Architectural Drawings

Architectural drawings are the heart and soul of a building project—virtually all other disciplines act in support of the architect's design, which is described in these drawings. To some extent this is due to the architect having overall control as the prime design professional and the uniqueness of building projects; however, in highway projects for example, where design standards for construction are common and projects are co-developed and managed by district and central DOT offices, the various engineers involved act more as equals.

There are as many approaches to design as there are architects—some conceive projects from the exterior and fit the functions within a shell; others determine the appropriate functional relationships of a building and develop the shell from them. No matter the origin of the design, the other disciplines take their cues from the architect's drawings, which are the most wide-ranging drawing set. Depending on the charges to the architect, the drawings can include master project planning and building design (frequently for multiple buildings) from basic systems or shells to complete buildings with interior details, furniture, and even fabric design.

I: Interiors

As just noted, architects might be given the responsibility to design a complete building or building shell for an owner. When the latter occurs, it is often because a developer or owner has anticipated that there will exist a demand for space within the building when it is complete and has undertaken to construct it on a speculative basis. Commercial real estate brokers monitor the construction and lease activity of buildings, and earn fees for facilitating lease agreements between building owners and tenants. Under certain lease agreements, tenants have the responsibility to design and construct their office space, within parameters established by the building owner, and will commission interior or building architects to produce the necessary design documents. It is these drawings, as well as drawings used in subsequent lease activity in the project, that are inserted in a drawing set under the interiors category.

Q: Equipment

Equipment drawings run the gamut of equipment that might be used in a project, from bank vaults, teller equipment, and ATMs to library, theater, videoconferencing, and commercial cooking, bakery, and laundry equipment. Some of the drawings required in this division can be complex, as, for example, in the case of bank vaults, which are subject to compliance with federal legislation governing their construction.

F: Fire Protection

To aid fire departments in their fire-fighting efforts, fire codes require comprehensive fire protection plans from project owners. Access and egress to the site; on-site street widths and radii; the location of hydrants, water mains, trees, overhead power lines, utility service disconnects;

and anything else that could affect the success of fire-fighting efforts are subject to review by fire districts.

Fire suppression systems, which are systems that actively fight fires (automatic sprinkler systems of a variety of types), as opposed to systems that simply detect or prevent fires, belong in this division as well. In addition to having some of the problems associated with other mechanical systems, fire suppression systems are complicated hydraulic systems whose performance is sensitive to minor changes in design. They are carefully reviewed in the design phase and are actively monitored during and after construction by the fire districts having jurisdiction in the community.

P: Plumbing

Plumbing systems are designed and described by mechanical engineers and recorded in plans, elevations, and sections, as well as in isometric schematics and fixture schedules. The basic parts of the system include drain waste and vent piping, hot and cold water supply, and fixtures.

D: Process

Process refers to systems that support the conversion of raw materials into a commercial product. Complicated forests of piping, controls, and storage facilities, process facilities are worthy of a distinct division in drawing sets. Refineries, canneries, and wineries are examples of process facilities—the latter being an example of projects for which building construction drawings and process facilities might be combined.

M: Mechanical

Mechanical drawings describe the location, size, and type of equipment for distributing, filtering, humidifying/dehumidifying, cooling and heating air, as well as the distribution and control systems required in a project.

E: Electrical

Electrical drawings describe the electrical service (utility-provided wiring, metering, main switches, and grounding), distribution (panelboards, switchgear, and wiring emanating from the boards), branchwork (circuitry), and devices used in a project. As with other drawings, the electrical engineer uses plans, sections, details, and schedules to describe the project.

T: Telecommunications Drawings

Changes resulting primarily from widespread computer use, as well as developments in telecommunications technology, have resulted in a dramatic increase in the attention given to telecommunications systems. The NCS has provided room for additional developments by creating a separate division for these systems.

R: Resource

Resource drawings consist of any drawings that are created prior to and sometimes during construction, as well as "measured" drawings—drawings that describe existing conditions that are used in the development of remodeling plans, among other types. As to subject matter, these drawings contain whatever information might be required for a remodeling or refurbishing project—structural, mechanical, and other plans are among the possibilities.

X: Other Disciplines

This category is a miscellaneous division. Any participant—an acoustical consultant, for example—could produce the necessary drawings for atypical kinds of work.

Z: Contractor Drawings

Shop or fabrication drawings are among the types of drawings that are the responsibility of the contractor, hence, the division "contractor drawings." Subcontractors or manufacturers use shop drawings to demonstrate to their shop personnel, the contractor, and to the design professional how an assembly or component—described in general terms by the architect or engineer—will be produced. Structural steel, trusses, fire suppression systems, and vertical transportation are examples of the kinds of work that are detailed in shop drawings. Shop drawings are the responsibility of the contractor; however, it is generally the architect who provides the list of work items that require shop drawings as a part of the submittal process. The drawings are reviewed initially by the contractor, and then are sent to the design professional, who reviews them for compliance with design intent. Although some controversy has arisen as to the timing (prior to permit approval) of certain submittals, as well as the liability associated with shop drawing approval, the design professional is interested in understanding generally how the contractor plans to execute portions of the work.

O: Operations

This category exists for the benefit of facilities management personnel, who have the responsibility for maintaining the facilities of a company or institution as well as for modifying facilities to suit changing needs. Drawings generated by facilities management employees or by design firms that describe proposed changes to the facility find a home in this category.

■ The International System of Units

Construction documents produced on behalf of the federal government and some state governments in the United States must comply with the International System of Units (SI). Additionally, the globalization of the economy is making an international standard—the metric system—increasingly useful to the design and construction industries in the United States (see sidebar).

SI and the Construction Industry

Building Modules

The basic building module in the United States is 4 inches, and the design and construction of most projects reflect its use. Building products and framing layouts, for example, are produced with this module in mind. Sheets of plywood are commonly manufactured in 4' × 8' sizes. Lay-in tiles for suspended ceiling systems come in 2' × 2' or 2' × 4' sizes. Gypsum wallboard is commonly available in 4' × 8' and 4' × 12' sizes. Wood light and light-gauge steel framing layouts are almost always 16" or 24" *on center* (abbreviated o.c., this phrase refers to the distance between the centers of members). Under the International System of Units, the module is 100 millimeters, or 3.937 inches. The dilemma caused by using both units simultaneously in drawings is apparent here.

Modular Construction

All construction is modular to varying degrees, but the term modular construction generally refers to construction in which specific units—4", 8", 16", 48", for example, or 100, 200, 400, 1200 mm—are used repeatedly to create a whole system or project. The module is used as the basis for planning as well as construction, and it constrains the finished dimensions of the project. The most common modular projects, and a good example of the modular approach, are masonry buildings. Modular construction is, theoretically at least, more efficient and less wasteful than nonmodular construction (fewer customized installations are required, resulting in lower labor costs and less wasted material). Not surprisingly, it is also less flexible, from both the design and construction standpoints.

Converting from U.S. Customary Units to SI

The agency that is responsible for shepherding metric into the mainstream of construction life in the United States is the National Institute of Buildings Sciences (NIBS), the same agency that is providing the forum for the National CAD Standard. Adjusting to SI influences a host of participants in design, construction, and manufacturing, for example:

Dimensioning Units

Dimensions in SI are limited to two units, millimeters and meters. Dimensions for structures are typically given in millimeters; site plans and civil drawings use meters.

THE INTERNATIONAL SYSTEM OF UNITS: BACKGROUND

In 1795, the French National Assembly adopted the metric system, first proposed by Gabriel Mouton, Vicar of Lyon, France, in 1670. Since its inception, the metric system has undergone a number of changes in response to its use by other countries and to developments in science and especially to industrialization and commerce. In 1960, the Eleventh General Conference on Weights and Measures (an international treaty organization) met and agreed upon an internationally accepted metric system, known universally as Le Systeme International d'Unites (SI), which translated from the French means The International System of Units. Among other accomplishments, the conference redefined the original term "metre," which was derived from the Greek word metron (measure), as meter. As used in the original system, metre was one 10-millionth of an imaginary line running from the Equator through Paris to the North Pole. The conference defined the meter as 1,650,763.73 wavelengths of the orange-red radiation of krypton 86 in a vacuum.

The Metric System in the United States

Though the U.S. Congress originally sanctioned the metric system in 1866, its use was limited in the United States until after the Eleventh General Conference on Weights and Measures was held in 1960. The U.S. National Bureau of Standards adopted the International System of Units for scientific use in 1964, but the transition from U.S. Customary Units did not get a significant boost until after the Omnibus Trade and Competitiveness Act of 1988 was enacted, making the conversion to SI mandatory for federal government agencies. And, as is common with federal legislation, states adopt their versions of it; consequently, SI is being used in some state projects as well.[4]

---

[4]*American Heritage Dictionary of the English Language,* Houghton Mifflin Company, 1978.

The centimeter, decimeter, and decameter have been abandoned because of their potential to confuse (decimal places would have to be used more frequently). Conventions for metric measurements have been established. Some common ones are:

- If a dimension is given in millimeters, the number has no decimal places. If it has decimal places, the measurement is in meters.
- Unit symbols are printed in upright type and lowercase, except for the liter (L) or units derived from a proper name (for example, the letter N, for newton, the unit of force named for Sir Isaac Newton).
- Unit names are printed in lowercase, even if they are derived from proper names (e.g., newton).
- The plural of unit symbols is not used (kgs), but it is used for unit names (kilograms).
- Numerals may be used with written unit names (10 square meters) in nontechnical writing, but in technical writing, symbols should be used in conjunction with numerals (10 m^2); if unit names are used, they should be written out (ten square meters).
- Symbol units and names should not be mixed; write "newton meters" or use the symbol Nm.
- Spaces instead of commas between numbers are used for any number over four digits, except in figures expressing money (e.g., 11 350 mm^2, rather than 11,350 mm^2).
- Periods after symbols are not used, except at the end of a sentence.
- The product of two or more units in symbolic form should be written with dots between the units, positioned above the line (kg·m·s^{-2}).
- Decimals, not fractions, are used to describe units that are not whole.
- For values less than one, a zero is used before the decimal point (0.12 g).

Drawings

Sheet sizes will conform to the International Standards Organization (ISO) designations, which are A0 (1189 × 841 mm, or 46.8" × 33.1"); A1 (841 × 594 mm, or 33.1" × 23.4"); A2 (594 × 420 mm, or 23.4" × 16.5"); A3 (420 × 297 mm, or 16.5" × 11.7"); and A4 (297 × 210 mm, or 11.7" × 8.3").

Drawing Scales

Drawing scales in SI are listed in straight ratios; for example, 1:200 means that one unit on the drawing represents 200 units of the object, whatever the unit is. So an object that actually measures 50 millimeters on the drawing represents 10,000 millimeters (10 meters) of the planned object. Table 3.2 compares the scales commonly used in construction.

TABLE 3.2 Comparison of Drawing Scale Ratios

Traditional U.S. Scales	Actual Ratio	Closest SI Scale
Architects' Scale		
3" = 1'-0"	1:4	1:5
1½" = 1'-0"	1:8	1:10
1" = 1'-0"	1:12	1:10
¾" = 1'-0"	1:16	1:20
½" = 1'-0"	1:24	1:20
⅜" = 1'-0"	1:32	1:33⅓
¼" = 1'-0"	1:48	1:50
⅛" = 1'-0"	1:96	1:100
1/16" = 1'-0"	1:192	1:200
Engineers' Scale		
1" = 10'	1:120	1:100
1" = 20'	1:240	1:200
1" = 30'	1:360	1:333
1" = 40'	1:480	1:500
1" = 50'	1:600	1:500
1" = 60'	1:720	1:800
1" = 100'	1:1200	1:1000

SI and Construction Products: General

The vast majority of construction products (over 90 percent) will not have to be changed to comply with the SI system; instead, they will undergo a "soft" conversion, that is, they will be relabeled. Modular products, however, such as concrete block, drywall, plywood, suspended ceilings, and raised floors will be produced in metric sizes (a "hard" conversion, in which products undergo a physical change). Products that are custom-fabricated for a project (cabinets, trusses, doors, and windows, for example), cut to fit on site, or that are not dimensionally sensitive (for example, fasteners, electrical components, plumbing fixtures, and HVAC equipment), or that otherwise remain unchanged, will be produced or labeled in SI units.

The spacing of structural and nonstructural members will be changed to reflect the new module; for instance, stud and rafter spacing will change from 16" and 24" o.c. to respectively, 400 mm and 600 mm (15.7" and 23.6"). A complete list of adjustments is available on the NIBS Web site at www.nibs.org.

▮ Jurisdictional Confusion: Drawing Interdependence

A variety of participants can be and frequently are involved in the design and installation of a single assem-bly or system in a project. Not all of the participants have contractual relationships with a single entity involved in the construction, making the coordination and control of a project more difficult. Owners may contract with certain design professionals—geotechnical engineers, civil engineers, and interior architects are examples—for portions of the work on the same project and then hire an architect to produce drawings and specifications for a building shell. Coordinating the work of the participants hired by the owner falls on the shoulders of the architect and prime contractor, who have no meaningful control over them. Because fees paid to design professionals are frequently inadequate and time constraints are severe, problems frequently arise. Adding to the problem is the nature of the process itself: Large numbers of widely disparate participants form a team designated to produce a single, complicated prototype in a relatively short period of time. The consequence of this situation is that drawings often leave the contractor with numerous questions.

Uncovering problems and resolving them proactively results from thorough searches in the different disciplines for information pertinent to a system. Information on a perimeter fence, for example, might be contained in civil drawings (the site plan and site details), architectural drawings (site plan and site details), structural drawings (foundation drawings and details), and in landscape drawings (plans and details). Aspects of mechanical equipment are often described in various views (plans, elevations, sections, details, equipment schedules) on the mechanical, structural, electrical, architectural, plumbing and landscape drawings. The mechanical engineer is typically in charge of this aspect of the work, whereas the structural engineer is concerned with the static and dynamic loads imposed on structural members by HVAC equipment, as well as ducts passing through them; and the electrical engineer must be sure to provide the appropriate high-voltage circuitry to the machinery. The controls for various kinds of equipment may be described in mechanical equipment schedules and installed by subcontractors who specialize in these systems. The architect evaluates how the HVAC equipment will affect habitability (noise pollution, for instance) and esthetics (how large pieces of machinery will be integrated into the architecture), and has primary responsibility for roof design, roof covering selection, flashing and HVAC equipment maintenance structures (access ladders, catwalks, and platforms). The engineer in charge of plumbing design must provide drainage and overflow lines from air handlers; and planting and hardscape plans can be affected by the location of compressor units and suction lines. Managers of the construction process learn to look for conflicts and resolve them before they become crises in the field.

Residential and Commercial Building Construction Drawings: A Comparison

There are four basic categories of project: residential building construction, commercial building construction, engineering construction, and industrial construction (refer back to Chapter 2). Two of those four types, residential and commercial building construction, constitute approximately three-fourths of the overall construction market. That proportion may change slightly as funds are invested in infrastructure projects, but it is likely that these two segments of the industry will continue to dominate the market. A comparison of the drawings produced in both sectors of the market is therefore worth making.

Residential Design

As noted in Chapter 2, some states do not require designers of detached single-family and duplex homes to be licensed, and consequently architects, individuals, builders, and building designers all produce the documents used in construction. Architects and the engineers in their employ design multiple-family projects (three- and four-family structures, townhouses, apartments, condominiums, and multistory residential projects) usually as a matter of state law.

Nature of the Work

While there are many variations in the documentation of residential projects,[5] the drawings for homes and duplexes bear numerous similarities to commercial building drawings but generally are less extensive, have fewer details, and are less prescriptive. Designers of buildings used by the public, particularly large numbers of people, are subject to myriad requirements having to do with minimum standards of health and safety, many of which do not apply to buildings occupied by a single family. Significant access and egress requirements (including by people with mobility challenges), panic hardware, signage requirements, and fire suppression systems are examples of constraints that do not apply to single-family and duplex projects.

As in comparable creative undertakings involving construction, architects or building designers conceive of solutions to their clients' design problems and then describe those solutions using drawings and text. Site and grading plans; architectural floor plans, roof plans, exterior elevations, building sections, wall sections, interior elevations, and architectural details; as well as foundation and floor framing plans and structural details are all used (see Chapter 4 for an explanation of the various views used in construction drawings). The simpler the project, the simpler the task of describing it. There are notable differences, however. For all their simplicity, small residential projects can be much more complicated than the typical commercial building project, particularly where building site issues and the geometry of the roof structures are concerned. A far greater number of homes are built in areas with severely sloping sites (the San Francisco Bay Area; the Hollywood Hills; Seattle; Portland, Oregon; and Boston are good examples) than are commercial building projects. The cost of creating a level building pad for a home on a steeply sloping site, if it can be done at all, is often prohibitive. Consequently, designers resolve the issue by requiring complex foundations comprising, for instance, stepped foundations over reinforced concrete piers, grade beams, and tie beams.

Simple, low-slope roofs, often hidden behind parapets, predominate in commercial building. Roof structures play a critical role in the aesthetic appearance of homes, and consequently pitched, ornamental structures are preferred in residential buildings. When those roof structures intersect at odd angles, building them can be challenging. Certain interior assemblies—curved or circular staircases, for example—can be elaborate in a residential building, particularly in more expensive homes. This complexity, combined with the tendency of building designers to be less prescriptive than architects who describe commercial buildings, can make residential building construction a frustrating exercise.

Documentation

The drawings for most small projects generally follow the sequence used in commercial building projects. In the drawing set, site drawings (which may be produced by the architect or building designer rather than by a civil engineer) precede the architectural drawings. Architectural drawings may consist of floor plans followed by a roof plan, exterior elevations, building sections (which may include interior and exterior elevations as a part of the section), and architectural details. Specifications in residential drawings, rather than existing in the separate book or binder form common in commercial construction, may be recorded on the drawings themselves. Some designers follow the same approach that structural drawings in commercial construction do: the first page or two is devoted to text and some drawings of generic details (standard details the designer includes on every project) followed by the building drawings (refer to the appendix for an inventory of information in a sample residential construction project). For all practical purposes, mechanical drawings in a residential construction set

---

[5]The influences on documentation include applicable state and local law, housing type (the two primary types being production housing and custom homes), developer or owner requirements, the individual preferences of the producers of the documentation, indigenous building practices, and cost.

are nonexistent, unless the project approaches commercial building size (10,000 or more square feet). Electrical drawings simply show the floor plan, the location and type of fixtures, and the location and the characteristics of the devices that control them (e.g., three-way switch, rheostat, number of switches in a "gang"). Specific information regarding key aspects of the drawings follows.

Site, Grading, and Utility Drawings

Every government entity has its idiosyncrasies, but many require little in the way of grading or utility drawings for a home that is constructed in a medium- to high-density subdivision—most of that work is produced by the civil engineer responsible for designing the subdivision and put in place by the earthwork contractor who creates the building pads and streets in the subdivision according to the engineer's design. Electrical power, the domestic water supply, and the sanitary sewer are installed in the streets or adjacent to them and are brought just inside the property line of each building lot. The constructor of the home on each lot connects the plumbing in the home to those utilities. Storm runoff on the lot is directed to gutters and storm drain inlets in the street or detention basins and bodies of water such as ponds or lakes in the subdivision via simple surface drainage mechanisms (slopes and swales). When other methods such as underground drain lines are desired, it is usually a landscape contractor who installs them as part of a landscape plan. Fire service underground is required in most commercial building projects because owners want to take advantage of the building square footage bonuses offered by the building code and lower insurance premiums that result when an automatic fire suppression system is used. Automatic fire suppression systems are used in residential construction (mostly for insurance reasons), but a separate underground system is not required; these systems use the domestic water supply. In some subdivisions, the improvements (streets, sidewalks, utilities) are installed but the home sites are not graded. In this circumstance site and grading plans, particularly on hilly sites, are required.

Architectural Drawings

The architectural drawings set for a residential building is very similar to that of a commercial building with one common exception: the walls frequently play a more integral role in the performance of the structural system in a residence. Businesses start, they grow, they move, and they disappear as well, and consequently commercial building owners require considerable flexibility within the building envelope to accommodate a changing clientele. The design of the structural system, then, anticipates those changes and carries out its role independently of the partitions that divide multiple businesses

in the shell. Many commercial building contractors construct a shell for an owner under one contract and install tenant improvements (TIs) under separate contract with the owner or lessee after the space has been designated (see Chapter 11 for a discussion of tenant improvements). While remodeling certainly takes place in residential projects, changes are not anticipated during the design development process like they are in a commercial building. In the drawings themselves, wall types, width, length, and orientation to each other are shown in floor plans. As in drawings for commercial buildings, architectural floor plans show room names, uses, or numbers; features such as sinks, counters, appliances, storage space, and plumbing fixtures; dimensions; and the numerous references to relevant drawings (major and minor building sections, wall sections) and legends for the various symbols used. As an alternative to employing a room finish schedule, reflected ceiling plan, or wall sections to identify ceiling heights, as commonly occurs in commercial buildings, the designer may list them directly on the floor plan, elevations, and/or sections.

Foundation Plans

The instructions to builders regarding foundations are generally the same in both residential and commercial building, although it is fair to say that commercial buildings exhibit a wider variety of foundation type and complexity than residences, which typically employ slabs-on-grade or conventional foundations (footing-stemwall systems with dimensioned or manufactured lumber floor structures). The biggest difference is in the detailing of the foundation system, which for a commercial building project of any consequence can represent hundreds of drawings.

Roof Structure Framing

As noted above, roof structures in residential projects are frequently more complicated than commercial building roof structures, due in no small part to their complex geometry. Pitched roofs (roofs with slopes greater than 3 in 12—three vertical inches to 12 horizontal) typically require the installation of many small pieces (e.g., shingles, shakes, slate, concrete roof tiles), whereas low-sloped roofs employ large sheet goods (built-up or single-ply roof coverings). From the builder's standpoint a pitched roof can be much more of a challenge, when access to the work, safety, and productivity are considered.

Exterior Elevations

Exterior elevations in both residential and commercial projects are very similar. They are pictures of the structure as it would appear to a bystander—views of the surface features of the structure from all sides, written descriptions of its physical features, with leaders pointing to specific elements, components, or assemblies.

Building and Wall Sections

Major building sections in both commercial and residential drawings are cuts through the entire project that expose the anatomy of main elements in the building. These drawings, depending on where in the structure they are taken, combine views of inhabitable spaces, the structure itself, and exterior surfaces. Wall sections, on the other hand, take smaller segments of the building and look at them in greater detail. Both architects and engineers make extensive use of wall sections. In the architect's case, the focus is on the general size, shape, location, and aesthetics of the system, its fire resistance capability, ventilation characteristics, and resistance to an invasion of the elements. In the case of structural engineers, wall sections are popular for identifying the structural elements required, the manner in which they are connected to one another, and the relationship of the wall system to larger structural elements. Considering that their professional responsibility is to make sure a structure performs as intended under a variety of static and dynamic loads, it is not surprising that structural engineers frequently ignore other aspects of the wall system (e.g., flashing).

Interior Elevations

Design professionals who produce drawings for commercial buildings frequently make use of interior elevations to describe the surface features of rooms in the structure. Particular attention is paid to public spaces such as lobbies, waiting rooms, restrooms, and specialty rooms such as locker rooms. In a hospital or healthcare project, for example, interior elevations of the typical hospital room and special rooms such as imaging or examination rooms are widely used (see Figure 11.5o). Interior elevations are used in residences as well, but not to the same extent.

Detail Drawings

The focus of detailing—describing the nature of smaller components and assemblies in the project—is the same in both residential and commercial building projects, although the specific subject matter may differ. Detailing of cladding and roof system components, a task that falls on the shoulders of architects, focuses on resistance to the penetration of the elements and sound.

For the engineer, how the cladding will perform with various loads applied to it (including thermal expansion and contraction) and the response of the structure to bearing the load of the cladding are the major concerns. Consequently, connections are of great interest. For projects that employ a system other than curtain walls, for example, a concrete tilt-up building (site pre-

cast concrete panels), the detailing of the panels is shared: architects focus on describing the aesthetic features of the panel, its resistance to the elements, floor and roof connections and their resistance to fire, and compliance with zoning codes among other things (refer to the figures in Chapter 9, in particular Figures 9.4g–n, for examples of the architect's work pertaining to cladding). Engineers focus on structural concerns involving the panel (how they are connected to one another, how floor and roof structures connect to the panels), and the structural elements all these assemblies are composed of.

Mechanical Drawings

Mechanical systems, consisting of piping (hot and cold water supply; drain, waste, and vent piping; hydronic systems; and fire suppression); pumps, valves, and other equipment; and heating, ventilating, and air conditioning (HVAC) systems and the equipment that controls them can be extremely complex in a large office building or hospital, due to the public nature of the buildings and the number and physical condition of the people occupying them. These systems are designed by mechanical engineering firms that operate as subconsultants to the prime design professional. In contrast, mechanical systems in residences serve a handful of occupants in a far simpler interior environment. Conditioned space in commercial buildings is subdivided into zones that various types of equipment serve, while in residential construction the entire home represents a single primary zone or perhaps two zones with individual rooms considered as subzones. Consequently, competent mechanical specialty contractors design and install HVAC systems on a design-build basis with few instructions from the design professional, with the possible exception of identifying the system type and location of large equipment (condensers and air handlers) and perhaps the key access points.

The design-build nature of HVAC or systems applies to plumbing systems in residential projects as well. Beyond identifying where the fixtures will be, the designer does little to describe the plumbing system. In commercial building projects, mechanical engineers produce complete plumbing drawings, including isometric drawings of drain, waste, and vent piping and hot and cold water supply (as well as other systems). Plumbing system drawings in residential construction may consist simply of a ghosted floor plan with fixtures emboldened on them.

Electrical Drawings

Electrical drawings in residential construction are similar in their character to the plumbing systems in that the designer identifies the location, number, and type of light or appliance and the switches that control them, the location and number of power outlets desired, the location of the service panel, and perhaps the type and color of switches, outlets, and covers. Beyond that, the installation is up to electrical specialty contractors who identify the number, type, and size and routing of the conductors. In contrast, electrical drawings in commercial buildings, though not as explicit as other drawings in the set, are far more prescriptive. (Nearly all information in electrical drawings is given in plan form and equipment lists and schedules. The third dimension—the vertical element between floor and ceiling—is ignored, leaving the mapping of the distribution system up to the specialty contractor.)

Summary Comment

What gets described in any drawing set is really the province of the individual or entity employing the design professional or building designer who resolves complex design issues on behalf of the client. A company that constructs projects for its own account using in-house personnel who know their company's preferences well (some apartment builders who develop, own, and operate their own housing complexes function this way) probably require less instruction from an architect than a construction company building an apartment complex for an investment group that has employed an entity to develop, design, and describe that project. It should come as no surprise that commonalities in residential and commercial building drawings exist, but variety in the information abounds.

Summary

Although there are a variety of formats for assembling drawings and for drawing content, most follow the same basic organization. The disciplines in each drawing set are organized more or less in the order of construction, and information within each discipline is organized from the general to the specific. In virtually all of the principal disciplines, plans precede elevations, which precede sections and details. It is a logical way to present information to the principal user of the drawings—construction personnel—and is a standard from which designers should not depart significantly.

CHAPTER 3 EXERCISES

1. Identify the contents of the project manual.
2. Explain the order of drawings within each discipline in a construction drawing set.
3. What is the basic building module in the United States? In Europe?
4. Explain the difference between a soft and hard conversion.
5. What is meant by the term *modular construction*?
6. Which dimensional units will be used in all drawings and for all projects done in SI?

4 Construction Drawings

Key Terms

Agreement
Approximate estimate
Architect's scale
Axial dimensioning
Axonometric projection
Boundary plane dimensioning
Cabinet projection
Conditions
Decimal scale
Detailed estimate
Dimetric projection
Direct costs
Elevations
Engineer's scale
First-angle multiview
     projection
Floor plans
Foundation plan
Indirect costs
Isometric projection
Longitudinal sections
Mechanical engineer's scale
Metric scale
Multiview drawings
Object
Oblique projection
Orthographic projection
Parallel projection
Perspective projection
Pictorial drawing
Plan views
Plane of projection
Projections
Projectors
Reflected ceiling plan

Sections
Site plan
Station point
Stipulated sum agreement
Technical drawing
Third-angle multiview projection
Transverse sections
Trimetric projection
Unit price agreement
U.S. Customary Units
Utility profile
Wall section

Key Concepts

- Successfully interpreting construction drawings requires an open mind, imagination, and an understanding of the goals of design professionals.
- A word frequently has several definitions; the correct one requires the reader to evaluate the choices in the context of the sentence. The same is true for graphic expressions—the "words" must be taken in context.

Objectives

- Determine the actual size of an object depicted on construction drawings using engineer's, architect's, and SI (metric) scales.
- List the various projections common to design and construction drawings and their appropriate applications.
- List and re-create the symbols most commonly used in construction drawings.
- Identify the various line types used in construction drawings.
- Produce scaled drawings of simple construction components.
- Perform quantity takeoffs for simple construction components in English and SI units.

Construction Graphics: A Practical Guide to Interpreting Working Drawings, Third Edition. Keith Bisharat.
© 2025 John Wiley & Sons Inc. Published 2025 by John Wiley & Sons Inc.

Scale, Scales, and Scaled Drawings

The word "scale" has many definitions. Those that pertain to construction graphics include: (1a) a system of ordered marks at fixed intervals used as a reference standard in measurement; (1b) an instrument or device bearing such marks; (2a) the proportion used in determining the relationship of a representation to that which it represents; (2b) a calibrated line, as on a map or architectural plan, to indicate such a proportion.[1]

Scale: The Concept

When translating from actual or planned object dimensions to a scaled construction drawing, the units of measure (the dimensions of the real object) are "compressed" into representative units. When a $\frac{1}{4}$" = 1'-0" scale is used, the reader is expected to understand that every $\frac{1}{4}$" in the drawing represents 1' in reality—1' is compressed into $\frac{1}{4}$". A 100'-long building, then, would be 25" long on the drawing—100', in reality, are represented by units of $\frac{1}{4}$" each (100' × $\frac{1}{4}$" = 25"). A very large object is being crammed into a very small area. Conversely, when translating from a drawing to reality, representative units are "expanded" into actual units. Since every $\frac{1}{4}$" on the drawing represents a foot in reality, the number of representative units must be determined in order to arrive at the actual length of the object. In the building just mentioned, 25" divided by the representative unit $\frac{1}{4}$" will give us the total number of representative units, which

must then be multiplied by what the unit actually represents (1'). So, 25" are divided by $\frac{1}{4}$" to get 100 units, each of which represents 1'. The answer is 100'. The beauty of the instrument known as a scale is that one is saved from doing the arithmetic simply by placing the proper scale onto the drawing and reading the dimensions on it. Figures 4.1a–c compare scales commonly used in construction.

Scales: The Instrument

Five different scales (the instrument) are commonly used in technical drawing (any drawing that expresses a technical idea), including the metric, decimal, architect's, engineer's, and mechanical engineer's scales. Most of the world (over 90 percent) employs the metric system, and metric scales (the proportions) in this system are expressed as ratios of some metric measurement, for example 1:5, (1 millimeter on the drawing represents 5 millimeters in reality), 1:10, 1:20, and so on. The decimal scale (used in machining and tolerancing details) reduces inches to decimal fractions of an inch (tenths and fiftieths of an inch). Architect's scales express measurements in inch-foot terms, for example $\frac{1}{4}$" = 1'-0" or $\frac{1}{2}$" = 1'-0". Engineer's scales are similar to architect's scales in that they also express a relationship of inches to feet; however, a whole inch in relationship to a foot or feet is used (for example, 1" = 1', 1" = 20', 1" = 100'). Mechanical engineers use metric, decimal, and mechanical engineer's scales, the latter being proportional expressions of an object that is ultimately described in inches and feet (full size, half-size, twice-size, three-times-size).

[1] *American Heritage Dictionary of the English Language*, Houghton Mifflin Company, 4th edition, 2000.

Figure 4.1a Architect's scales are distinguished by the division of the individual foot into inches for each scale.

ENGINEER'S

Figure 4.1b Engineer's scales express measurements in terms of an inch in relation to feet.

SI

Figure 4.1c In the International System of Units, scale is expressed in straight ratios.

Scale in Construction Drawings

In the construction industry, the three common scales used are the architect's, engineer's and metric scales. Federal and some state government construction drawings use metric scales. Since the conversion to metric is not complete in the United States, metric drawings and drawings using U.S. Customary units both exist. Combining both units on the same set of drawings, however, leads to confusion and error, and should be discouraged.

Architect's Scales

Architects are commissioned to design very large objects (buildings and other large projects) as well as the minute details within them (e.g., trim, fixtures, ceiling, and floor patterns). The architect's scale (the instrument) must be versatile enough to show objects in full size as well as greatly reduced sizes. The typical triangular scale has one full size and 10 overlapping reduced-size scales (the proportions), enabling the architect to depict an object, in feet and inches, in its full size to ¹/₁₂₈ of its full size. If one were to overlook subdividing the ends of the architect's scale into inches—which all 10 reduced-size scales do—depicting a large object at ¹/₁₉₂ of its full size is possible using the full scale, which is 12" divided into 16 subdivisions per inch. What distinguishes the architect's scale from others is that the ratio is expressed in terms of a fraction or multiples of an inch to a single foot (e.g., ¼" = 1'-0", ½ " = 1'-0", 1 ½" = 1'-0" 3" = 1'-0"), as opposed to a straight ratio or an inch to feet, as with, respectively, the metric and the civil engineer's scale.

With the exception of the civil sheets in building construction projects, all of the drawings, including structural, mechanical, and electrical engineering drawings, are produced in architectural scales; the architect, after all, is the prime design professional on these projects. One of the advantages to producing drawings in SI is that only one scale (the instrument) is used.

Engineer's Scales

Civil engineers, who typically produce the infrastructure drawings on building construction projects and most of the drawings in civil construction projects, have traditionally used scales (the proportions) that describe distances in feet and tenths of feet, and their scales (the instrument) reflect that tradition. One whole inch on a civil engineer's scale (also called a *chain scale* because of its use by surveyors) represents 10', 20', 30', 40', 50', or 60', (or fractions or multiples of those distances, e.g., 1" = 5' 100', 200', 500', or 1000'). Interestingly, for utility profiles, some highway sections, and sections of drainage systems, engineers will use two different scales in the same drawing, one for the vertical and another for horizontal distances. Slopes on drainage systems, particularly very large pipe, can be very gradual—as little as a half percent (6" in 100'), or less, and the pipe itself may not stand out on the drawing. Consider a 100' section of a 24" storm drain on a 1" = 20' scale drawing: a half-percent slope would be imperceptible to the user of the plan (one end of the 5"-long line on the profile would be ¹/₄₀ of an inch lower than the other). Lines indicating the top and bottom of the piping would be ¹/₁₀" apart, or about ³/₃₂"—worthless for all intents and purposes. Using a different vertical scale results in dramatically exaggerated vertical elements. A 1" = 1' scale for the vertical, for example, would result in one end of that same 5" line being ½" lower than the other, with parallel lines 2" apart—a display much more easily perceived by the eye. Additionally, the accuracy of earth-quantity takeoffs performed manually by the average end-area method depends to a large extent on the scale of the sections (the larger the scale of the drawing itself, the more accurate it is graphically). One method of selecting the two scales for sitework drawings is to choose a vertical scale that is 10 percent of the horizontal scale, so if the horizontal scale is 1" = 100', then the vertical scale would be 1" = 10'.

Metric Scales

Conceptually, metric scales are simpler than architect's and engineer's scales, since they do not combine different units (inches and feet). Metric scales use straight ratios, wherein one unit on the scale represents 10, 20, 30, 40, 50, 100, 200, or more units (millimeters or meters) in reality (refer back to Figures 4.1a–c).

■ Determining Dimensions without a Scale

Occasionally, a design professional will neglect to identify the scale of a drawing, and in some depictions (usually presentation or conceptual drawings), the omission of a scale is intentional. Floor plans on model homes, for example, often depict the layout but do not show the dimensions of rooms. It is nevertheless possible to estimate the sizes of nearly any object depicted on a drawing without any scale being listed, as long as one is reasonably confident that the drawing is proportional. By using an object in the drawing that is a known size—a bathtub, entry door, or material such as concrete block or brick, for example—one can create a reliable scale with which to measure other objects in the same drawing. Masonry varies to some extent with the manufacturer and the type, but concrete masonry units, (CMU) are normally 7⅝" high, 7⅝" wide, and 15⅝" long (190 mm × 190 × 390 mm under a hard conversion to SI), and modular brick masonry is 2¼" high, 3⅝" wide, and 7⅝" long (57 mm × 90 × 190 mm, hard conversion). When the typical mortar joint (³/₈" or 10 mm) is added to the actual dimensions of CMU or modular brick, dimensions that reflect the building module of 4", or 100 mm, are achieved. For CMU, the nominal dimensions are 8" × 8" × 16"; for modular brick they are 2 ²/₃" × 4" × 8" (a three-course wythe of brick is 8" high, 4" thick, and 8" long: the same wythe in SI is 200 mm × 100 mm × 200 mm). These are the dimensions that are most useful in determining dimensions without a scale. Transcribing the edges of a 3'-0" entry door (a very common residential entry door size) onto a piece of paper, cutting it on the mark, then folding the paper in two (three times) will result in creases at 4" intervals. The first fold creates a crease at 16", the second at 8" and 24", and the third at 4", 12", 20", and 28".

■ Projection Types

Background

In the fifteenth century, Italian architects Filippo Brunelleschi and Leon Battista Alberti developed the idea of projecting an object onto planes; in fact, the rediscovery and refinement of perspective drawing technique—the method of drawing that best duplicates reality, originating with the Greeks and Romans, but lost in the intervening years—is widely attributed to these two Renaissance figures. Others who followed, including Leonardo da Vinci, refined and recorded drawing techniques; and in the early nineteenth century, the French professor Gaspard Monge articulated his theories of descriptive geometry—and developed technical drawing as we now know it—in a treatise entitled "La Geometrie Descriptive."

Projections

There are two basic projection types, perspective and parallel. The term *projection* refers to the transcription of lines of sight, called projectors, from an imaginary or real object onto a plane. The points at which the projectors intersect this plane are then connected to create a two-dimensional image of the object as seen in a particular view (see Figure 4.2). Figure 4.3 comprises the "family" of projections used in technical drawing. Of this group, the drawings commonly used in design and constructions include third-angle multiview, axonometric, and oblique projections (see Figure 4.4).

Elements of Projections

All projections have four components (refer back to Figure 4.2): the station point (the observer's eye), the object (whatever the observer is looking at), the projection plane (the drawing), and the projectors (lines of sight).

Perspective Projection
One-Point (A), Two-Point (B), and Three-Point (C)

Parallel Projection
Multiview (D), Axonometric (E), and Oblique (F)

Key
1 Station Point
2 Projectors
3 Plane of Projection
4 Object

View through Picture Plane (top)
Plan View (bottom)
* Multiview
** Oblique

Figure 4.2 Parallel and perspective projections.

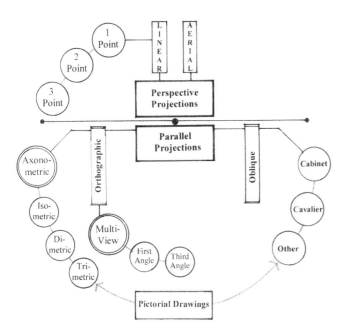

Figure 4.3 The family of commonly used technical drawings.

Third-Angle Projection

Modern technical drawing is characterized by the use of six contiguous planes (a transparent box) surrounding an object, on which six basic views are projected (top, front, bottom, rear, left, right), then "unfolded" and arranged in quadrants on a drawing sheet (see Figure 4.5).

For simpler objects, in which two or three views are all that is required to adequately describe an object, nonessential views can be eliminated. The common projections—first and third—vary according to the placement of the object relative to the viewer and the planes of the transparent box. In a first-angle projection, the object is placed between the viewer and three of the six planes of the box, and the respective views are projected onto a plane much like a shadow is cast on a wall by an object. The top, front, and right side views are recorded in, respectively, the lower right, upper right, and upper left quadrants on the drawing sheet. In a third-angle projection the top, front, and right side planes separate the object from the viewer. The main difference between the two is in how the views are conceived and then "unfolded." In third-angle projection, the top, front, and right side views are recorded in the upper left, lower left, and lower right quadrants. In the United States, first-angle projections were common until the early twentieth century, at which time the third-angle projection—thought to be a more logical depiction of an object—became the standard in technical drawing. First-angle projections continue to be used elsewhere in the world. Figure 4.6 compares first- and third-angle multiview drawings.

The distinguishing characteristic of the third-angle projection—the placement of the plane of projection between the station point and object—is consistent in the hundreds of pictures that are required to adequately describe a medium-sized building construction project. Compared with other graphic techniques (axonometric drawings for instance), multiview drawings allow complicated designs to be easily and quickly produced and measured with simple instruments. Their principal disadvantage is that a considerable amount of "cognitive construction," or constructive imagination, is required to understand the object. Foreshortening, for example, presents real difficulties to the viewer of a multiview drawing, since there are many ways to interpret a drawing. Figure 4.7 is an example of a single drawing that may be interpreted in a variety of ways, and it illustrates why additional views are required to accurately describe an object.

■ Views: Plans, Elevations, and Sections

Construction drawings consist of numerous mulitview drawings of a building or a structure's features. The three views used are plan views, sections, and elevations.

Plan Views

Plan views are used to show any element, component, or assembly from above. For example, site plans show the building or structure situated on the project site, as well as ancillary structures, parking, and frequently, landscaping or hardscape features. Foundation plans are structural drawing depicting the foundation designed to support the structure. Roof plans are bird's-eye views of the roof of a building, including such features as mechanical equipment, the walls that screen it, canopies, crickets, ridges, and slope direction.

Floor plans, which depict the number, width, and location of a building's walls, are actually horizontal cuts (sections) through the entire building, taken about four feet above the floor, viewed from above. This convention puts features that transcend horizontal planes—stairs, for example—or features that exist in proximity to the elevation of the cut such as soffits, chases, and hanging cabinets, in their proper context.

One interesting challenge facing architects is how to describe the ceilings of a building project. Depicting them as one would see them upon completion—standing on a floor looking up—causes orientation difficulties: a light fixture over the northeast corner of a room in plan view would be oriented in the southeast corner of a view of the ceiling. The solution to this dilemma is the *reflected ceiling plan*, a view of the ceiling looking down on a mirrored floor. Using this method, the light fixture just mentioned would be shown in its proper location above the northeast corner, in a different horizontal plane. Figures 4.8a–c are examples of the different views commonly used in construction drawings.

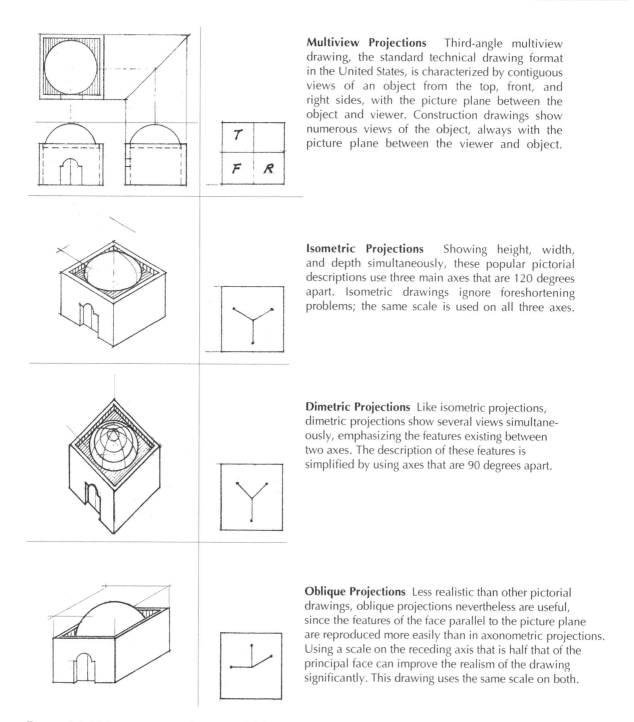

Multiview Projections Third-angle multiview drawing, the standard technical drawing format in the United States, is characterized by contiguous views of an object from the top, front, and right sides, with the picture plane between the object and viewer. Construction drawings show numerous views of the object, always with the picture plane between the viewer and object.

Isometric Projections Showing height, width, and depth simultaneously, these popular pictorial descriptions use three main axes that are 120 degrees apart. Isometric drawings ignore foreshortening problems; the same scale is used on all three axes.

Dimetric Projections Like isometric projections, dimetric projections show several views simultaneously, emphasizing the features existing between two axes. The description of these features is simplified by using axes that are 90 degrees apart.

Oblique Projections Less realistic than other pictorial drawings, oblique projections nevertheless are useful, since the features of the face parallel to the picture plane are reproduced more easily than in axonometric projections. Using a scale on the receding axis that is half that of the principal face can improve the realism of the drawing significantly. This drawing uses the same scale on both.

Figure 4.4 Multiview, isometric, dimetric, and oblique drawings.

Sections

Sections are cuts—most commonly vertical cuts—designed to reveal the insides of critical parts of the project. They are dissections of the project's "anatomy." Cuts through the entire structure, usually drawn in the same scale as the floor plan, are called building sections—either transverse (perpendicular to the long axis of the building,

also referred to as cross sections) or longitudinal (parallel to the long axis of the building). Wall sections show parts of wall assemblies and generally depict construction within a couple feet of the wall itself. Sections through even smaller parts of the project, for example, foundations, beam-to-column connections, floor systems, roof systems and coverings, and through window and door heads and jambs (top and side pieces) can be, and are, used.

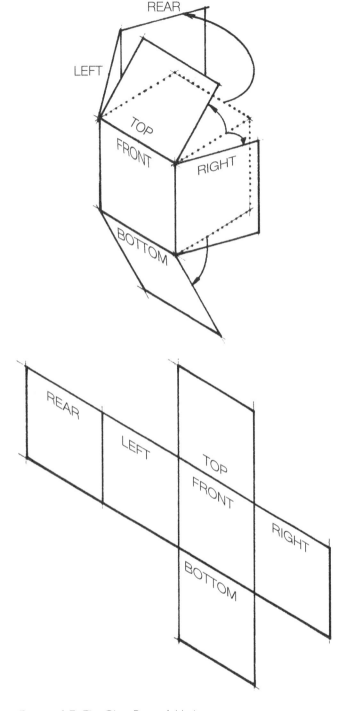

Figure 4.5 The Glass Box unfolded.

The design professional needs to show only as much critical information as is required to properly describe the project. It stands to reason that more complicated, oddly shaped, or unusual components or assemblies require more views than simple ones, but since theoretically there is no limit to what could be shown, the design professional must strike a balance between too much and too little information.

Elevations

Elevations are views of surface features of the entire project or any part of it, including small connections. Exterior elevations show the exterior surfaces of a building, and interior elevations show finished surfaces inside the building, for instance, the walls of a food preparation area. Elevations are commonly used to describe column-to-beam, bracing connections, framing systems, or various wall structures as well, frequently in association with sections and plan views of the object.

▪ Pictorial Drawings

There are projections used in construction drawings known generally as *pictorial drawings*, that imitate views of an object as conceived by the mind and eyes. Axonometric and oblique drawings—both parallel projections—make up the collection of pictorial drawings used in construction drawings illustrated in Figures 4.9a–d.

These drawings are popular because, as pictures go, they are fairly simple to produce freehand or with simple instruments, and they convey ideas quickly. Their drawback is that they are not entirely realistic, like the perspective drawing is. But perspective drawings, although most similar to the images that the eye normally sees, are complicated to produce and, due to the convergence of the projectors in one, two, or three vanishing points, are not particularly helpful for determining the true dimensions of an object. It goes without saying that being able to determine dimensions accurately, especially for the constructor, is critical.

Axonometric Projections and Drawings

Axonometric projections—isometric, dimetric, and trimetric—are pictorial projections in which the object's principal edges and surfaces are inclined relative to the plane of projection. The advantages of axonometric drawings are that they show height, width, and depth accurately in a single view, they are reasonably realistic, and they convey ideas quickly. The principal disadvantage of axonometric projections is that they are more difficult and time-consuming to produce than multiview drawings, in part due to the adjustments to scale that have to be made. The edges of a cube drawn in isometric, for example, are about 80 percent of their true length—a result of rotating the object relative to the picture plane. In dimetric projections, two different scales are required; and in trimetric projections, three. The differences in scale that exist in axonometric "projections" are ignored in axonometric "drawings," since these drawings are mainly used to convey ideas quickly, and using the same scale on all three principal axes sacrifices little in pictorial value.

Figure 4.6 First- and third-angle multiview drawings compared. (Drawing by Tom Krejci, after Keith Bisharat.)

Figure 4.7 Numerous interpretations of an object are possible with just one plan view.

Figure 4.8a Floor plans are horizontal sections through the structure. Roof plans are views looking down on the roof from above.

Figure 4.8b There are two common section types; longitudinal and transverse, or cross sections.

Figure 4.8c Elevations depict either exterior or interior surfaces of large and small elements, systems, and assemblies. From time to time elevations are used to describe framing systems as well.

Isometric projections (and drawings) derive their name from the orientation of the three main axes in the drawing, which are 120 degrees apart ("isometric" means equal measure). Two of the three main axes in dimetric projections are equal; and in trimetric projections, none of the angles between the axes is the same. Selecting one of the three axonometric types depends on the aspects of an object that the drafter wants to emphasize to the user. In construction drawings, isometric drawings are commonly used in water supply and wastewater schematics (plumbing isometrics), for certain details (e.g., light standards or fixtures, decorative concrete masonry units), for flashing and roofing details, and in any other situation in which a pictorial display is more effective than a multiview drawing. Architects, to reveal spatial relationships within a living space (which is frequently harder to capture in a simple plan view), use one very popular dimetric drawing type—the plan oblique.

D3 **PARAPET WALL ISOMETRIC**
SCALE: 1 1/2":1'-0"

Figure 4.9a Isometric drawings are popular for a variety of components in construction drawings, including flashing . . . (Drawing courtesy of Comstock Johnson Architects.)

STRETCHER BLOCK 1A

LEGEND:
A = 16" Long Block
B = 8" Long Block

NOTE:
Figure 1A or 1B are applicable to the construction of all control joints in the 8" hollow block walls except where steel or conc. cols occur. See Architectural. Elevations for req'd. locations.

Figure 4.9b . . . and structural components.

TABLE C

CEILING BRACING
SCALE: NOT TO SCALE

Figure 4.9c Dimetric, or plan oblique, drawings accentuate a different aspect of a component from isometric drawings. (Drawing courtesy of Comstock Johnson Architects.)

Oblique Projections

Oblique projections are pictorial drawings in which two of the three principal axes are 90 degrees apart, with a third (receding) axis at any angle desired by the creator. They are especially appealing when one main face—placed parallel to the plane of projection—is ornate. Casework, for example is a good subject for an oblique projection. Since the axes in the main face are at 90 degrees, more difficult geometric constructions in an object—circles, for example—are very easily replicated with standard drawing instruments. By comparison, in any face of an object depicted in an isometric drawing, a circle becomes an ellipse, which requires considerably more work to produce than a simple circle. Additionally, certain objects, largely because of their proportions, are better depicted in an oblique view than in one of the axonometric views. The main disadvantage to oblique projections is that, as pictorial drawings go, the oblique is the least realistic. In order to increase the realism of the drawing, the receding axis can be adjusted to a scale other than that used on the two main axes. Figure 4.10 compares a cavalier projection (which uses the same scale on the receding axis as is used on the main axes) and a cabinet projection (which uses a scale that is half that of the scale used on the two main axes).

Architects commonly use two oblique drawings in the course of their studies of a design and in presentations to owners. Plan and elevation oblique drawings are quickly

PLANTER STRIP DRIVEWAY
CITY DETAIL C.C. 48

Figure 4.9d Trimetric drawings are used also, though not to the extent that isometric drawings are (from the City of Sacramento Standard Specifications).

CAVALIER PROJECTION
FULL SCALE ON ALL AXES

CABINET PROJECTION
FULL SCALE ON MAIN FACE,
HALF SCALE ON RECEDING AXIS

Figure 4.10 Oblique drawings can be made more realistic by reducing the scale on the receding axis. On the left, the same scale on both axes is used, while on the right, the scale on the receding axis is half that of the principal axes.

and easily produced and are sufficiently realistic to be useful in decision making. The plan oblique, though technically a dimetric or trimetric drawing depending on the angle of rotation chosen, derives its name from two of its main axes being 90 degrees apart. The plan oblique uses a typical floor plan (walls in a structure are commonly 90 degrees to one another) rotated to 30 degrees, 45 degrees, or 60 degrees above horizontal. The corners, or intersections of the walls, form the vertical axes. The effect is similar to a bird's-eye view of a structure with the roof removed.

The "elevation oblique" uses a normal elevation view and adds receding elements to increase the realism of the projection. The angle of the receding axis is chosen to emphasize a secondary plane (one side or the top of the object). Figure 4.11 compares plan and elevation oblique drawings.

PLAN OBLIQUE (DIMETRIC)

ELEVATION OBLIQUE

Figure 4.11 Comparison of plan and elevation oblique drawings.

Dimensioning: How Big Is It? Where Does It Go?

The purpose of dimensioning is to communicate the precise size of a structure (in all three spatial dimensions) and its location on the site. The size and location of individual pieces, components, assemblies, and systems within the structure itself must be located as well. Effective dimensioning technique reflects the sequence of construction trades and, in the interest of producing uncluttered plans, offers only critical information. Even in the simplest of buildings there is an exceptional amount of information on drawings, and the design professional will communicate better and cause fewer complications for the numerous users of the drawings if none of the information is superfluous.

Dimensioning Conventions

There are two basic approaches to dimensioning: identifying the centers of components (axial dimensioning) or identifying their boundaries (boundary plane dimensioning). The choice to use one or the other (or some combination of the two) is dictated by the type of construction and the component being described. Modular construction, in which wall lengths and openings in walls are multiples of the basic module, lends itself well to boundary plane dimensioning, as does wood light framing. The structural system for a high-rise, in contrast, is laid out according to a grid system that identifies the centers (axes) of the structural elements themselves. The changes in column and beam size from the bottom to the top floors in a tall building are so significant that axial plane dimensioning is the only reasonable method to use for the structure. The centers of columns must align to avoid undesirable eccentric loads on the elements in the system below them. In the vertical direction, consistent information as to the distance between systems and elements (floor to floor, for instance) is required.

Combining both boundary plane and axial dimensioning methods in the same drawing set is common. An architect might be certain where a window should be placed in a wall, but not of its exact dimensions. In this instance, axial dimensioning works well. The dimensions of the component (in this case, the window) may vary slightly from manufacturer to manufacturer, and by identifying the center of the component, the architect can control the esthetics of the assembly, the constructor can comply with the contract requirements by selecting one of several potential suppliers of the component, and the owner receives the best outcome—value derived from the competitive process. The boundaries of a column wrap or a masonry structure within a structural steel frame on the other hand might be more critical than the centerline of the column, even though those boundaries may be located according to the centerline of the column. Occasionally, a subgrid system is used to identify such elements. The boundaries themselves can be rough (for example, to the face of concrete [f.o.c], face of stud [f.o.s.]) or finished boundaries, whichever the designer deems most appropriate.

Constructing Dimension Strings

Architectural Graphic Standards recommends that dimensions strings be developed as follows: The string closest to the structure, located several inches away from it on the drawing, includes walls, partitions, centers of windows and doors, offsets, and so on. The second dimension line, 3/8" to 1/2" farther away from the first string, identifies walls and partitions. The third string (spaced the same as the first two) may identify the dimensions of the structural grid, if one is used, or the overall dimensions. The grid identifiers—circles attached to the end of a line with numbers or a letter within them—should be placed just inside the overall dimension string. Repetitive dimensions are handled with notes accompanying the dimension strings ("DO," for example, stands for ditto). Dimensions that are less than one foot are expressed in inches; those over a foot are expressed in feet and inches. The lines on which the numbers are located are called *dimension lines*. *Extension lines* are those that cross the dimension lines at right angles and terminate very close to the outside face of the object being described. The intersection of the two lines is typically identified by broad, short hatch marks at 45 degrees to the dimension line, by dimension lines with opposing arrowheads at intersections, or by dots over the intersection, all of which are called *terminators*. Dimension strings that are oriented vertically on the page should be set up to read from the right. The dimensions themselves should be placed above the dimension line in all cases. Figure 4.12 is a summary of dimensioning conventions.

Good dimensioning includes:

- Critical information only
- Consistency in the methodology
- Clarity (including where in the dimension string the builder might make adjustments—not all dimensions are critical)
- Closure (dimension strings should add up)

■ Lettering, Notes, and Leaders

While manual lettering is highly individualistic, standards are followed. Among the desirable characteristics of effective letterers are:

- Legibility
- Uniformity of style, height, proportion, inclination, and intensity
- Consistency—all upper- or lowercase letters (the former being the most desirable)

Figure 4.12 Dimensioning conventions.

- Speed
- Adequate spacing between letters and words
- Slightly larger shapes in the lower half of the letters, which anchors the letter to the line

One of the great benefits of computer technology is that repetitive tasks such as lettering can be quickly and flawlessly performed. Indeed, the entire list of desirable lettering characteristics is incorporated in the word processing function of drafting software. Though people justifiably lament the loss of "character" represented by manually produced drawings and type, the benefits of computers are considerable. The following guidelines apply to manually produced notes:

- Consistency results if horizontal guidelines are used to contain the letters in a line of printing. Generally, these guidelines are twice as high as the space between the lines; for example, if 1/4"- (6 mm-) high lettering is desired, the space between lines is 1/8" (3 mm). Title lines are frequently 1/4" (6 mm) high, with 1/8" (3 mm) spaces between lines; subtitles are 3/16" (5 mm) with 3/32" (2 mm) spaces between lines; and other lettering is 1/8" (3 mm) high with 1/16" (2 mm) spaces between lines.
- Using a triangle as a guide for vertical strokes enhances the consistency of lettering, especially if the triangle glides along a straight edge such as a parallel rule or T-square.

- Spaces between letters vary with the letter being printed—a T and an A can be printed closer together than a D and an O, for example—but all spacing between letters is ultimately a simple choice: What looks good? Spaces between words, in contrast, are roughly equivalent to the width of a letter. The smallest acceptable lettering size is ⅛" (for manual notation).

Notes and Leaders

Although the specifications are the proper place for most text pertaining to the quality of work, materials, and proper installation, supplementing drawings with text is helpful and necessary. The Construction Specifications Institute (CSI), the leading organization for construction specifications guidelines, encourages design professionals to use generic terms in notes; for example, "gypsum wallboard," not "sheetrock," is the proper term for the material commonly applied to walls in buildings. (Sheetrock is a brand name.) Abbreviations can be confusing, and thus should be avoided in notes. Terminology between drawings and specifications should be consistent.

Leaders refer to the thin lines with arrowheads that connect a note to the object being described. A few guidelines pertain to leaders:

- Leaders should not be confused with the numerous other lines that appear in drawings. As such, angled or curved lines are used to avoid confusion (see Figure 4.13).
- Crossing dimension lines and other leaders is to be avoided if possible.
- Notes located on the left side of an object should have leaders that connect the end of the note to the object being described.
- Notes located on the right side of the object should have leaders that connect the beginning of the note to the object being described.

Lines

It is worth noting that lines do not exist in nature; they are our best attempt to replicate the juxtapositions of light, color, and reflection in the world that surrounds us. As such, they are the "alphabet" used in the visual language we call graphics. A few simple lines cleverly combined become a "word," which can be combined with other words to form a "sentence"—a complete graphic thought that stands on its own. Lines are distinguishable from one another by three characteristics: width, intensity, and continuity.

Line Width

Most of the lines used in construction drawings can be grouped into five widths: fine, thin, medium, wide, and extra wide, as shown in Table 4.1.

Figure 4.13 The line family.

TABLE 4.1 Fundamental Lines in Construction Drawings*

	Fine	Thin	Medium	Wide	Extra Wide
Width (in./mm)	0.007/0.18	0.0.010/0.25	0.014/0.35	0.020/0.50	0.028/0.70
Pencil Lead Range, if Manually Produced	3H & harder, or light pressure	H, 2H	F, HB, B	2B and softer, or more pressure, and more than one pass	2B and softer, more pressure and several passes
Application	Poché, hatching, layout or construction lines, and text guidelines	Dimension and extension lines, leaders, hidden lines, center lines, break lines, matrix grid lines	Minor object lines, lettering text	Major object lines, cut lines, section cutting planes, titles	Large titles, special-emphasis object lines (profiles), object lines in large-scale details, ground lines for elevations, building footprint

*This matrix of lines reflects the National CAD Standard conventions for lines.

Fine lines are frequently used to accentuate part of an object (poché and hatching) and for layout (construction) and text guidelines, both of which are meant to disappear in the drawing reproduction process. Thin lines are used for dimension and extension lines, leaders, hidden and centerlines, and long break lines. Object lines, text, terminators, and matrix grid lines are medium-width lines. Medium, wide, and extra-wide lines are all used as object lines, depending on the scale of the drawing and the relative importance of the parts of the object being described. Wide lines are used also for cutting plane lines and titles, special-emphasis object lines (for instance, lines that are used to profile an object), and drawing block borders. Extra-wide lines are commonly used for borders and large lettering, as special-emphasis object lines in large-scale drawings, as title block borders, and as sheet borders.

Intensity

The intensity of a line—its response to light—generally reflects the emphasis the drafter chooses to place on it. Critical lines are crisp and dense, regardless of width. Less critical lines—lines on the periphery of an object, for example—may be less intense and well defined; in fact, line intensity is used to express depth in a drawing: lines close to the viewer tend to be bolder (and broader), and those farther away are lighter and less dense.

Continuity

Varying the width and intensity of lines is one way to add to the alphabet; however, it is difficult to create a whole language from seven "letters" (five widths and two intensities). Consequently, lines are broken into various patterns, each of which has its own meaning. Dashed and dotted lines, as well as combinations of dashes and dots in an otherwise continuous line, add still more variation to the line alphabet. The following is a sampling of line conventions:

- Dashed lines, depending on the length of the dashes, describe hidden objects, objects that are not in the same plane as the subject of the drawing, or existing contours on a project site.
- Lines with a dot between long dashes, or a short dash between long dashes, represent centerlines.
- Lines with two short dashes between long dashes commonly represent property lines.
- Extra-wide lines composed of a short dash between medium-length dashes are *match lines*, lines that identify where one portion of the project joins an adjacent portion (where adequate space to show the entire project on a single page does not exist).

While conventions exist, there is some variation in how lines are used. Fortunately for the builder, most design professional offices include a legend of lines to help interpret project drawings. Figure 4.14 is collection of lines commonly used in construction drawings.

Good line work is like good speech: The words in a sentence are in the correct place, they are distinguishable from one another, and the syllables are properly enunciated. It behooves the construction professional to learn the various lines and employ them properly in the execution of drawings and sketches. Figure 4.15 is a comparison of good and not-so-good line work.

Symbols

Symbols are used extensively in construction drawings because they communicate pertinent information promptly. A variety exist in four categories:

- *Reference symbols.* As their name suggests, reference symbols refer the user to other drawings in the set that contain pertinent, often more detailed, information. Building and wall sections and elevation symbols (interior and exterior) are examples of reference symbols.
- *Material symbols.* Myriad construction materials can be easily and quickly associated with a system or assembly by using material symbols. Although used most frequently in sections (cuts through an element, component, assembly, or system), material symbols are used in plan views and elevations as well.
- *Object symbols.* Object symbols include abstract representations of elements or components used in construction—for example, the switches that con-

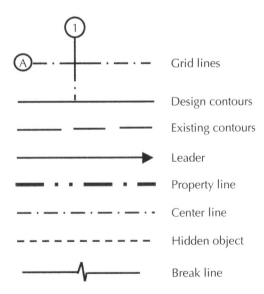

Figure 4.14 Common lines in construction drawings.

Figure 4.15 Good (left) and better (right) line work compared.

trol lighting in a room—as well as symbols with text that identify doors. Object symbols are normally not scale-specific; that is, they do not reflect the true scale of the object being described.

- *Template symbols.* Template symbols derive their name from the plastic templates that were used in manual drafting to draw bathroom and kitchen fixtures, cabinets, and many other components used in construction projects. They were made in different scales and were intended to facilitate accurate and consistent descriptions of those items in manual drafting. Computers have libraries of such items, as well as wall, door, and window assemblies to name a few, which serve the same purpose.

It is worth noting that symbols must be interpreted in context. The symbols for return and supply air, for example, are the same as the symbols for, respectively, blocking and continuous wood. The symbol for continuous wood is the same as the symbol used for fluorescent fixtures; certain fence symbols on civil drawings and the symbol for welded wire fabric are the same; and the datum point used in many drawings to denote elevations is the same as irrigation system remote control valves, points of connection on landscape drawings. Figure 4.16

is a very brief list of symbols that are used repeatedly in construction drawings.

■ Schedules

Schedules—matrices of various products in construction—are used to identify, sort, and organize objects and spaces in a construction project. The common object schedules include column and footing schedules; door, window, louver, and finish schedules; and equipment, fixture, and furniture schedules. Schedules are also used to identify spaces in a building, including rooms, entrance lobbies, balconies, terraces, elevators, escalators, and stairs. It goes without saying that consistency in numbering spaces is essential to clear communication. The Uniform Drawing System, CSI's contribution to the National CAD Standard, offers the following guidelines to numbering spaces and objects:

- The room that is closest to the most prominent access to the floor (the lobby, for instance) should be the first room of that floor. Working clockwise, the next-closest room becomes the second in the sequence of rooms on that floor.

Reference Symbols

Object Symbols (non-scale specific)

Material Symbols

Figure 4.16 Common symbols from the National CAD Standard.

- Stairs or escalators with the largest egress capacity should be the first in the sequence of stairs. Subsequent stairs are identified as they are encountered in a clockwise sweep of the building.
- Doors are identified by room number, followed by an alpha character (where more than one door exists) assigned to the door closest to the corridor access.
- Windows have their own numeric identifier.

What Constitutes Good Graphics?

The culmination of the design professional's efforts is primarily the drawings and specifications, although there are many other important documents in a construction contract. The principal user of the drawings is the contractor, so it follows that the designer should give the contractor's needs the highest priority when producing construction documents. Indeed, designers that fail to do so may add considerable cost and time to their clients' projects, a consequence that few owners are willing to tolerate. There are immediate costs—the quality of a set of documents is often evaluated by contractors and assigned a value prior to submitting a bid—and longer-term costs, which manifest themselves in the excessive project administration costs and dispute resolution, which may involve attorneys, expert witnesses, and the courts. It is worth noting that there are battle-hardened contractors who see considerable opportunity in poorly conceived and executed drawings and specifications, and pursue projects that exhibit those traits.

So what is it that contractors need? Construction documents are instructions as to what is to be built, and as such should be relevant, concise, unambiguous, complete, easily and quickly assimilated, and properly coordinated. The people who determine a project's cost and manage its construction typically have very little time, relatively speaking, to familiarize themselves with the contract drawings and specifications. It is not unusual for contractors to have from three to six weeks to absorb the information recorded on a hundred or more drawing sheets and a thousand pages of specifications—information that they are seeing for the very first time—then commit their entire worth to a construction plan and cost estimate for a stipulated sum contract (one in which the cost to the owner is more or less fixed). An architect, by contrast, may have lived with the same project for months, maybe years, and consequently possesses extensive unrecorded knowledge of numerous and varied details. Consequently, it is very difficult for design professionals to see their completed work from the contractor's viewpoint, that is, for the first time. Add to this the differences in professional orientation (design professionals tend to focus on the building as a product, while contractors focus on process) and it is not difficult to see

why conflicts arise. The most useful drawings to constructors will exhibit the following characteristics:

- Prompt access to pertinent information
- An effective mapping methodology (audit trail, or cross-referencing system)
- Consistency (compliance with drafting conventions, uniform location of subject matter, and proper coordination of the design consultants' work)
- Unambiguous, complete, accurate information that is clearly communicated and easily absorbed
- Thorough details that exhibit practical construction knowledge, and for more complicated details, axonometric or oblique projections
- Concise drawing notes that use generic terminology
- Respect for the separate but crucial role the specifications play
- Clear, concise, relevant specifications (when those specifications are recorded on the drawing set)

Mapping

Designers produce drawings in a deductive process, that is, from the general to the specific, starting with programming (discovering project parameters), then working through schematic design (in which assemblies and systems are selected), design development, and, finally, to construction drawings, in which the smallest details are worked out. Builders, particularly specialty contractors, use drawings as reference books, and do not necessarily familiarize themselves with the drawings in the same sequence as they are produced. They may devote enough time to identifying the designer's choices regarding basic project requirements (site requirements, foundation and framing systems, cladding, and mechanical and electrical systems), then will explore the documents in detail according to a hierarchy of costs from the most to the least expensive systems in a project. In a concrete tilt-up project, for example, the items of work having the greatest monetary impact are the sitework, concrete (slab-on-grade and walls), roof structure, and roof covering. Consequently, of immediate interest are the following:

- Soil characteristics and the extent of required grading
- Planned utilities and the location and elevation of existing utilities
- Foundation requirements, particularly the connection between walls and foundation
- Unusual wall features
- Connection requirements between the roof structure and walls, particularly where drag panels (shear panels) are concerned
- Column/beam connections
- Roof structure drainage and diaphragm nailing requirements
- Unusual roof covering requirements

The information listed is, for the most part, found in the details of a drawing set, so it stands to reason that identifying the path to and from details and their origin, or context, would be very helpful to the constructor. One very effective, simple method of directing the constructor to details and their origin is a referencing methodology known as the *split-bubble* referencing method, in which the upper half of a detail bubble lists the detail number, and the lower half, divided into quarter circles or simply by a short vertical line, lists the source page(s) on the left and the detail page on the right. This enables the constructor to use the drawings like an index in a reference book: a subject of interest can be found in the index, and the page(s) on which the topic is discussed are listed. This saves the builder from searching chapter by chapter, page by page to find the topic of interest. Figures 7.5–7.8, 8.6a–8.6i, and 11.5a–11.5f are good examples of this indexing method.

Consistency

Consistency in a drawing set contributes significantly to the rate at which the information in drawings is absorbed. Identifying the standards used (drafting, building code, and others) and complying with them is a basic requirement. Locating critical information in the same place from drawing set to drawing set is also essential. The Engineers' Joint Contract Documents Committee (EJCDC), the American Institute of Architects (AIA), and the Construction Specifications Institute (CSI) cooperate in the publication of a document that identifies the proper location of subject matter in construction contracts. Called Document 1910–16 by EJCDC, and A521 by the AIA, this 12-page document is a comprehensive checklist of the location of critical construction information.

Coordination of Design Consultants' Work

As noted in an earlier chapter, the role that the architect plays on the design side is the same as the role the builder plays on the construction side. Each has overall responsibility for the work and each contracts with specialists for portions of the work. The coordination of the work of specialists, then, falls on the shoulders of the design professional for design work, and the prime (or original) contractor for construction. Common areas of discrepancy were mentioned in Chapter 2; prudent professionals on both sides of a construction effort would note problem areas and mitigate the causes of them in future projects.

Unambiguous Information

The definition of "ambiguous" is "susceptible to multiple interpretations," and "the presence of two or more possible meanings, usually because of faulty expres-

sion."[2] Little needs to be said of the great value of limiting information to a single interpretation, particularly where construction is concerned; however, examples of design work (graphic and written) that was not reviewed in this light abound. Consistent use of terms in drawing notes, use of varying line weights and intensity, the consistent use of standard symbols, and the selection of the proper view(s) help reduce the likelihood of multiple interpretations of a drawing.

Details, More Details, and Constructability

Determining which details to include in a drawing set is no doubt a challenge. The project architect (for building projects) is forced to balance the firm's profitability against the graphic requirements of a project. That said, constructors rue the lack of adequate numbers of details in drawing sets, the difficulty in finding pertinent information in details without undue "chasing," and lack of field experience of many design professionals. It is worth mentioning that constructability is not limited to whether a component or assembly can be built; it is more an evaluation as to the effect on the construction process that a particular detail might have. Convincing owners to budget adequate fees for the design professional would help the situation tremendously, but construction knowledge and experience on the part of the individual who draws the details are essential. In the absence of knowledge, experience, or time, consultation with a builder early in the design process is prudent. Constructors often provide preconstruction services at very reasonable rates, if they charge for them at all. The trade-off for the constructor is a more profitable construction effort.

Concise Drawing Notes

Much of the design professional's work can be accomplished through pictures and symbols alone; however, notes accompanying drawings are inevitable. When notes are used, it is most helpful when they are placed in proximity to the element they describe, when generic terminology is used, and when the terminology between drawings and specifications is the same. Following is a comparison of generic and specific terms.

Generic *(desirable)*	Specific *(less desirable)*
CMU (concrete masonry unit)	Basalite block
Porous fill	Gravel
Gypsum wallboard	Sheetrock
Expansion bolt	Redhead
Framing clips	Simpson clips; A 35s

[2]*American Heritage Dictionary of the English Language*, Houghton Mifflin Company, 1978.

Plastic laminate	Formica
Acrylic sheet	Plexiglas
Headed stud anchors	Nelson studs
Cement-asbestos pipe	Transite pipe

Figure 4.17 shows the connection of steel vessels in an equipment room to a concrete slab. In detail B, the engineer calls for "wedge anchors," while in detail C, a Hilti Kwik Bolt II (a brand of expansion bolt) is noted. Construction dictionaries define wedge anchors as the

Figure 4.17 These two details list expansion anchors in lieu of wedge anchors in two different ways.

hardware devices that anchor post-tensioning tendons in post-tensioned concrete. "Expansion bolts" and "expansion anchors" are the more common term for the fasteners described here. It is worth noting that the graphic depiction of the anchors varies also. The builder is left to ponder what the engineer actually wants. Another example of this appears in 11.5i, note 8.

Graphics and Specifications

As noted in an earlier chapter, drawings are particularly effective for describing complex shapes and spatial relationships, whereas specifications perform the critical function of describing material quality, the quality of work, handling and storage requirements, preparation, and installation procedures—things that are difficult, if not impossible, to draw. Each document must be reviewed in light of the other; they are complicated, extensive, interdependent documents. In an environment of severe project deadlines and remarkable editing and reproduction technology, there is great appeal to solving the daunting task of producing project specifications by assembling them from preexisting work. While this is conceptually not a problem, the reality can be quite different. The consequence of taking a cut-and-paste approach without analyzing the relevance of the information is, not surprisingly, specifications that are not necessarily pertinent to the job at hand. The generic specifications that frequently appear on the drawings themselves in some disciplines are an example. For instance, the first sheets of structural drawings, dubbed "S-1" sheets by some engineers, may include standard wood framing details, details addressing penetrations of pipe through foundations, and material requirements for framing lumber, steel and concrete—details that frequently conflict with the specifications for the project described in the drawings. CSI's tenet, "say it once in the right place" is an appropriate and effective guideline.

■ Interpreting Working Drawings

Background

One of the more compelling but often tedious and frustrating tasks in construction is extracting the information one needs to build a project profitably, within the established schedule, budget, and quality using a two dimensional representation of the project. There is an enormous amount of information presented by design professionals on every page of the typical project (see Inventory of Information in a Residential Construction Drawing Set and "A Journey Through a Commercial Building Drawing Set" in the appendix), and it takes a substantial knowledge base and considerable time and effort to grasp even a simple building project in its entirety (especially if one

is preparing to do a detailed estimate). A working knowledge of business and the law; architecture and structural, mechanical, and electrical engineering; familiarity with numerous construction documents; and knowledge of hundreds of different materials, processes, and the operations required to integrate them into a project are all required to fully appreciate (and evaluate) the information in a drawing set. And the challenges are not limited to those inherent in the description method (using a wide variety of projection types, views of a component from several observation points, myriad line types, hundreds of different symbols in four categories, abbreviations, and words). It simply takes time to absorb it all. Knowing one is about to embark on an exercise in memorizing 10,000 elements of something would cause one to accept from the outset that a great deal of time, effort, determination, and repetition is required for a successful outcome.

Retaining Information Extracted from a Drawing Set Review

Digital modeling (so-called 3D modeling) will no doubt help builders understand the design professional's goals quickly, but there is considerable benefit, from the learning and retention standpoints, of struggling through a conventional set of drawings and trying to conceptualize what is to be built. To understand a component described in a drawing, one must first understand what the drawing depicts in each of at least three views. One is fortunate when one can see all views of an object side by side (or three facets simultaneously, as in isometric drawings), but parts of the object may be shown on different pages in the same discipline as well as other disciplines. Once the elements of a component are understood in each view, the correlation between views has to be resolved in the imagination until such time as the mind's eye has a clear picture of the component (rarely do drawings show just an element) in all three dimensions. This struggle is a valuable cognitive exercise that may not occur when the source document is a 3 D model, simply because the information takes less exercise to grasp.

Drawing Set Reviews

Probably more than anything else, understanding drawings involves a commitment to fully exploring a drawing set, carefully observing the information presented, determining its relevance, comparing it to the information in the specifications, identifying and resolving apparent discrepancies, answering as many "why" questions as one can muster, and taking good notes. Most of the information one needs is there. After all, it is the design professional's responsibility to adequately describe the project, and while omissions, errors, and discrepancies certainly occur, there is no intention to mislead or omit critical information. As with other professionals who conduct business, architects and engineers are doing their best to operate profitably, which means putting some sort of limit on the time they expend developing project documents. Additionally, the iterative nature of design—the discovery of the client's problems, an analysis of them, the development of solutions, the communication of those solutions to involved parties, and further problem clarification and solution development—creates a rhythm that never really ceases.

Conceiving and describing the space that will fulfill a client's needs is a profoundly complex effort that is sometimes fraught with doubt because it results in a permanent, very visible outcome that is conceived, explained, and initially reviewed entirely in the imagination. Renderings, though often realistic, are still just two-dimensional depictions of what someone's imagination indicates the project will be. Even a superb, accurate scale model, digital or otherwise, is a model—not the actual object. Detailed working models work well for resolving certain kinds of practical problems—two components occupying the same space, for example. But for depicting space (and frequently very large spaces), they are limited. What effect will the building have on the passer-by? What will a visitor's first impression of the business be, based on the design of the public entrance? Will the space reflect the spirit of the owner's business? Will it be conducive to a productive workforce? These and hundreds of other difficult questions arise and must be answered during the design process. Often enough, the result of the effort leaves something to be desired in the final product. In production housing development, this conceptualization challenge is met by constructing model homes (see "Identifying Expectations" side panel in Chapter 14).

Professional Responsibility

It should be pointed out here that reviewing a set of construction drawings and ignoring the varying professional obligations and proclivities of the participating architect, landscape architect, and structural, mechanical, and electrical engineer and other consultants makes one vulnerable to overlooking problem areas. For example, structural engineers hired as consultants for a commercial building project are charged with the task of supporting the architect's design, but they are sometimes so focused on their responsibilities (designing an economical structural system that will effectively resist the numerous forces imposed on it), they fail to see how their solutions affect the design. This applies to all the other sub consultants and it works in the other direction too: an architect should be aware of how a design and changes to it affect sub consultants' work. The list of potential clashes in a project is an extensive one. It should also be noted that the builder's own predispositions often play a role in how the information is interpreted.

Exploring the Drawings

What does "fully exploring a drawing set" mean, then? As noted elsewhere in this text, there are numerous sources of information on a single component, assembly, or system within the same discipline and drawing set (see for example, Figures 8.5c–8.5h) and it behooves the user of the drawings to review and understand all of the notes and references pertinent to a detail. A full component or system evaluation may involve reviewing tens of drawing sheets and numerous details within them. There may be several details within the structural drawings (see Figures 8.7b–8.7d and 10.4p and q) and on the architectural sheets pertaining to the same component, and it is not unusual to see a component depicted on sheets in numerous disciplines. For example, a retaining wall on a site might be shown on the grading plan, the architectural site plan, structural details, and the landscape drawings, and there might be several relevant details on each of those sheets. Determining the finish requirements of a room often requires a review of the architectural floor plan, building or wall sections in the architectural sheets, the reflected ceiling plan, finish schedule, structural floor plan, building and wall sections on the structural sheets, and structural detail sheets (not to mention the specifications). Much like police detective work and forensic analysis are reputed to be, every lead must be pursued until it bears fruit or proves to be fallow.

Generally the information in a drawing set proceeds from the general to the specific (refer back to Chapter 3), which is an effective way to introduce new information to a user; paint the big picture first, then home in on the details of the numerous relationships between elements, components, and assemblies. It makes sense to review drawings in a similar sequence; that is, starting with the site plan, floor plans, elevations and major building sections to gain an understanding of the project generally, then exploring other drawings as circumstances or tasks dictate. However one approaches the work, it makes sense to go about it methodically and leave evidence as to where one has been. Failing to do so increases the potential of missing critical information.

Getting Started

One indisputable reality of modern business is that time is a precious commodity; no one can afford to dawdle. One therefore must make the most of the time available to complete a task.

As with any other information gathering effort, interpreting construction drawings requires a clear purpose. What needs to be accomplished? What has to happen first? What information is required to accomplish the task? What information is available and what, if any, has to be produced? How does the information tie to existing information or knowledge? How will the information be used? What information should be recorded, who will use it, and how should it be presented? How much time is there to complete the work? Having these questions in mind aids in a productive review of drawings just like identifying the purpose of reading helps one to extract essential information from text. It also helps to get into the habit of leaving a good audit trail: it is very easy to get confused after looking at a drawing set for hours, particularly if the drawings are not well organized. During the review process one can annotate drawings, color code them, take notes in a separate document (say a binder that one can add to or delete information from) and mark the drawings up. Interruptions in a construction company office are common, so a trail helps in reorienting the user after being distracted. In short, a drawing set review should be carefully planned and contemporaneously recorded.

Some Caveats

There are a number of caveats to bear in mind in a drawing review:

Building in the Mind's Eye

1. Get a feel for the project by becoming familiar with the project type, size, location, site constraints, and major systems (e.g., underground utilities, framing, cladding, roof covering, HVAC). This may require reviewing many drawings briefly, starting with title pages and moving to civil and architectural site plans, followed by architectural floor plans, roof plan, exterior elevations, building sections then the foundation and floor plans on the structural sheets, and imagining what the space is like. Comparing the new information with known and familiar information is helpful.

2. It is safe to say that people live much of their lives without often reflecting on a number of mundane things. How high is a step? How big is a car? How tall is 10 feet? What is a room 150 feet by 150 feet with a 30 foot ceiling like? How long is a city block? Fortunately, these things are easy to explore with simple tools: knowledge of one's height, vertical reach, stride, arm span, waist height, span from thumb to little finger, common door sizes, masonry, and floor and countertop tiles for example, are all useful in exploring the built environment.

3. It is difficult to imagine the true magnitude of a project represented by a set of drawings that fits on a desk. What are the dimensions of the entire project? What is the size of the building footprint? A 500,000 square foot single story warehouse, for instance, covers 11⅛ acres, or about the same space 60 single-family detached homes in a medium density residential development do. A 500,000 square foot Class A high-rise office building in an urban area might rise to forty-five stories, including the parking.

4. Once the project begins to take shape, it helps to see it in the context of building challenges. How will I get people, large equipment, and materials to the site? Where will I store materials? Where will I locate the field office? What should the sequence of building be? How do I get people to the places the work has to be done? This and many other practical questions must be asked and answered during the drawing review process. There is a similarity between architects and constructors that is worth mentioning here: architects must imagine projects that do not yet exist and describe them accurately in two dimensions (even a 3D computer model is two dimensional). Constructors must learn what the design professional has described, then use just as acute an imagination to conceive of the myriad requirements that are never described in the design professional's documents: the impact of excavations; type and number of temporary structures such as shoring, falsework, staging, material hoists and personnel lifts, temporary roads; and the equipment and labor required to make it all work. Whereas the architect imagines the whole project, the builder imagines the whole project under construction.

5. It is critical, at least for builders, to recognize that understanding what is being described in construction documents is a key milestone, but only an early one in the quest to construct the project.

Practical Considerations

6. Make sure you understand the scale of the drawings under review—drawings that have been reduced to 50% of their original in the copying process are common.

7. Beware the unit type. Drawings using SI (metric) employ only two units of measurement, the millimeter and the meter, and consequently calculations involving components given in millimeters can be huge (a cubic meter is a **billion** cubic millimeters!).

8. Projects share characteristics, so what emerges, with time and experience, are the atypical requirements of each project.

9. Be patient—there is an enormous amount of information even in simple drawing sets and it takes time to absorb it all (see Appendixes C and D for an inventory of information in a simple drawings set).

Background

As noted above, once the goals set forth by the design professional are clear, that is, when one understands what is required by the documents, the task of building a project in the mind's eye (described repeatedly in this text as "cognitive construction," a process that often begins during the review) begins in earnest. In addition to understanding the project goals as described in the documents and the context of the project (the building on its site, the site in its immediate environment, the environment in the region), cognitive construction requires an acknowledgment of all of the "missing" realities that are the builder's responsibility. The estimator of a construction project, to be effective, must at least be aware of how hundreds of operations—mobilization; demobilization; excavation, grading, and paving; formwork installation and removal; concrete placement, finishing, and curing; structural steel erection, and myriad others—are executed. The entire forecast (the estimate) is based on something that has not happened yet, except in the estimator's imagination. Materials are easy enough to quantify when one has a good set of drawings to work from: what is not easy is determining labor costs which, owing to their volatility, hold the greatest threat to the profitability of general and specialty contractor alike. Most prime, or original contractors[3] subcontract substantial portions of a project to specialty contractors for a variety of reasons. The reasons include:

- "Purchasing" specialized equipment, personnel, expertise, and management experience which, when procured in other ways by the contractor, can be prohibitively expensive. Skilled specialty workers and those that manage their work tend to become very proficient at what they do, and it normally costs less, even with profit included, to hire a specialty contractor than it would if the prime contractor were to execute the work with its own forces.
- Mobilizing rapidly for a project.
- Sharing risk.
- Sharing a financial burden (in the typical building construction project, subcontractors are paid approximately 10 days after the prime contractor has been paid for the sub's work by the owner, so the prime contractor is relieved of the burden of carrying labor, equipment, and material costs for that particular specialty—the specialty contractor bears that burden).

Consequently, the depth into which a general building contractor's drawing review goes varies with the management program established for the project. Construction cost estimates are forecasts of the probable costs of a project, but there are different types with different purposes.[4] Among them are:

Detailed estimates In a detailed estimate, every element, the cost of the labor and equipment required to install it (**direct costs**), project overhead (management personnel and other **indirect costs,** including bonds and insurance), general office overhead (General and Administrative, or G&A expenses), and profit must be accurately determined. Both stipulated sum and unit cost estimates[5] fall into this category.

Approximate estimates Also referred to as budget, preliminary, order of magnitude, and conceptual estimates (among other names), these forecasts are based on limited design information (e.g., preliminary design drawings) to which unit such as cost/square foot, cost/cubic foot, or cost/system are assigned, or for projects in which just sketches or no graphic information of any kind exists. In the latter estimates, costs/bed (hospital), cost/unit (apartments or condominiums), cost/room (hotels), and cost/space (parking garages) are common units. Such estimates require an accurate historical database and keen judgment as to how and to what extent a proposed project differs from those that made up the database.

Data Gathering for an Estimate

There are many ways to collect and sort data for an estimate, and while similarities in approach exist (company management may mandate a particular approach), estimates vary considerably in how they are done. That variation can often be traced to the strategy developed by upper-level management to procure the project. Someone has to figure out how to approach the estimating task and provide the necessary structure to insure a successful outcome, as much as that is possible. What is the nature of the project? How much, if any, of the work will be performed with a company's own forces? How much will be subcontracted?

For in-house work, the operations to be monitored, reported on, and accounted for, and the detail the tasks get broken down into need to be determined. How will productivity be measured? It is possible, for example,

---

[3] A synonym for prime contractor used in the Business and Professions Code of the State of California. *Black's Law Dictionary*, 6th edition, Henry Campbell Black, West Publishing Co., 1990.

[4] Definitions for different estimate types vary widely, but the essential differences amount to the level of accuracy required, from counting every nut, bolt, and washer to approximating the cost of a project based on whole systems that are assumed rather than specified by the design professional.

[5] Stipulated sum agreements, also known as lump sum and fixed price agreements, are popular for projects that involve numerous work items and for which the quantities are readily determinable. This agreement type establishes a fixed value for the work, and language in the agreement and conditions is focused on containing costs (refer back to Chapter 2 for a description of these two contract components). In California stipulated sum contracts, while often problematic, are still required on most public works building projects. Although any project can be based on unit prices, that is, one can determine a cost per unit of work, a unit price agreement here refers to projects involving fewer work items, where the work is repetitive, and for which quantities are difficult to determine. The contractor determines a value for units of work (lineal feet of pile, tons of paving, cubic yards of earth) using the owner's quantities, the units are counted as they are produced, and payment is made accordingly. The value for any work over or under the owner's estimate of quantities is normally liquidated prior to work commencing. State transportation departments, for example, commonly employ this agreement type for highway construction projects.

to reduce labor-paced tasks—the work of people—into rudimentary movements for the purpose of determining and managing productivity (and thereby arriving at a cost). Frederick Winslow Taylor, the founder of Scientific Management, did just that with his time and motion studies in the early twentieth century. But is that level of detail necessary in the project at hand?

Knowing the answer to some of these questions provides clues as to how to accumulate and organize the material quantities for a take-off. For example, in a cast-in-place post-tensioned parking garage project, the beam forms and deck (which become littered with paper, sawdust, tie-wire remnants, fasteners, tendon sheathing, and other debris during the reinforcing steel installation) must be cleaned prior to concrete placement. The quantity of space to be cleaned must be determined, so that a cost can be assigned to that operation. What is the appropriate unit for the labor cost? Square feet of slab, with beams calculated as slabs? Separate the slabs and beams and develop two units reflecting the differences in how the work is done? Use the volume of concrete to be placed, or the lineal feet of beam form? Toss it in with general labor costs? In a million-square-foot multistory parking garage (some 3,000 spaces), or in a 40-story high-rise, this last solution does not appear to be prudent. For certain operations—furnishing, placing, and finishing concrete for instance—two different units are used, the cubic yard, which is used to purchase concrete (the "buy" unit for this material) and measure the productivity of a placing crew, and square footage, the unit often used to determine cost and measure productivity in a finishing operation or to cure concrete. There certainly exist circumstances in which yet another productivity unit might be more useful, for example, a lineal footage of slab rather than area of slab unit. After all, it is much easier for a foreman to pace off the length of something than it is to measure and calculate area, and it is much more effective to point to a specific place on a set of forms as the daily goal than it is to tell the crew, "We need to place and finish 11,400 square feet of slab today." For work with one's own forces, it all boils down to what management chooses to implement, monitor, and account for. Like so many other decisions in construction, these things are matters of judgment, but a drawing review and the subsequent organization of information, from take-off to complete estimate, must be done in the context of the larger picture of managing the construction process.

toward a single drawing standard are being made, there will be architects and owners who choose not to adopt the standard format for any number of reasons. Builders will be most effective when they view the design professional as a partner and review contract requirements with flexibility, imagination, and an effort to understand the design professional's goals.

Construction drawings are produced for the builder's use by the design professional (who represents the owner), and as such should be clearly presented, unambiguous, relevant, and complete instructions as to what should be built. To be most useful they should be properly organized and coordinated, and the information should be readily available without undue effort. If these fundamental traits are exhibited in a drawing set, the owner's investment in the construction effort, from the standpoint of both time and money, will be maximized. Well-executed and well-cordinated drawings contribute significantly to an efficient construction program.

CHAPTER 4 EXERCISES

1. Transcribe the following architectural scales in the manner preferred in civil drawings, where the scale is expressed in terms of whole inches and feet: $1/8" = 1'-0"$; $1/4" = 1'-0"$; $3/4" = 1'-0"$; $11/2" = 1'-0"$.

2. A building 13 000 long on a set of drawings is how long on the drawing sheet at 1:20 scale? What would the dimensions of the drawing of a building 100 meters long by 60 meters wide be at 1:100 scale?

3. What type of drawing uses one scale for the horizontal and another scale for the vertical?

4. List the elements of projections.

5. List and describe the principal projection types and their benefits and drawbacks.

6. List the characteristics of good dimensioning.

7. What are axonometric drawings, and in what situations are they appropriate?

8. How are the scales used in construction drawings distinguishable from one another?

9. What is the distinguishing characteristic of third-angle projection?

10. What is the purpose of drawing sections?

11. Fundamentally, what is good drawing?

▨ Summary

Communication in any professional discipline is a two-way street. Translating construction documents imposes demands on the constructor as well. Though efforts

5 Sketching

Key Terms

Arc
Chord
Destination point
Ellipse
Major and minor axes
Rhombus

Key Concepts

- Effective sketching involves imagination, good observation skills, knowledge and use of drafting conventions, respect for the hierarchy of lines, a sense of proportion, and speed.
- Sketches are frequently approximations, and are used to convey ideas quickly; strict adherence to technical drawing standards is not required.

Objectives

- Produce vertical, horizontal, inclined, and curved lines accurately and quickly using freehand drawing techniques.
- Find the center of a square or rectangle using diagonal lines.
- Draw circles, arcs, and ellipses using freehand drawing techniques.
- Divide lines into equal pieces using a simple lined instrument such as a ruler or scale.
- Draw perpendicular lines on long line segments.
- Convert pictorial drawings of objects into multiview drawings, and from multiview to pictorial drawings.
- Format and compose properly proportioned sketches of construction components and assemblies.

Construction Graphics: A Practical Guide to Interpreting Working Drawings,
Third Edition. Keith Bisharat.
© 2025 John Wiley & Sons Inc. Published 2025 by John Wiley & Sons Inc.

Introduction

The key to formulating and presenting good questions for the design professional and to defining and communicating problems and solutions for oneself, subcontractors, and the skilled labor force that actually does the work is good sketching skill. Preliminary sketches—quickly produced and reproduced in a "thinking with a pencil" process—also help to organize one's thoughts and reduce errors on formal drawings, when such drawings are required.

Effective sketching requires clear mental images developed through the eyes or the mind's eye. Thoughtful and active observation, imagination, and plenty of exercise using both are therefore essential to developing effective sketching ability.

It also helps to remind oneself that the reasons for sketching vary—one may want simply to approximate reality, to suggest an idea, or to produce a preliminary drawing. Properly composed and executed, sketches are every bit as effective as drawings produced with drafting instruments or a computer.

Drawing Instruments and Media

Sketches can be (and frequently are) produced in the field on pieces of lumber, scrap gyp board, or cardboard using a carpenter's pencil or crayon. Rudimentary in nature, these sketches focus on very basic information and may consist of only a few simple lines. Books and articles featuring noteworthy building projects frequently include the architect's original conceptual sketch, drawn on a napkin or the back of an envelope. For more formal sketches, ordinary papers, such as copy paper or paper bound in tablets, are perfectly adequate. Tablets of grid paper in 4, 5, 8, and 10 squares per inch, corresponding to architect's and engineer's scales, and paper created specifically for isometric drawing, are widely available. Sketchbooks (hard bound blank paper) and daily diaries, which many project managers and superintendents maintain, can also be used. The only caveat where sketching media are concerned is that the surface must have sufficient texture to record a decent line; slick paper does not produce satisfactory sketches.

> "Learning to draw is really learning to see—to see correctly—and that means a good deal more than merely looking with the eye."
>
> —Kimon Nicolaides, *The Natural Way to Draw* (Houghton Mifflin, 1969)

By far the most versatile sketching instrument is an ordinary number 2 pencil. Lightly held and applied, this pencil is adequate for layout, and when applied with varying inclination and pressure and with various point thicknesses, it can produce the whole range of lines commonly used in technical drawings—extension, dimension, leader, object, section, and border lines. Pencil lead holders—for a time the principal drawing instrument in the design professional's office—can be used the same way. Both of these instruments require sharpeners, which come in a variety of designs, motorized and not. Mechanical pencils, being of a single lead diameter, are designed never to need sharpening; however, therein lies their limitation: a single line width.

Although the plastic and pink pearl erasers are effective on most papers, erasers are frequently ignored altogether in sketching; many people simply opt for drawing over lines. Proper planning and resisting the temptation to draw prematurely frequently obviates the need for an eraser.

Manner and Style of the Presentation

As in most technical drawing, identifying the user of the drawing is an essential preliminary step. Sketches produced for one's own use—just like notes taken at a lecture—need not comply with standards of any kind; they merely need to be understandable to the drafter. In contrast, sketches produced for skilled workers, management personnel, and design professionals should reflect all of the desirable characteristics of good drawing: diligent, consistent use of drafting conventions; line use that reflects the hierarchy of thicknesses and weights; clarity in the presentation and view(s); accuracy; and the inclusion of necessary information only. The goal is to present unambiguous information that is easily and promptly absorbed by the user.

Views

Any of the standard views—plans, elevations, sections, or pictorial views such as isometric drawings—are acceptable for sketches. As noted in a previous chapter, multiview drawings are easy to produce, but they require the viewer to exercise the constructive imagination. Isometric or oblique drawings, being closer to the reality that the mind and eyes conceive, require less effort to understand but require a bit more drawing skill and discipline.

Drawing Lines

The majority of lines produced in construction drawings are straight lines of varying lengths, types, and thicknesses. It is well worth the effort to learn how to quickly and accurately produce straight lines, which requires little more than some discipline and a fair amount of practice.

Horizontal, Vertical, and Parallel Lines

There are as many methods of producing straight lines by hand as there are hands to draw them, and ultimately you will have to develop the technique that works best for you. However, abiding by several maxims is helpful:

- Your hand position should be relaxed and comfortable, as shown in Figures 5.1a and b).
- Horizontal lines should be drawn from left to right if you are right-handed; left-handed drafters may feel more confident drawing horizontal lines from right to left (see Figure 5.2).
- Vertical lines should be drawn from top to bottom (Figure 5.2).
- Inclined lines can be drawn from top to bottom or bottom to top, whichever direction gives you the greatest confidence (Figure 5.2).

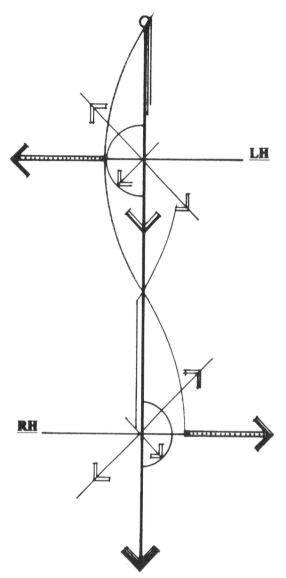

Figure 5.2 Draw lines in a natural direction.

Figure 5.1a Hand position – drawing lines. (Photo by the author.)

Figure 5.1b Hand position – lettering. (Photo by the author.)

- The drawing medium can usually be moved so that inclined lines can be made either horizontal or vertical (see Figure 5.3).
- Most lines begin as layout lines (very lightly drawn, thin lines, also called *construction lines*).
- Lines drawn to destination points result in straighter lines overall. Your eyes should be focused on the destination as the line is being drawn (see Figure 5.4a).
- Long straight lines drawn freehand toward the center of the paper tend to curve, hence it is best to conceive of and execute a single long line using a series of shorter line segments as illustrated in Figure 5.4b.
- Long straight lines can be drawn quickly and effectively using visual guides (for example, the edge of the paper) or by using your finger as a guide, as shown in Figures 5.4c and d).

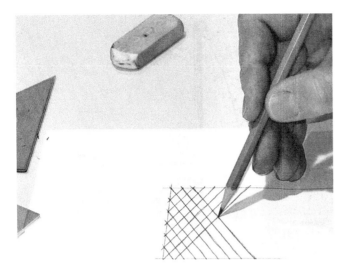

Figure 5.3 Orient the medium to produce consistent lines. (Photo by the author.)

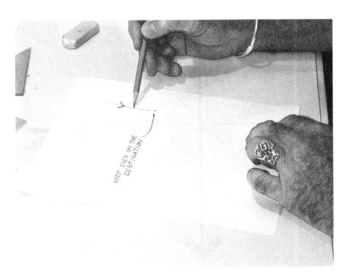

Figure 5.4a Draw to a destination. (Photo by the author.)

Figure 5.4b Draw long lines using segments.

Circles, Arcs, and Ellipses

There are several handy methods for drawing circles, arcs, and ellipses; but constructing any of these shapes requires that their centers be located first. The simplest way to to do that is to draw an encompassing square, rectangle, or rhombus (equilateral parallelogram), then draw diagonals from corner to corner. The intersection of the diagonals is the center of the shape. Figure 5.5

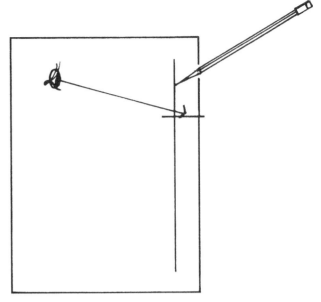

Figure 5.4c Maintaining a visual image of the margin ahead of the pencil results in reasonably straight, parallel lines.

Figure 5.4d Using your finger as a guide.

demonstrates several techniques for drawing an encompassing shape, locating centers, and identifying critical radii.

Circles and Arcs

The accuracy and circle size are the principal determinants for which method of drawing circles to use (see Figures 5.6a–d and 5.7a–d). The simplest, most efficient method for very small circles is simply to print an "O." Larger circles can be accurately drawn using a freehand arm movement:

1. Determine the proper diameter and move your arm and hand several times in a circular shape with the pencil lead held above the paper.

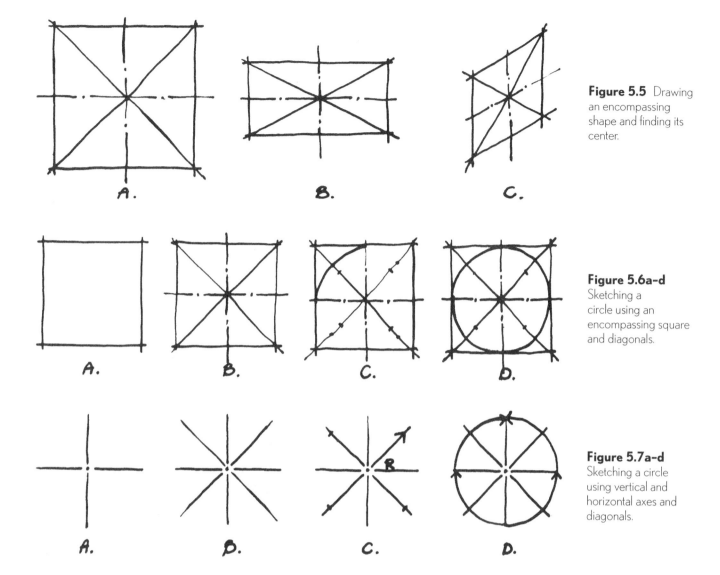

Figure 5.5 Drawing an encompassing shape and finding its center.

A.   B.   C.

Figure 5.6a–d Sketching a circle using an encompassing square and diagonals.

A.   B.   C.   D.

Figure 5.7a–d Sketching a circle using vertical and horizontal axes and diagonals.

A.   B.   C.   D.

2. When the size and shape of the circle are correct, lower the pencil onto the paper and lightly draw several circles. Allowing the pinky finger to glide on the surface of the paper may help to stabilize the strokes.

3. Select the best of the circles and darken it.

The freehand method works well when the diameter of the circle is between one and four inches. Another useful method for drawing circles is to use an encompassing square with diagonals. The edges of the square define the tangents of the circle. Using vertical and horizontal axes and diagonals without an encompassing square is another effective method, but without an encompassing shape, it is more difficult to envision the circle before it is drawn. Figures 5.6 and 5.7 demonstrate two techniques for drawing circles.

Larger circles can be produced quickly and accurately using the hand as a compass, or by using two pencils in the hand, one at the center of the circle and the

other at the circle edge (shown in Figures 5.8a and b, respectively).

The former method is effective with 2" to 8" diameter circles; the latter is effective with circles varying in diameter from 1" to 6". Of the two hand compass methods, the two-pencil method is consistently more accurate.

A variety of simple tools can be used as circle-drawing aids, including pieces of string, paper clips, and thick pieces of paper. Figure 5.9 demonstrates circle construction using a trammel.

Since arcs are parts of circles, the techniques for drawing them are the same as those used to sketch circles (see Figure 5.10).

Ellipses

When round objects or voids are shown in axonometric drawings, they become ellipses, which consist of two pairs of arcs on major and minor axes. There are several useful techniques for sketching ellipses.

Figure 5.8a Sketching a circle using the hand as a compass. (Photo by the author.)

Figure 5.8b Sketching a circle using two pencils. (Photo by the author.)

Figure 5.9 Sketching a circle using a trammel. (Photo by the author.)

The freehand arm technique for drawing ellipses is the same as for drawing circles. The arm and hand move several times in an elliptical shape with the pencil lead hovering above the paper. When the size and shape of the ellipse are correct, the pencil is lowered lightly onto the paper, and several revolutions are drawn on the sheet. The drafter then selects the best and darkens the line (see Figure 5.11).

When an ellipse is required in an axonometric drawing, the easiest method is to draw an encompassing rhombus with its sides parallel to the main axes. The major axis of the ellipse is always perpendicular to the centerline of the ellipse, and the minor axis is perpendicular to the major axis.

1. Draw a rhombus that encompasses the ellipse, using the diameter of the circle as the guiding dimension (Figure 5.12a).
2. Sketch diagonals from corner to corner to find the center of the ellipse (Figure 5.12b). These lines form the major and minor axes of the ellipse.
3. Draw lines from the midpoints of adjacent sides of the parallelogram through the center to the midpoint of the opposite side (Figure 5.12c).
4. Locate the tangent of the arcs on the diagonals (they are about 71 percent, or a fraction over two-thirds of the length of the diagonal, measuring from the center of the rhombus), and lightly sketch each from midpoint to midpoint on the sides of the rhombus, completing first one corner, then its opposite (Figure 5.12d).
5. Once satisfied with the result, "burn in" (darken) the ellipse.

The four-center method for drawing ellipses also requires the construction of an encompassing rhombus, whether the drawing is an isometric, dimetric, or trimetric drawing.

1. Draw a perpendicular line from the midpoint of each of the sides of the rhombus until it intersects with the perpendicular lines drawn from adjacent sides. The four intersections form the centers of four circles with two different radii (Figures 5.12e–h).
2. Draw quarter-circle arcs from the centers to form the ellipse (the center points of both of the longer radii are on the minor axis; the center points of the shorter radii are on the major axis).

If the receding axis of the rhombus is 45 degrees above horizontal, two of the centers will fall outside of the rhombus; in an isometric drawing, with receding axes 30 degrees above horizontal, two of the centers will occur at corners of the rhombus (at both ends of the minor axis); in drawings with a receding axis at 15

Figure 5.10 Sketching arcs.

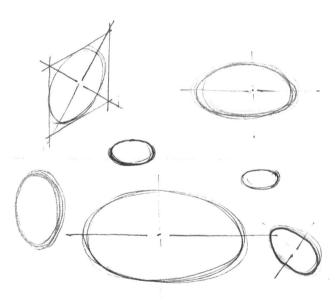

Figure 5.11 Drawing ellipses with free arm movement.

degrees, all four centers fall within the encompassing rhombus (see Figure 5.13).

Another method that works for circles and ellipses both is to draw an encompassing square or parallelogram, then divide its sides into four equal spaces and connect the first and third segments on each side with the corners on the opposite side. The result is a series of points of tangency that are then connected with arcs to form the complete circle or ellipse. Figure 5.14 demonstrates this technique.

■ Other Useful Geometric Construction Techniques

Dividing a Line into Equal Parts Using a Ruler or Scale

Being able to divide a line into equal or proportional parts is a very useful skill, for example, when interpolating points of elevation on a survey grid or contour map. Although computer programs that transform survey grids into topographic maps are now common, it still helps to be able to perform this task manually. A ruler or scale of any type, or any instrument with consistent units, perhaps a triangle or protractor, and a pencil are helpful; and if one can tolerate some imprecision, the work can also be performed with just a pencil and paper.

There are three good approaches for dividing lines graphically (Figure 5.15a–c). The first method is effective because vertical lines are easier to "eyeball" (judge with the naked eye) than lines at other angles.

1. Draw a perpendicular line through one end point of the line being subdivided.

2. On a ruler or scale, select the number of equal or proportional units to divide the line into. (Note: For

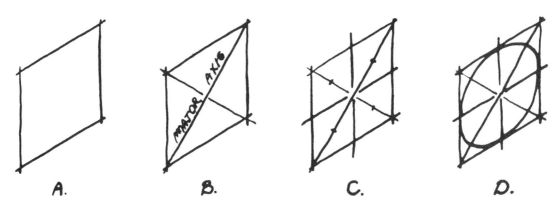

Figure 5.12a–d Sketching an ellipse using major and minor axes.

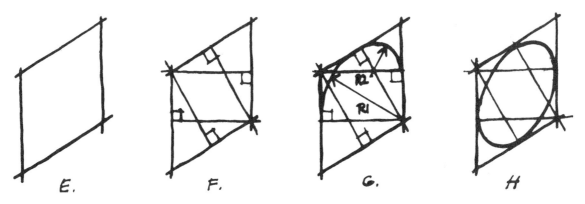

Figure 5.12e–h Drawing an ellipse using the four-center method.

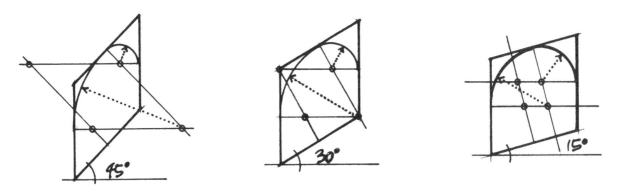

Figure 5.13 The locations of arc centers in axonometric drawings vary according to the angles of the axes.

this method to work, the sum of the units on the ruler or scale must be greater than the length of the line. For example, a 7½"-long line is easily subdivided into six equal spaces by using a standard ruler and six 2"-long units.

3. Place one end of the ruler (the zero point) on one end of the line.

4. Rotate the ruler up or down until the 12" mark (using the preceding example) intersects the vertical line drawn through the other end.

5. Mark the proper intervals, in this case 2", 4", 6", 8", and 10", on the angled line.

6. Draw vertical lines from each mark to the line being subdivided.

The intersections of the vertical lines with the line being subdivided identify the ends of each equal segment (Figure 5.15a).

The second method is similar to the first, except that no vertical line is drawn through one end, and it is not necessary to use a ruler or scale that is longer than the line being subdivided (Figure 5.15b):

1. Place the zero point of the ruler or scale on one end of the line.

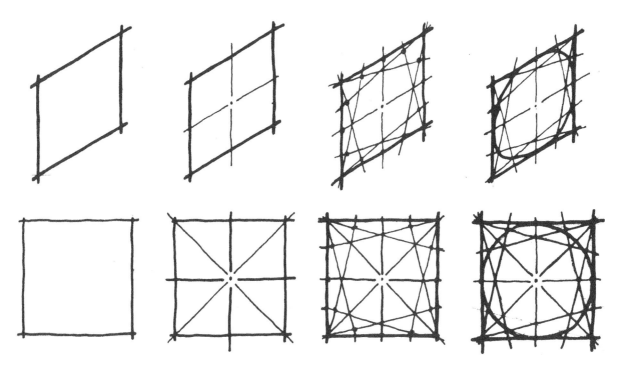

Figure 5.14 Drawing ellipses and circles using 12 points of tangency.

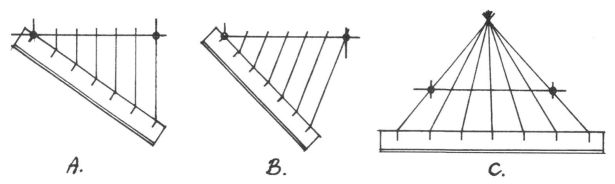

Figure 5.15a–c Subdividing a line graphically—three methods.

2. Rotate the ruler or scale up or down at any angle.

3. Draw a line from the ending unit on the ruler or scale through the other end of the line being subdivided. The ending unit could be the 6" or 12" mark on the ruler (or some other unit that is a multiple of six), assuming one desires six equal pieces.

4. Draw lines through the intermediate units parallel to the first line drawn. To produce six equal spaces, for example, it would be necessary to draw parallel lines connecting the 5", 4", 3", 2", and 1", or 10", 8", 6", 4", and 2" marks on the ruler to the line being subdivided.

The result will be six equal spaces on the subdivided line.

The third method involves drawing a line parallel to and longer or shorter than the line being subdivided (Figure 5.15c). In the example described next, the second line is longer.

1. Draw a line parallel to and longer than the line being subdivided, either above or below it. The midpoints of the lines need not line up vertically.

2. Draw lines from the ends of the longer line through the ends of the shorter line until they intersect.

3. From that intersection, draw lines through the subdivided line to the desired unit on a ruler next to the longer line.

The result will be equal or proportional spaces on the line being subdivided.

Drawing Perpendicular Lines in Large-Scale Sketches

Drawing radii from the ends of a line performs two useful functions: It identifies the midpoint of the line and it provides the targets to accurately and quickly draw a line

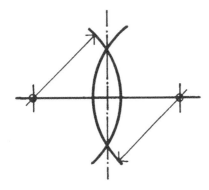

Figure 5.16 Finding line midpoints and drawing perpendicular lines using arcs.

perpendicular to the first (see Figure 5.16). This is especially helpful on large-scale sketches, in which finding the midpoint of a longer line is more difficult. The technique is simple:

1. Using a string, paper clip, or trammel,[1] draw two semicircles using the end points of the line as the radii. The length of the radii must be equal, and greater than half the length of the line.
2. Draw enough of a circle that the ends of the arc cross back over the estimated midpoint of the line.
3. Draw a line connecting the intersections of the arcs.
4. The resulting vertical line will be perpendicular to the original line and will pass through its midpoint.

Translating from Pictorial to Multiview Drawings and Back

Translating pictorial drawings into multiview projections and back develops a good imagination, which is the single most important tool required to produce and interpret graphic expressions. Despite significant improvements in computer technology, design professionals persist in describing construction systems, components, and elements using multiview projections (contiguous two-dimensional views of an object). Until that changes, being able to see an object under construction and as a finished product, based on a two-dimensional depiction of it, will continue to be a critical skill.

Translating Third-Angle Multiview Drawings to Isometric Drawings: Planning, Layout, and Execution

Multiview projections are by far the most common drawing techniques for construction projects. Translating

these views into pictorial drawings to better explain them is a common requirement. Here is one approach:

1. Identify the overall dimensions of the object.
2. Establish the axes of the isometric drawing (one vertical, one on each side of the vertical at 30 degrees above horizontal).
3. Using a hard lead, or a sketching pencil with very light touch, create the box in which all the parts of the object will fit and properly orient it on the drawing medium.
4. Step off the dimensions of plane breaks along or parallel to the axes.
5. All dimensions must be taken parallel to the main edges of the enclosing block, and can begin at any corner.
6. Lines that are not parallel to the principal axes cannot be easily measured, therefore locate the ends of the angled lines at their intersections with lines that are parallel with the axes.
7. Once the ends are located, connect them.
8. Complete the drawing using light lines.
9. Be sure to check your work. Does it make sense?
10. Once satisfied that the drawing is accurate, "burn in" the object lines.
11. Apply whichever other features—shading, cross-hatching, or symbols—next, after first determining that they are necessary.
12. Draw any enclosing lines—profiling or border lines — last.

Figure 5.17 demonstrates this sequence of steps.

Translating Isometric Drawings to Multiview Drawings: Planning, Layout, and Execution

Not surprisingly, translating isometric drawings to third-angle multiview drawings is similar to the process just described, except that the outcome is three drawings, rather than one.[2]

1. Identify the overall dimensions of the object.
2. Determine the overall dimensions of each critical face (top, front, and right side), measuring along the principal axes of the isometric drawing.

---

[1]Hand circle drawing techniques are less accurate in this application due largely to the existence of two circle centers (one at each end of the line). It is difficult to maintain the exact same radius when the hand has to move from one end of the line to the other.

[2]At least three views of an object—the top, front, and one side—are required to describe it effectively. In the United States, the preferred technical drawing method is the third-angle multiview, which shows the object from the top, front, and right side, *with the plane of projection between the viewer and the object*. Construction drawings are conceptually the same; however, more than three views of a component or element are frequently required.

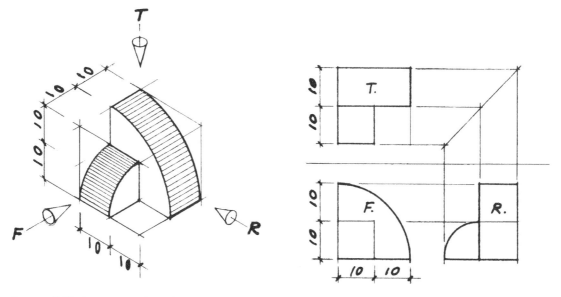

Figure 5.17 Steps to translating from a multiview to an isometric drawing.

3. Arrange the top, front, and right side views in the proper format (top view in upper left quadrant of the sheet, front in lower left quadrant, and the right side in the lower right quadrant), allowing for the object itself and border spaces (around the object as well as around the sheet).

4. Using a hard lead, or a sketching pencil with very light touch, create squares or rectangles that will encompass all the elements of the object that are featured in each of the faces (top, front, or right side) and properly orient them on the sheet.

5. Step off the dimensions of plane breaks along or parallel to the axes and transfer the dimensions to the proper view on the multiview drawing.

6. All measurements must be parallel to the main edges of the enclosing block, and can begin at any corner.

7. Lines that are not parallel to the principal axes cannot be easily measured, therefore locate the ends of angled lines at their intersections with lines that are parallel with the axes.

8. Once the ends are located, connect them.

9. Complete the drawing using light lines.

10. Be sure to check your work. Does it make sense?

11. Once satisfied that the drawing is correct, "burn-in" the object lines.

12. Apply other features—symbols, shading, crosshatching—after first determining that they are necessary.

13. Draw any enclosing lines—border lines, profiling—last.

Figure 5.18 demonstrates these steps.

Figure 5.18 Steps to translating from an isometric to a multiview drawing.

Formatting and Composing Properly Proportioned Sketches

Reducing an object to its fundamental shape facilitates the production of a quick and accurate sketch. Here is one approach to sketching objects freehand:

1. Determine the view(s) (plan, section, elevation, isometric, diametric, oblique, or perspective).

2. Identify the basic geometry and proportions of the object and outline it using simple forms (most objects, particularly in construction, can be fit into squares, rectangles, triangles, and cubes).[3]

3. Determine whether you will add to or subtract from the forms (enclose the object in a transparent box and carve it out of the box, or assemble a series of basic shapes).

4. Using a light touch, outline the object within the basic form chosen.

5. Evaluate the accuracy of the outlines. Do all the parts make sense?

6. Articulate critical lines using a hierarchy of line thicknesses and intensity (border and profile lines are thicker and more intense; object lines are of medium thickness and intensity; leaders, dimension, and extension lines are lighter still; and surface lines— lines that describe the texture or material of a surface — if used, are the lightest).

7. Add minor details, such as shadows and shading, but only if these details improve the readability of the object.

Figure 5.19 demonstrates the two approaches.

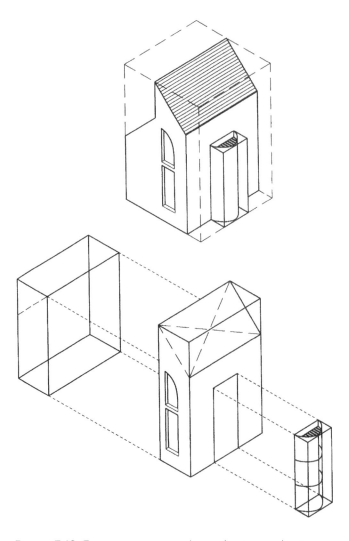

Figure 5.19 Two common approaches to drawing an object are carving from a large piece and assembling basic shapes. (Drawing by Tom Krejci, after Keith Bisharat.)

Summary

Sketching is a widely applicable skill in the construction and design professions. It helps all the participants in the construction process to explore and understand the problems that are endemic to the construction process, and it is essential to the development and communication of the solutions to those problems. Graphics-oriented participants, primarily design professionals, as well as tradespeople and people for whom English is a second language respond well to problems and solutions that are graphically displayed. Sketches, often approximations of reality or ideas, can be used to quickly convey ideas or to outline a more detailed drawing. An understanding of basic sketching techniques, conformance with graphic conventions, attention to proportion, and a little practice will result in more effective communication skills. In an industry in which the most frequently cited skill is the ability to communicate—to listen, talk, write, and draw well—sketching is indispensable.

CHAPTER 5 EXERCISES

1. Using a blank 8½" × 11" sketchbook, draw a borderline ³⁄₈" (10 mm) in from each edge of the paper. Then divide the area within the border into 32 equal spaces using the diagonal technique for finding the center of the rectangle(s). Within each space, practice drawing horizontal, vertical, and inclined lines.

2. Repeat the preceding exercise, but draw arcs and circles within 16 of the spaces. In the remaining 16 spaces, draw circles and arcs of a variety of sizes.

---

[3]Proportions are often easy to determine simply by comparing parts of the object to one another and estimating the ratios of components, or by knowing the dimensions of the object.

3. Using the format just described, draw rectangles within the spaces in the first and third columns, each with a 3/16" (5 mm) space around it. In the second and fourth columns, neatly list each of the 16 most common material symbols in a drawing set that was actually used in a project, and re-create the symbol in the space to the left of the list.

4. Select several details, reproduce them with all equal line weights, evaluate them, and then redraw them to reflect the hierarchy of lines.

5. Select a detail from a construction drawing set and sketch it from four different positions in isometric (above left, above right, below left, below right).

6 Site Construction

Key Terms

Best management practices (BMP's)
Contour interval
Contour line
Crib wall
Drop inlet
Gabions
Geocomposite
Geogrids
Geomembrane
Geonet
Geotechnical engineer
Geotechnical report
Geotextile
Hachure
Index contour
Intermediate contour
Lifts
Thrust block
Wattle

Key Concepts

- Site construction is a source of considerable controversy in construction, in large part due to the uncertainty of subsurface conditions (especially soil permeability and groundwater), the complexity of some geologic environments, variability of soils on the same site, and the difficulty of determining the costs of the work.
- The purpose of subsurface site investigations is to determine with reasonable accuracy what the engineering characteristics of the earth underlying a site are, so that a reliable plan for preparing it for its intended role can be developed. A certain amount of uncertainty is inherent in this work, partly because geotechnical engineers can only forecast soil behavior.
- Soils and rock are critical construction materials that one is, for the most part, forced to work with. One cannot specify the appropriate soil—as in the case of concrete or structural steel—with fresh, new ingredients of a particular character.

Objectives

- Identify the projection types, lines, and symbols commonly used in grading, utility, Storm Water Pollution Prevention Plans, and landscape drawings.
- Explain the shortcomings of depicting the earth's physical features in two dimensions.
- Use graphic and mathematical interpolation techniques to find various elevations between two known points.
- Reproduce drawings of some of the earth's features.
- Produce and interpret site sections.
- Manually perform earth quantity takeoffs using grid averaging and the average end-area methods.

Construction Graphics: A Practical Guide to Interpreting Working Drawings, Third Edition. Keith Bisharat.
© 2025 John Wiley & Sons Inc. Published 2025 by John Wiley & Sons Inc.

▪ Site Construction: Preparation and Improvements

Until humankind begins to inhabit space as a solution to housing and other pressing needs, its projects will be attached to the earth. Consequently, geotechnical and civil engineers will continue to explore the existing soil conditions at proposed construction sites and make recommendations as to how the earth should be modified to accommodate new projects.

It is the geotechnical engineer's responsibility to examine the earth (on land and under bodies of water) and develop the guidelines for preparing soil to perform an identified role, say, to support a building. Removing, remediating, relocating, treating, and compacting the earth, and groundwater issues are all normally addressed in geotechnical reports for construction sites. Logs of soil test borings, their locations, and summaries of findings may appear on the drawings themselves, for example in highway and bridge construction drawings.

It is generally the civil engineer's responsibility to survey a site, verify property lines, locate existing improvements, and determine what its final condition and appearance will be when the project is complete. Although the architect (and sometimes the landscape architect) normally designs and describes the surface site improvements that are required on virtually every building construction project of any size (parking areas, property fences, and site concrete, for instance), the civil engineer must incorporate these improvements in the grading and paving drawings. The engineer's directions are recorded in sitework specifications and grading and paving plans. Civil engineers are also responsible for designing underground utility systems (sewer, storm drain, domestic water supply, and fire service) on the building site. The power companies in whose jurisdiction the project is being constructed may design and install the gas supply and electrical underground systems to the planned building structures, or they may be designed and described by a mechanical engineer (for gas supply) or an electrical engineer (for electrical power) and installed by an underground utility (gas lines) or electrical subcontractor (electrical supply).

Increasingly, civil engineers are developing and producing the storm water pollution prevention plans (SWPPS) that are required of project owners by the Federal Clean Water Act. These plans identify temporary pollution prevention measures to be carried out and the compliance procedures followed on a construction site, and are recorded in text and graphic form.

Landscape architects are hired by the A/E to develop ornamental landscaping plans, which includes selecting and locating trees, shrubs, ground cover, and grasses, as well as designing the irrigation system required to support them. Hardscape—vernacular for trellises, fences, screens, pedestrian bridges, water features, dry stream beds, patios, walkways, seating, and the like—may also be designed and detailed by the landscape architect.

Licensing requirements for builders is a matter of state legal jurisdiction, and not all states require licenses for construction work. The federal government does not have a licensing requirement, although other requirements are imposed on builders performing work for it.[1] That said, some states regulate construction work through state license boards, which may establish separate license requirements for various kinds of work. In California, for example, general engineering contractors (A license classification) construct highways, bridges, dams, underground work, and the like. General building contractors (B license holders) construct buildings running the gamut from houses to high-rises. Specialty contractors (C license holders) perform a wide variety of specialty work including landscaping, plumbing, HVAC, concrete, masonry, drywall, plastering, painting, and floor coverings, to name a few of the 30 or so specialties. These specialty contractors may perform work on sites controlled by A and B license holders, for example, a concrete specialty contractor may install the headwall for a culvert on a site or the foundation and walls for a concrete tilt-up project.

Executing the recommendations of the geotechnical engineer in treating and handling soil and putting it where the civil engineer's drawings dictate is often performed by general engineering contractors (also known as earthwork contractors). Utility, or underground contractors, excavate, install, backfill, and compact the underground utilities described by engineers in utility drawings. Landscape contractors normally take responsibility for the installation of irrigation systems, irrigation system controls, planting, and associated improvements such as mow strips and planter boundaries under a specialty construction license. These contractors may also construct the hardscape mentioned if required to by the general contractor. They also maintain the landscaping for a period of time following installation to assure that the plants and grasses are properly adjusted.

▪ What to Expect in the Drawings

As noted, site construction drawings include graphic and written instructions to the contractor, a description of existing conditions and improvements, the final topography of the site, utility construction, required erosion control measures, and landscape drawings. Although this book focuses on graphics, the interdependent nature of construction documents is an underlying theme; therefore, textual documents pertinent to site construction are listed as well.

---

[1]A map illustrating the states in the U.S. that require contractors' licenses can be found at www.contractors-license.org.

Written Information

The geotechnical report describes (among other things) the past uses of a site, the characteristics of the earth encountered on it, the nature of groundwater conditions, how the soil is stratified (revealed in soil borings), the location of the borings that produced the soil samples, the prescription for properly treating the soil during the course of construction (remediating it, drying it out, mixing it with other soil, treating it with water or chemicals, compacting it, and perhaps disposing of it) and its expected engineering characteristics after treatment. Some highway departments include the results of soil investigations on the drawings themselves.

Specifications are one of five fundamental components in construction contracts. Generally, they describe the quality of work, materials and products to be used in a project. The Construction Specifications Institute (CSI) has developed the most widely used construction products indexing system in the construction industry. Called MasterFormat, it reflects how construction materials are specified, purchased, and installed. Although MasterFormat is a complete format for developing project manuals (including introductory information, bidding requirements, and contracting requirements), it is primarily a way of organizing qualitative information relating to construction work, materials, and products. It is currently segregated into 49 divisions:

- Division 1 is administrative in nature, and is a collection of generic guidelines on payment, quality, temporary facilities and controls, execution and protection of the work, and facility commissioning and decommissioning, to name a few.
- Divisions 2 through 49 focus primarily on the quality of work (acceptable tolerances, preparation guidelines, finishes, etc.), materials (grades of steel or wood, cement and mortar types, and mixing guidelines, for example) and components in the project (listed by acceptable manufacturers or otherwise quality rated).

Each division is segregated into numbered line items describing types of work, material, or products; for example, 31 22 13 is the identification for rough grading. The first pair of numbers identifies one of the 49 MasterFormat divisions (Earthwork), the second a subcategory in Earth Moving (itself a subcategory of Earthwork) called Grading, and the third identifies the grading as Rough, which is the specific work type that is the subject of the specification). The specification itself is divided into three parts:

- Part 1, General Requirements, lists the administrative, procedural, and temporary requirements related to Division 1 that are unique to that material or product; for example, how a specific piece of equipment will be handled, stored, and started-up, and what its maintenance and warranty requirements are.

- Part 2, Products, is a description of the required quality of a construction material, product or piece of equipment, its fabrication tolerances, required finishes, and so on.
- Part 3, Execution, describes installation or application requirements of a product or material, including the required skill level of the installer, preparation, cleaning and protection requirements, repair or restoration requirements, and required protection measures, among other things.

Graphic Information

Demolition drawings show existing improvements and give the constructor instructions as to what should be removed and, increasingly, what should to be recycled. The nation's rising real estate costs, diminishing resources, solid waste problems, increasing construction costs, and persistent concerns about the environment have given rise to the "green building" movement, characterized by the evaluation of projects on a life-cycle basis, energy efficiency in buildings, and extensive recycling efforts. Consequently, demands are increasing that, for example, old asphaltic concrete parking lots be ground up and reused, structures be deconstructed and materials resold, and myriad other recycling efforts be made.

Grading plans juxtapose existing and planned contours (contours are the principal tool for describing the earth's features two dimensionally) and contain a variety of details, including street and walkway sections, driveway and parking lot details, number and type of parking spaces to be constructed, handicap access ramps, and similar features. Occasionally, designers will superimpose existing improvements, vegetation, and demolition instructions over the planned improvements, resulting in drawings that are very difficult to read and are prone to misinterpretation. Figure 6.1 is excerpted from a drawing that tries to accomplish far too much on a single sheet.

Utility drawings for building construction projects normally describe the sanitary sewer, storm drain, domestic water, fire service, and gas and electrical systems. Among the critical bits of information are drain inlet locations, sizes and configurations, grate or rim elevations, and flow line elevations for storm and sanitary sewer systems at access vaults (manholes), the access vaults themselves, pipe sizes and slopes, locations of clean-outs for sewer systems, valving for pressurized systems such as domestic water and fire systems, fire hydrant locations and landscape irrigation supply requirements. Immediately after the rough grading is completed,[2] the trenches for these systems are excavated and the pipe installed, backfilled, and

---

[2] This is the common sequence. When very deep fill is required on a site (over 20'), a site contractor may choose to construct the storm drainage and sanitary sewer systems as the fill is being placed, to avoid the very high costs of deep excavating, extensive trench shoring, and accidents.

Figure 6.1 This excerpt from bid drawings shows how difficult it is to interpret drawings that try to do too much on a single sheet. The objects to be removed (building, septic tank, leach field, concrete walk, stairs, AC pads, fences), left in place (certain trees), and those newly created are all shown on the same sheet.

compacted in lifts (layers) with special bedding materials such as sand, crushed rock, and native soil. Pressurized lines, like domestic water and fire service lines, require thrust blocks (crudely cast masses of concrete used to stabilize the pipe where it changes direction or branches). Reinforced concrete pipe, corrugated, and other plastic pipe are common for storm drain systems, and vitrified clay pipe is common for sanitary sewer systems. Steel and (increasingly) various plastics are used for domestic water, fire service, and gas lines, and electrical conduit. It follows that one would see these materials and the bedding materials in graphic form on utility drawings.

Storm water pollution prevention plans (SWPPPs) derive from the Federal Clean Water Act, comprehensive federal legislation designed to reduce the contamination of natural bodies of water by human activity, including building construction. Enforcement of the legislation is the responsibility of the states, which approach the problem in different ways. At least one state has relegated enforcement of the Clean Water Act to regional water quality control boards, which require owners of construction projects to file pollution prevention plans prior to the commencement of construction work, and to monitor construction activity until the project is completed. Most of the work involved in complying with the act, as far as the construction industry is concerned, has to do with erosion control, sediment management, and the containment of chemical pollutants coming from construction materials and products that are stored and used on site. While it is the owner's responsibility to develop the SWPPP, the task is often handed to a civil engineer who provides contractors with drawings identifying how the owner intends to comply with the requirements, or the task is given directly to the contractor. Among the *best management practices* (BMPs)[3] described in these plans are silt fencing, straw bale and straw wattle filters, drain inlet filters, rock sacks, plastic sheathing draped over slopes, hydroseeding, specialized construction entrances, equipment wash-down pits, and material storage areas. Some states meet the failure to comply with the Clean Water Act with substantial penalties.

Landscape drawings normally present the planting scheme for a project first, followed by the irrigation system drawings, then details showing the prescribed way to plant and stake trees, prepare and plant shrubs, mount system controls, install control valves and moisture sensors, and install mow strip, to name a few. Landscape contractors (not underground utility contractors) are the contractors who normally install these systems. A variety of symbols exist for identifying trees, shrubs, ground cover and grasses; and, normally, the title pages of landscape drawings

(or civil sheets devoted to landscaping) or the planting plan include a planting legend that explains the symbols used. Concrete, reinforcing steel, earth, rock, gravel, cobble, boulders, gabions, crib walls, piping, and filter fabric all find their way into site construction drawings.

Graphic Expression in Site Construction Drawings

Projection Types

Most site construction is described in plan views— views looking down on a specific area from above; topographic maps and grading plans, site plans, utility plans, SWPPP plans, planting, and irrigation plans are examples. Vertical sections are also used, to describe highways, streets, walks, large underground drainage systems, and details of various improvements such as catch basins. Pictorial views such as isometric and dimetric drawings are used also, primarily in details of site improvements such as driveways or on SWPPP drawings, to describe various BMPs such as wattles and straw bale filters.

Applicable Line Types

A list of lines that are commonly used in site construction drawings appears in Figure 6.2. Figures 6.3 and 6.4 demonstrate how some of the lines listed above are used.

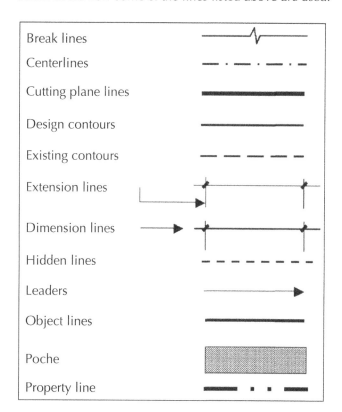

Break lines	
Centerlines	
Cutting plane lines	
Design contours	
Existing contours	
Extension lines	
Dimension lines	
Hidden lines	
Leaders	
Object lines	
Poche	
Property line	

Figure 6.2 Common lines used in site construction drawings.

---

[3]Best management practices, or BMPs, often developed with input from state agencies, are the most effective erosion control, sediment management, and chemical containment techniques currently known to those involved in Clean Water Act compliance.

Figure 6.3 Site plan for a remodel, showing a variety of line types. Note that the adjustments to the existing grade around the structure are minor, with the bulk of the work taking place north (top of the page) of the structure.

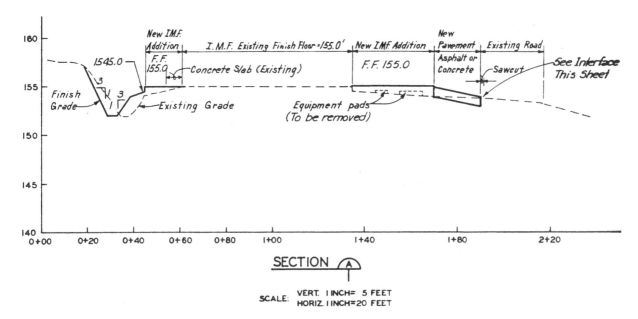

Figure 6.4 This is section A taken from the project in Figure 6.3. Note the use of two different scales, one for vertical dimensions and one for horizontal dimensions. Using two different scales exaggerates gentle slopes that otherwise would be indiscernible.

Symbols

Figures 6.5a–c comprise symbols that are commonly used in site construction drawings.

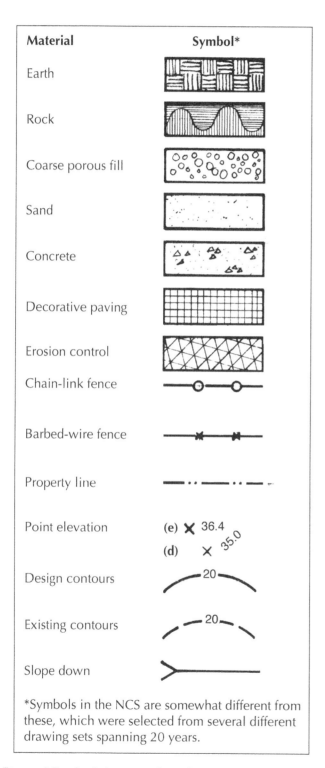

Material	Symbol*
Earth	
Rock	
Coarse porous fill	
Sand	
Concrete	
Decorative paving	
Erosion control	
Chain-link fence	
Barbed-wire fence	
Property line	
Point elevation	(e) ✗ 36.4 (d) ✗ 35.0
Design contours	20
Existing contours	20
Slope down	

*Symbols in the NCS are somewhat different from these, which were selected from several different drawing sets spanning 20 years.

Figure 6.5a Symbols commonly used in site plans.

Material	Symbol*
Sanitary sewer line	8"S
Sanitary sewer vault	(e) ○ (d) ●
Storm-drain line	18"D
Storm-drain vault	(e) ▢ (d) ▣
Gas line	G
Domestic water	DW
Fire service	FS
Fire hydrant	▫
Gate valve	◁▷
Electrical conduit	
Electrical pull box	▢

Utility Drawing

*Symbols in the NCS are somewhat different from these, which were selected from several different drawing sets spanning 20 years.
(e) existing
(d) designed

Figure 6.5b The family of symbols commonly used in utility drawings.

Material	Symbol*
Lawn	
Ground cover	
Shrubs	
(various sizes)	
Trees	
Main irrigation line	
Sprinkler heads	
(lawn & shrub)	

Landscape Drawings

*Symbols in the NCS are somewhat different from these, which were selected from several different drawing sets spanning 20 years.

Figure 6.5c Landscape drawings use a few simple symbols.

Drawing Subject Matter and Orientation

Topographic Maps

Topography, derived from Greek, means "to draw a place." Topographic mapping is the principal method for describing the earth's features two-dimensionally. The main tool used in topographic mapping is the contour line—a line connecting points of equal elevation. Following is a list of graphic conventions that are observed in topographic mapping:

- *Subject matter.* Topographic maps show bodies of water (lakes, streams, canals), the physical characteristics of the earth (ridges, mounds, valleys, swales, and the like), and certain human creations (boundaries, existing improvements, roads, and so on).
- *Orientation.* Topographic maps are plan views of the earth, that is, views looking down from directly above the mapped area.
- *Graphic tools.* The principal graphic tool in topographic maps is the contour line.
  - Contour lines exist in nature. A lake, for example, that has experienced different water levels over a period of time creates contour lines. When viewed from above, they are the same as contour lines on drawings, except that the intervals are not as consistent as those on a drawing.

- Contour lines are essentially the edges of horizontal planes passing through the earth, like layers of cardboard stacked upon one another.
- The vertical distance between contour lines is called the *contour interval.* Using the layered cardboard analogy again, the contour interval would equal the thickness of the cardboard (see Figures 6.6a–c).
- Contour lines represents whole numbered elevations.
- Every fifth contour line is called an *index contour,* which is displayed as an emboldened line interrupted by an elevation notation.
- Lines between index contours are referred to as intermediate contour lines, and are not commonly labeled with elevations.
- Supplementary contour lines are contour lines used in areas where the intermediate interval is not small enough to display more detailed features of a site. Light lines are used to distinguish the supplementary contours from the intermediate contour lines. It makes little sense to produce an entire map with close intervals when only a very limited area on the map might require it.
- Contour lines do not touch each other, except when describing certain geographic features such as cliffs. Contour lines are, for the most part, parallel lines.
- Contour lines close on themselves—they do not simply end (except at the map's edges).
- Contour lines do not split, except in very rare instances, for example, when a contour line happens to coincide with the base of a knife-edged ridge or the edges of a similarly shaped valley, or on bridge deck contours in the centerline of the highway.
- Contour lines that describe streams are V-shaped, with the point of the V pointing upstream.
- The convergence of two streams results in an M shape, with the points of each M pointing upstream.

The basis for selecting a given contour interval varies according to several factors, namely the purpose for which the topographic map is being created, the physical size of the area being mapped, the map sheet size, the variations in elevation of land features, the desired accuracy, and the budget allocated to surveying the site. Maps commonly have 10', 50', and 100' contour intervals, while building construction sites frequently use 1', 2', or 5' intervals.

Grading Plans

The main purpose of grading plans is to describe the existing features (topography) of a site and how the site should look upon completion. The principal technique for doing this is to superimpose the planned topography over the existing topography. To distinguish between

Figure 6.6a Contour lines and a few symbols are the instruments of expression in topographic maps.

Figure 6.6b Contour lines are analogous to the edges of layers of cardboard stacked upon one another. When viewed in plan, the cardboard model replicates the topographic map shown in Figure 6.6a. (Photo by the author.)

Figure 6.6c This model aptly describes contour lines: at the contour line itself, the elevation changes abruptly. The thickness of the cardboard is the corollary to the contour interval. (Photo by the author. Model constructed by David and William Gill.)

existing and design contours, design professionals use two different lines: dashed lines of medium thickness represent existing contours; solid medium lines represent planned contours. Existing contours are frequently irregular, whereas planned contours are often much more regular, straight, and parallel (manifesting the imposition of humans' sense of order over nature's). The difference between the two, tempered by the structural earthwork requirements set forth in the geotechnical report, is what represents the engineering contractor's work.

The scale selected for any drawing is a function of the size of the object, its complexity, the desired level of detail, and the size of the drawing medium. Site plans for highways are frequently drawn in small scales, so that large portions of the work can be recorded on a single drawing. Building construction projects use somewhat larger scales—1" = 20' (1:240) and 1" = 30' (1:360) are very common scales for grading and utility plans. Site plans frequently contain sections—vertical cuts—that are views of features such as building pads, roads, ditches, and walks. Having these details on the site plan is helpful to drawing users, who otherwise would have to search through detail sheets to find the pertinent information.

The development of site plans varies from design professional to design professional, but a common method is for the architect to develop the site plan from the engineer's survey or from a legal description of the property.

This drawing includes the property lines, easements, building footprint, parking areas, entrances, exits, fencing, and open space. Once the site plan is developed, it is sent to the civil engineer, who develops the grading plan for the project.

Utility Drawings

Utility drawings contain essentially the same information as grading plans, except that the focus is the underground utilities rather than the topography. Under-ground utilities are typically installed after rough grading is completed, so knowing the planned contours and the elevations of the storm drain and sanitary sewer systems is critical. The depths of the existing infrastructure frequently dictate the elevations at which these systems are set. They depend on gravity to propel storm runoff and effluent to their discharge or collection points, so pipe diameter and slope—relatives in these systems—are critical. Twenty- and thirty-foot-deep systems are not unusual. Normally, sewer systems are buried the deepest, followed by storm drains, fire service, domestic water, and gas and electrical. The latter four utilities are frequently in proximity to one another—their depths are a function of code requirements that specify minimum coverage of about 3'-6" (1067 mm), which means that design professionals and contractors alike need to be vigilant in avoiding conflicts within the 4'-thick plane in which the systems exist. Their proximity to structural elements such as drilled piers can be an issue, too.

Details of components and elements in underground systems are also called for on utility drawings. Sections through underground systems, called *utility profiles*, reveal the piping as well as access points and other relevant information. Underground power drawings also contain details showing, for example, the proper separation of conduit and coverage requirements in trenches. Figures 6.7a–d comprise relevant site construction details taken from several different projects.

Figure 6.7a Site construction details describe a variety of systems and components, above and below ground.

Storm Water Pollution Prevention Plans

Drawings associated with SWPPS are relatively simple, consisting of a site plan with the locations of silt fences, wattles, detention ponds, access roads, vehicle washdown and storage areas, and material storage areas. Since this drawing type is fairly new, the symbols for the various BMPs are, for the most part, of the design professional's own making.

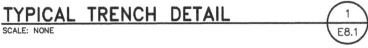

Figure 6.7b This is a section through a trench for underground electric utility service.

25 mm SLOT ON INSIDE
FLANGE OF ENCLOSURE

INSTALL EYE BOLT
WITH 41mm INSIDE
DIAMETER

PCC FOUNDATION
PCC PAD

INSTALL #5 PULL BOX

216 mm

300 mm

510 mm

100 mm

200 mm

1.2 m

100 mm

0.6 m

40 mm TYPE 3 CONDUIT

NOTE:
ANCHOR BOLTS IN FOUNDATION
SHALL BE 10 mm DIAMETER X
155 mm LONG X 53 mm OFFSET
(4 REQUIRED)

DETAIL A
NOT TO SCALE

TRAFFIC MONITORING STATION

Figure 6.7c This drawing is a simple isometric of a traffic monitoring station, part of the instructions accompanying a bridge improvement drawing set.

TRAFFIC GRATE & FRAME

18'
MIN.

9" THICK(MIN.) CONCRETE COLLAR REINFORCED
WITH # 4 HOOPS AND CLOSED TIES SPACED 6' O.C.
NOT REQD IN NON-TRAFFIC AREAS.

FINISHED GRADE

SUBGRADE

DEPTH TO FLOW LINE, Z

PRECAST OR CAST-IN PLACE CONC. WALL.

INSIDE
DIMENSION,D

DEPTH OF EARTH COVER

GROUT

TO STORM DRAIN

6' DIA.(MIN.)

CONCRETE FILL

CAST-IN-PLACE CONC. BASE SLAB (6' MIN.)

Z (FEET)			D (INCHES)
Z <		2.0 =	18
2.0 <	Z < 3.0 =		24
3.0 <	Z < 5.0 =		30
5.0 <	Z < 7.0 =		36
Z >		7.0 =	48

STANDARD DROP INLET
CITY TYPE I

NOTE:
1. MINIMUM WALL THICKNESS IS
    6 INCHES FOR CAST-IN-PLACE
    CONCRETE.

2. PERMISSABLE TO USE A CONE SECTION WITH
    24' FRAME AND GRATE ON PRECAST MANHOLES
    BUT DIAMETER TO DEPTH RATIO PER TABLE
    ABOVE MUST BE MAINTAINED.

Figure 6.7d Sections of underground utilities, in this case a drain inlet, are common in underground utility drawings.

Landscape Drawings

Large commercial buildings frequently have limited landscape areas; most of the real estate is devoted to the building footprint, driveways, and parking areas. Nevertheless, hundreds of trees, plants, and shrubs—and the irrigation system to sustain them—may be called for. The methods of identifying them vary; however, using several different tree and shrub symbols distinguished by the size of the plant is common. Ground cover and grasses tend to be identified by poché (shading) of varying types. The scale of planting and irrigation plans is commonly the same as the site plan, which means the engineer's scale is used, except when drawings are in metric units, in which case the metric scale is used. Interestingly, landscape details are frequently not drawn to scale like most other details—they are simply schematic descriptions of landscaping elements and components.

▒ Translating the Drawings into Work

Once the determination to bid a project has been made, contractors set about determining, with reasonable accuracy, what the design professional is directing them to do. As noted previously, these directions are given primarily through drawings and specifications. For earthwork contractors, the geotechnical report plays a far more significant role than with most other specialty contractors, since it contains the prescription for the handling and treatment of the earth on site. The earthwork contractor prepares the site for all subsequent improvement work, including underground utilities and foundations, so it is essential to understand what currently exists on the site, what fin-

ished state the designer has in mind, and what requirements regarding earthwork and groundwater issues will be insisted upon by the geotechnical engineer.

Sitework involves a variety of tasks, among them:

- Removing soil (cutting) from locations on the site—under buildings and parking lots, for example—temporarily stockpiling it, and reinstalling it in lifts under controlled conditions, or placing it in low areas to raise them (filling).
- Installing and maintaining an excavation retention system (soldier beams and lagging, tied-back sheet piling, and sloped excavations, for example) and site access.
- Dewatering, as may be required (drawing water out of the earth).
- Removing earth from a site and hauling to another location, as with an urban office project with several levels of parking below the street.
- Importing earth from one location (referred to as a *borrow pit* in highway work) and hauling it to a site that requires fill.

As much as possible, engineers "balance" a site to avoid unnecessary costs (see sidebar on page 90).

Determining what equipment should be used and identifying the tasks, sequence, and time required to execute the work are the responsibility of the earthwork contractor. Being able to see a site under construction while estimating its costs is an unusual, and essential, ability. The photographs in Figure 6.8a–d show what the estimators conceived as the solution to the excavation of a two-city-block site months before the work started. A shallower site with no groundwater issues and straightforward access allowed for a much simpler approach on the project shown in Figures 6.9a–c.

Figure 6.8a This photograph shows one collection point in the dewatering system of a large urban office building site. The large tank is a detention basin that captures sediment before the water is expelled into the existing stormwater system. (Photo by the author.)

Figure 6.8b The excavation plan called for teams of excavators with long shovel arms working with a dozer and loaders. End-dumps (in the background) were used to haul the soil off site. (Photo by the author.)

Figure 6.8c The full team effort is visible in this view. The walers (horizontal members) attached to the tiebacks that are keeping the sheet pile wall in place are visible in the background. (Photo by the author).

Figure 6.8d Deep excavations are monitored regularly to detect any change in the buildings and improvements surrounding an urban site. In this photograph, the inclinometer, which measures movement in the sheet pile wall, is visible between the first two guardrail posts. A crack is developing in the soil behind the wall. (Photo by the author).

Figure 6.9a A shallow cut in this project allowed for a simple access road. (Photo by the author.)

Figure 6.9b Though the soil in this project had similar characteristics to the one depicted in Figure 6.8a-c, the difference in the magnitude of the quantities was significant. (Photo by the author.)

Figure 6.9c The prescription for soil treatment was simple on this project; it required periodic watering and compaction only. (Photo by the author.)

The elements that affect excavation costs are: the characteristics of the soil; groundwater; magnitude of the work; access to, size, and location of the site; the disposition of excavated material; and the availability of equipment. The project shown in Figures 6.10a–b, for example, while relatively simple from the construction standpoint, was made immeasurably more difficult by the very confined workspace—a pedestrian access on the left, a railroad track behind it, a small business to the right, and a busy street in front of it. The A/E's drawings do not show projects in context—conceiving them that way is up to the contractor.

As noted in a previous chapter, construction work must be correlated to units of cost and time. One of the most common units for sitework is cubic yards (or cubic meters, for drawings done in SI), which are correlated to production times for various types of equipment. A sampling of units commonly used in site construction is shown in Table 6.1.

Determining Earth Quantities: Two Methods

The earth quantity takeoff process is ultimately designed to determine a value for implementing the plan for a site.

Figure 6.10a This very small site presented the builder with a considerable challenge: It was over 20' deep with very little workspace around it. (Photo by the author.)

Figure 6.10b Limited access may result in the need for special equipment for what are ordinarily common operations, in this case backfilling a sump with porous fill. The intake chamber that is partially completed in Figure 6.10a is visible in this photograph, between the pickup truck and the streetlight base. (Photo by the author.)

TABLE 6.1 Sample Unit Designations for Materials and Work—Site Construction

	Day	Acre	EA each	LF lineal foot	CLF hundred lineal feet	VLF vertical lineal foot	CF cubic foot	CY cubic yard	SY square yard	SF square foot	MBF thousand board feet*	Ton	
Demolition			X	X			X	X		X			
Structure Moving	X									X			
Site Clearing		X	X										
Stripping								X					
Ripping								X					
Shores											X		
Sheet Piling										X	X	X	Grading and Paving
Cribbing and walers										X			
Excavate, backfill, compact								X					
Haul								X					
Slope/Erosion Control				X					X				
Gabions			X					X				X	
Rip rap								X				X	
Pavement base								X	X			X	
Paving									X			X	
Extruded Curbs				X									
Piped Utilities				X									
Valves			X										Utilities
Catch basins			X			X							
Irrigation			X	X									
Soil Preparation							X	X		X			Landscape
Planting			X							X			

*One board foot = 144 cubic inches using the nominal dimensions of the boards.
Metric drawings require different units: liters (liquids), kilograms (weight), lineal meters (length), cubic meters (volume), square meters (area).

Balancing a Site

"Balancing a site" refers to the practice of using existing soils and rock (including excavation spoils), as well as recyclable materials on site, to achieve the prescribed topography. Soil and rock are removed from high spots and transported to low spots on the same site, so that importing and exporting soil, which adds significantly to cost, is avoided. The elevation of existing utilities may influence site balancing, since one cannot excavate below a point where sufficient slope for critical drainage systems can be created.

Although computer technology has improved the estimating process considerably, earth quantities are sometimes difficult to determine accurately—the task is not like counting light fixtures in a reflected ceiling plan or numbers of plumbing fixtures of a specific type—and determining the cost of actually doing the work is challenging, especially if groundwater or unsuitable soils are encountered. The unit price contract is acknowledgment of this difficulty—a price for the units themselves (tons, cubic yards) is agreed upon, not the number of units. Nevertheless, earth quantity takeoffs are performed every day in thousands of earthwork contractors' offices by a variety of methods.

Identifying how much material must be excavated, treated, and replaced is one of the initial steps in determining sitework costs. Generally, the earthwork contractor must calculate the difference in cubic yards between the existing contours and the design contours, taking into consideration any special requirements the geotechnical engineer may have, for example, to over excavate the area underneath a proposed building and backfill it under a stringent soil prescription. In substructure excavation projects such as parking garages, and in tunnel and utility work, tasks that are not shown on drawings can be especially critical, for example, the amount of overexcavation that may be required to provide working room, access to the work and the site by excavating and hauling equipment, and safety issues (shoring, benching, personnel restraints). All construction drawings show the finished product, not the processes that have to take place to make them a reality.

There are many approaches to quantifying earthwork, but grid averaging and the average-end-area method are two common methods. The two methods of interpolation are used to find critical elevations on drawings when they are not given. Once the elevations are known, the volume of earth (the quantities) can be determined. Cubic yards of earth are essential to analyzing the productivity of large machinery and, ultimately, costs.

Grid Averaging

Grid averaging refers to the imposition of a grid of squares, rectangles, or triangles (or combinations of these shapes) over a site. The boundary of the grid circumscribes the area to be graded. Existing elevations are determined at each intersection and are compared to the design elevations, and quantities of cut and fill are tabulated in a matrix. The quantities of cut and fill are then recorded for the entire site. The spacing of the grid is a function of the terrain and the desired level of accuracy. Grid averaging is useful on sites where slopes are mild and the design does not differ radically from the existing topography. Where the existing grade and planned grade converge—known as the *daylight line*—the grid is often broken up into subgrids. Figure 6.11 is the source document for the commentary that follows.

In section 1, the entire area (other than a very small triangle) is fill: to calculate the amount, the fill depths at each grid intersection are added, then divided by 4 to arrive at an average depth for the area. The area of the grid is then multiplied by the average depth to arrive at a volume figure. For section 1 of the grid, the computation would look like this:

$$(.8 + .5 + 0 + .6) \div 4 \times (20 \text{ m} \times 20 \text{ m}) = 190 \text{ m}^3$$

The small triangle at intersection B2 is ignored, largely because of its size and the gentle slope of the surrounding terrain. Making it a five-point figure rather than four-point figure (although it would account for the triangle) would result in a noticeably different depth calculation; therefore, section 1 was treated as a four-sided figure.

In the sections where both cut and fill are required (sections 2 and 4 in Figure 6.11), the daylight line (on either side of which cut and fill separate) is determined first. The areas of these two subsections are then calculated and multiplied by the average depth of cut or fill. For example, the cut portion of grid 2 could either be subdivided into two triangles or treated as a trapezoid to quantify the area of the subsection, then multiplied by the average depth. Depth for the each subsection could also be determined by interpolating the elevation at the centroids of these areas. The volumes of cut or fill in subsections are then added to the other sections to determine the total amount of cut or fill required. The daylight line can be determined by graphic means, by juxtaposing the existing and planned grades in section, or by mathematical means. One formula for determining where the daylight line occurs is:

$$X_f = (f/f + c) \times d, \quad \text{or} \quad X_c = (c/f + c) \times d$$

where X_f = distance to the daylight line from the fill intersection
X_c = distance to the daylight line from the cut intersection
f = the amount of fill
c = the amount of cut
d = the distance between the two intersections[4]

---

[4]Prepared by Donald W. Nostrant.

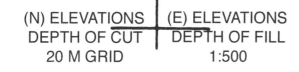

(N) ELEVATIONS	(E) ELEVATIONS
DEPTH OF CUT	DEPTH OF FILL
20 M GRID	1:500

Figure 6.11 Grid averaging is one method of determining the volume of earth in a given area. Here the number in the upper left quadrant of each intersection is the design elevation; in the upper right is the existing elevation. The depths of cut and fill are recorded in the lower left and lower right quadrants, respectively. Elevations are given in meters.

Mathematical and Graphic Interpolation

Placing a grid over a site inevitably results in grid intersections falling between contour lines, so both the existing and planned elevations for each intersection need to be determined. One of several mathematical and graphic interpolation techniques can be used to determine the elevation of the grid intersections between contour lines.

Interpolation—the process of approximating a given function by using its values at a discrete set of points—is the technique used to determine various elevation points between two known elevations. Its success as a technique is based on the assumption that the lines between points of elevation are straight (the slope is uniform). For example, the midpoint (horizontally) between the two elevations in Figure 6.12 is a point half the distance vertically.

By knowing two of three factors—the slope (or gradient), the horizontal distance between elevation points, or the difference in elevation—you can determine the

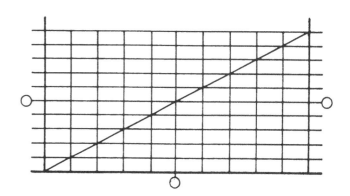

Figure 6.12 The underlying assumption in all interpolation is that the line between points is a straight one. A point halfway vertically is also halfway horizontally.

locations of any elevations falling between the two points. The formula for determining slope, or gradient is:

$$G = D_e/D_h$$

where

   G = gradient
   D_e = elevation difference between the two points
   D_h = horizontal distance between the two points

When the horizontal distance (D_h) and gradient (G) are known, you can calculate the difference in elevation (D_e):

$$D_e = D_h \times G$$

When the gradient and difference in elevation are known, you can calculate the horizontal distance between points:

$$D_h = D_e/G$$

If the horizontal distance between the two elevations mentioned is 100', for instance, you can calculate the slope; a 16' rise (or drop) in 100' is a 16% slope (16'/100' = (.16 × 100%) = 16%). So, if you want to determine what the elevation of a point 50' from either elevation is, you multiply the horizontal distance from one of the two points by the percentage of slope: 50' × 16% = 8', which you either add to the lower elevation, if measured from the low point, or subtract from the higher elevation if you measured from the high point (176' + 8' = 184', or 192' – 8' = 184'). By the same token, you can locate where on the horizontal line another elevation lies, for example, the index contour 180. This elevation is 4' higher than elevation 176 or 12' lower than elevation 192'. The total difference in elevation is 16': If you divide the difference in elevation between 176 and 180 (4') by .16 (the slope), you get 25'—the horizontal distance to elevation point 180, measured from point 176 (4'/.16 = 25').

Another way to approach this problem is to compare the difference in elevations between a known point and one you are trying to find with the overall difference in elevation between two known points. Suppose you are trying to find the 2' contours occurring between intersections on a survey grid, one at elevation 176.4 and the other at 188.2. The contours you are trying to locate are 178, 180, 182, 184, 186, and 188. The overall difference in elevation is 11.8' (188.2 – 176.4 = 11.8). You can begin at either point, but try 176.4 to start. You are looking for the distance to the first 2' contour on the line, or elevation 178. The difference between the starting point elevation and the first contour is 1.6' (178 – 176.4 = 1.6). By dividing this increment by the overall difference in elevation, you can determine what percentage of the vertical distance is represented by these two elevations (1.6/11.8 = .14, or 14%). You can apply what you learned from Figure 6.12 to this problem: If you are 14 percent of the way up to elevation 188.2,

then you are 14 percent of the way over to it as well. If the horizontal distance between the known points is 100', then you will find the contour you are looking for 14' away from the grid intersection at elevation 176.4 (.14, or 14% × 100' = 14'). On an actual survey grid or grading plan, you would want to use the proper scale to locate contour 178.

The next contour to find is 180'. The difference between 178' and 180' is 2'. Dividing again by the overall difference in elevation, you determine the increment to be .17 (2'/11.8' = .17, or 17%). Contour interval 180 is 17 percent higher than 178, and 17' away from it as well (17% × 100' = 17'). Since the increment you evaluated was between 178 and 180, you would want to measure 17' from point 178, not from the starting point. A word of caution here: When measurements are transcribed to a drawing from successive points, the margin for error increases—a minor error in each measurement can have a significant effect on the overall accuracy of the drawing. You can avoid inaccuracy by measuring from the starting point every time; however, you must be sure to evaluate the correct increment. To find elevation contour 180 from the starting elevation—176.4—you would take the difference vertically between the two points (3.6') and divide by the overall difference in elevation (11.8') to determine what percentage of the overall elevation difference is represented by these two points (3.6'/11.8' = .31, or 31%). The horizontal distance from the starting elevation to elevation 180 is the same percentage as the vertical measurement—so .31, or 31% × 100' = 31'. The sum of the horizontal distances in the first two calculations should be the same as the third calculation, if the points are in the same place, and indeed it is: 14' + 17' = 31'.

There are really only three calculations required to find all the even contours between elevations 176.4 and 188.2: one to determine the distance from the starting point to the first even contour, one for the even contours, and one for the last even contour and the ending point. Figure 6.13 demonstrates how the even contours between 176.4 and 188.2 are mathematically determined.

Graphic interpolation can be much less time-consuming, since no calculations are required. The basis for this approach is the division of the line into equal parts (refer back to Figures 5.15a–c). The engineer's scale and the metric scale are best for graphic interpolation. Figure 6.14 shows how graphic interpolation can be used to find even contours between the two elevations just listed.

Once the elevations at the grid intersections have been identified and recorded, the grid averaging technique of determining earth quantities can be applied. You can also produce an entire topographic map from a survey grid by one of these techniques: Once you have identified the points of elevation between grid intersections, you simply connect the dots. After all, the definition of a contour is a line connecting points of equal elevation.

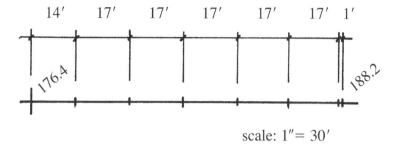

scale: 1″= 30′

Find even 2' contours between the two
elevations

Overall elevation difference = 11.8'

176.4 to 178 = 1.6'/11.8' = .14 x 100' =   14'
178 to 180 = 2'/11.8' = .17 x 100' =   17'
180 to 182 = 2'/11.8' = .17 x 100' =   17'
182 to 184 = 2'/11.8' = .17 x 100' =   17'
184 to 186 = 2'/11.8' = .17 x 100' =   17'
186 to 188 = 2'/11.8' = .17 x 100' =   17'
188 to 188.2 = .2'/11.8' = .01 x 100' =   1'
                                        100'

Figure 6.13 Mathematical interpolation takes different forms. This method is quick and accurate, especially when there are few elevation contours to find between points.

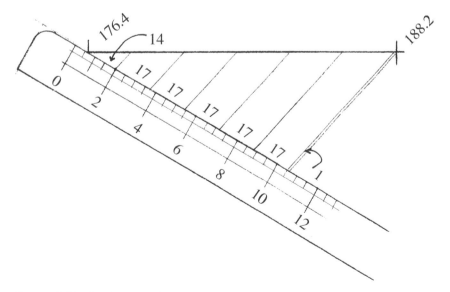

Figure 6.14 Graphic interpolation is quick, reasonably accurate, and requires no calculations. It is useful when there are numerous contours to find between points. In this example, the method shown in Figure 5.15b is applied.

Average-End-Area Method of Determining Earth Quantities

Another effective way of determining earth quantities is by the average-end-area method, which is the basis for some computer programs used to calculate earth quantities. While most accurate when applied to earthwork that is regular—for example, in highway construction—in most cases it is sufficient for other applications.

In this method, site sections (vertical cuts) are taken at intervals called *stations*, their areas are calculated and averaged with adjacent sections, and the averaged ends are multiplied by the distance between stations to arrive at a volume for that portion of the site. The formula follows:

$$V = [A_1 + A_2] / 2 \times D$$

where

V is the volume
A_1 is the area of the starting station
A_2 is the area of the adjacent station
D is the distance between them

Although this is not a sitework estimating book, it is worth mentioning here that quantities of earth are customarily calculated in cubic feet (or meters) and converted to bank cubic yards (bcy or bcm), loose cubic yards (lcy or lcm), and compacted cubic yards (ccy or ccm). The three different units take into consideration the swelling that in situ earth undergoes when it is excavated (requiring a conversion from bank cubic yards to loose cubic yards for hauling) as well as the amount that loose earth shrinks when it is compacted under prescribed conditions (requiring a conversion from loose cubic yards to compacted cubic yards). The process of determining earth quantities by the average-end-area method, when done manually, follows this basic outline:

1. Determine specifically what you are being asked to do.
2. Determine what you have to work with and what information needs to be developed.
3. Define the graphic parameters of the work, in all three dimensions.

Figure 6.15a This is the existing topography in the area where "Imaginary Road" will be built. The centerline of the road bisects both hills.

4. Select the appropriate scales, vertically and horizontally (as noted in Chapter 4, site sections are frequently drawn to two scales, primarily to exaggerate slight slopes—.5 percent to 5 percent slopes) and to fit large areas onto a single drawing sheet).

5. Determine the number and frequency of sections to be taken,[5] and identify them by station point. This is frequently done for you on highway drawings, which use station points as reference markers (like a grid in drawings of buildings); however, you may add sections where significant changes in either planned or existing contours occur.

6. Plot the existing and design profiles, bearing in mind requirements such as overexcavation.

7. Calculate the difference in areas, in square feet (or square millimeters or meters) between existing (or overexcavated) profiles and design profiles at each station—this represents the work.

[5]See Appendix G on taking a site section.

8. Using the customary format for earth quantities shown in Table 6.2, record the quantities of cut and fill and convert to cubic yards (or cubic meters if they are the proper unit).

The average-end-area method of determining earth quantities is especially effective when the "shape" of the earth is fairly regular, such as in highway construction.

Figure 6.15a–d show how the average-end-area method may be applied to a simple road project. There is just enough information to determine the quantity of earth required on this site. The imaginary road parameters follow:

1. Contours: Existing contours (solid lines in Figure 6.15a) and highway centerline are shown in Figure 6.15b; the contour interval is 2', starting at 200'.

2. The road has two 10' lanes and 3' shoulders.

Figure 6.15b Roadway design contours, like the existing contours in 6.15a, are shown as solid lines in this drawing. Station points were selected on the basis of terrain changes.

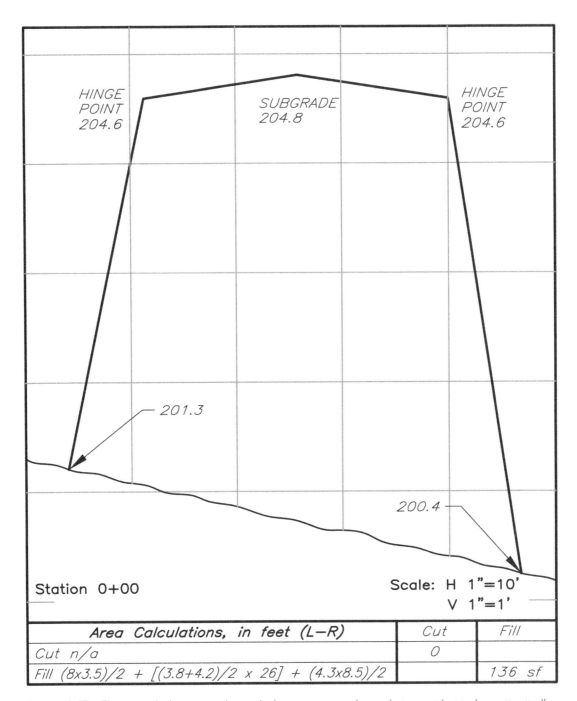

Figure 6.15c The area calculations correlate to the basic geometric shapes that are evident in the section itself. This part of the road calls for fill. (Drawing by Tom Krejci, after Keith Bisharat.)

3. The roadway cross slope is 2 percent, measured from the centerline.

4. The surface course elevation at station 0 + 00 is 206.

5. The pavement section is 4" asphaltic concrete (AC) over 10" aggregate base (AB).

6. All slopes are 2:1 (two horizontal, one vertical).

7. Ditches occur where necessary.

8. The drawing has been reduced; the scale is therefore approximately 1" = 57'.

This exercise requires that design contours be developed and superimposed over the drawing, starting with the road centerline. Though they are not shown dashed, the new contours are shown in Figure 6.15b, as are the station points and the profiles of the road and existing terrain. Figures 6.15c and 6.15d are examples of sections taken at two station points: the starting station (0 + 00) and station 1 + 20 (120' from the starting station), with area calculations shown at the bottom of the section. Table 6.2 is a summary of

HINGE
POINT
205.77

SUBGRADE
206.03

204.6

Station 1+20

Scale: H 1"=10'
V 1"=1'

Area Calculations, in feet (L—R)	Cut	Fill
Cut (7.5x.25)/2 + (6x1.2)/2	5 sf	
Fill (2x1.1)/2 + [(1.1+.35)/2 x 13] + (.35x5.5)/2		11 sf

Figure 6.15d At this station point, some cutting is required. (Drawing by Tom Krejci, after Keith Bisharat.)

the earth quantity for Imaginary Road in the standard format for the average-end-area method.

The wonder of computer technology and its value to the visualization process are evident in Figures 6.16a–c, which is a wire frame model of the Imaginary Road site before and after the construction of the road. This ever-improving technology is making a great contribution to dispute resolution and a reduction in errors: What once resided in peoples' imagination is stored, retrieved, and subjected to thousands of manipulations in the course of the design and construction of a project.

TABLE 6.2 Imaginary Road Earth Quantities

Station	Distance (ft)	Cut			Fill		
		Area (sf)	Ave. area (sf)	Vol. (cf)	Area (sf)	Ave. area (sf)	Vol. (cf)
0 + 00		0			136		
	57		10	570		76	4304
0 + 57		19			15		
	11		64	704		8	88
0 + 68		109			0		
	28		99	2772		0	0
0 + 96		88			0		
	24		47	1128		6	144
1 + 20		5			11		
	80		3	240		110	8800
2 + 00		0			209		
	88		0	—		174	15312
2 + 88		0			138		
	24		43	1032		69	1656
3 + 12		85			0		
	48		44	2112		12	576
3 + 60		2			23		
	25		1	25		76	1900
3 + 85		0			129		
				8583 cf			32,780 cf
				318 cy			1214 cy

The standardized display for earth quantity take-offs determined by the average-end area method lists the sections by station number, the distance between stations, and the area of cut and/or fill at each section. The area of cut and/or fill in adjacent stations is averaged, listed, and multiplied by the distance between stations.

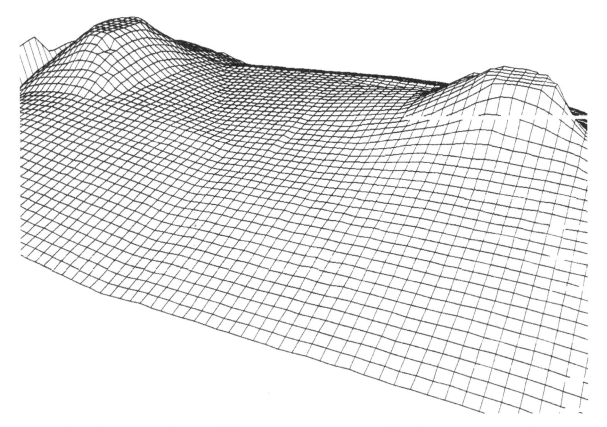

Figure 6.16a This drawing of the site in 6.15a was generated on a computer. The value of computer technology is evident in this and the next two drawings. Can you identify any evidence of incorrect data input in this drawing? (Drawing courtesy of Nicole Wertz.)

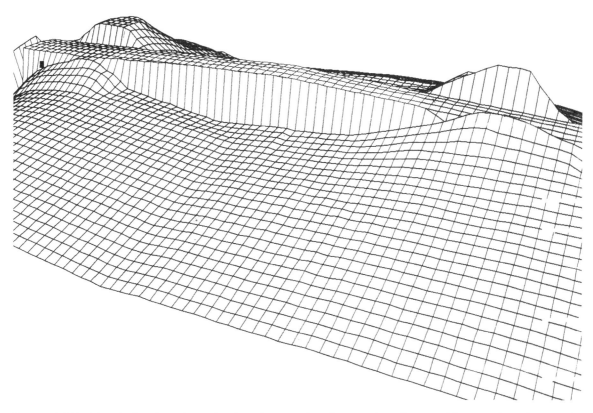

Figure 6.16b Note the literal translation that the software makes of contour lines: the tops of the mounds are lopped off. (Drawing courtesy of Nicole Wertz.)

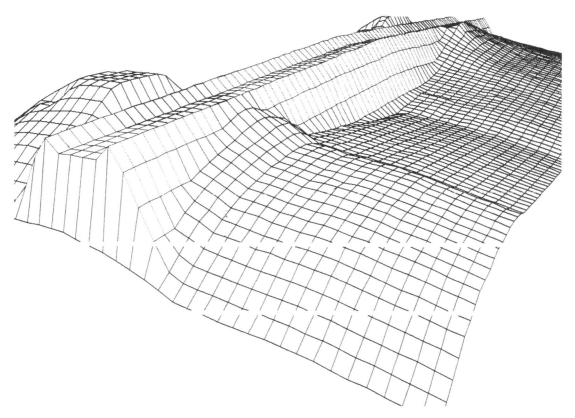

Job Name:CM 22 ROAD

Figure 6.16c The value of computers to visualizing a project is obvious in this and the preceding drawing. (Drawing courtesy of Nicole Wertz.)

Summary

Site construction presents challenges to project owners and contractors as a result primarily of soil properties and water. It is very difficult for the contractor to precisely determine what to expect in sitework, therefore it is inherently riskier than other types of construction work. The variety of issues (determining quantities and the magnitude of the work accurately, selecting the proper equipment, determining the effect that soil properties, groundwater, site location, and access to the work are likely to have on profitability) that get resolved effectively months prior to construction, based on geotechnical reports, specifications, and drawings is a testament to the skill and judgment of site construction contractors.

CHAPTER 6 EXERCISES

1. How do the professional roles of civil engineers, geotechnical engineers, and landscape architects differ where site drawings are concerned?
2. Where would one find information pertaining to sanitary sewer, storm drain, fire service, and domestic water systems?
3. What is the underlying purpose of developing storm water pollution prevention plans, and what is their primary focus?
4. What key tool is used in topographic maps?
5. Name five "rules" pertaining to contour lines.
6. List the cost generators in excavation work.
7. What are the steps to manually determining a quantity of earth using the average-end-area method?
8. Using an old cardboard box, create a model of a site using a topographic map. An eighth-inch layer of cardboard is about the right thickness for contour intervals of 1' foot, using a 1" = 10' engineer's scale. Using the same topographic map, select a section and superimpose a pad over a portion of it. Draw the new contour lines and then construct a cardboard model of the site.
9. Using the photograph in Figure 6.17, draw an axonometric sketch of the walkway that will result from casting the concrete into the forms shown.

Figure 6.17 ADA-compliant street corner.

7 Foundation Systems

Key Terms

Angle of repose
Benching
Buck
Caissons
Cohesionless soil
Cohesive soil
Continuous footing
Dampproofing
Dead loads
Differential settlement
Floating foundation
Foundation
Geotextiles
Grade beam
Hydrostatic pressure
Imbedment
Keyway
Lateral loads
Layback
Live loads
Mat foundation
Operations
Overturning
Parging
Pile caps
Piles
Slab-on-grade foundation
Spread or column footing
Substructure
Superstructure
Surcharge
Temperature steel
Tensioning tendons
Tie beam
Waterproofing

Key Concepts

• Understanding the goals and responsibilities of the design professional facilitates the interpretation of construction drawings: One can anticipate what the drawings are likely to show. The typical construction contract imposes the responsibility on the constructor to build according to the drawings and specifications provided by the owner (through the A/E), so, to varying extents the design dictates the construction process and sequence.

Objectives

• Identify the projection types commonly used to describe foundation systems.
• List and draw the symbols commonly used in foundation drawings.
• List the common components of a foundation system.
• Reproduce drawings of foundation systems.
• Describe how to construct a simple foundation.
• Perform a material quantity takeoff of a simple foundation.

Construction Graphics: A Practical Guide to Interpreting Working Drawings,
Third Edition. Keith Bisharat.
© 2025 John Wiley & Sons Inc. Published 2025 by John Wiley & Sons Inc.

Purpose of Foundations

Fundamentally, the purpose of a foundation is to transfer the various loads imposed on it to the earth. To fulfill its basic requirements, a foundation must be safe against structural failure, settle uniformly, and be economically and technically feasible.

What to Expect in the Drawings

Foundations for many structures are divisible into two basic categories: deep and shallow foundations. Deep foundations include end-bearing or friction pile foundations, caissons, mat, and floating foundations. Deep foundations are commonly used to support tall buildings, bridges, or structures overlying incompetent soil. Shallow foundations comprise spread or column footings, continuous footings, and slabs-on-grade, and are used on smaller structures and low-rise buildings sitting on competent soil.

Pile foundations are commonly constructed with precast, prestressed concrete, wood, or steel piles of varying sizes, lengths, and groupings, or with cast-in-place reinforced concrete caissons. Cast-in-place reinforced concrete pile caps join pile heads together, and transfer the loads from the columns that sit on them to the piles below. The piles are normally designed to share the load equally. Cast-in-place reinforced concrete tie or grade beams frequently tie groups of pile caps or caissons together.

Shallow foundations, in contrast, are often composed of cast-in-place reinforced concrete or, in the presence of expansive soil, post-tensioned cast-in-place reinforced concrete slabs-on-grade. Continuous reinforced footings connected to steel-reinforced, grouted concrete masonry units (CMU) are popular as well. Figure 7.1 shows several common foundation systems.

The list of forces that foundations resist varies for the type of project. Commonly, however, foundations resist vertical loads (e.g., gravitational loads, uplift due to hydrostatic loading or seismic events), multidirectional lateral loads (hydrostatic pressure on the substructure, seismic events), and various reactions to the loading (uplift due to the lateral loading of a tall structure, bending in vertical and horizontal structural members, and shear forces, for example). With the foregoing in mind, what should someone reviewing drawings of foundations expect to see? In a cast-in-place reinforced concrete system, reinforcing steel of various lengths, diameters, and numbers is located where stresses, particularly tensile stresses, act. The manner of connecting structural elements to one another—where columns of a steel superstructure connect to pile caps, for example—is of interest, since the overall performance of the structure is dependent upon the adequacy of its weakest link. Connections within the

elements are noteworthy as well. The structural engineer, for example, in addition to calling for epoxied reinforcing bars in pile heads might require that a given length of the tensioning tendons in precast prestressed concrete piles be exposed and tied into the reinforcing steel cage in the pile cap. The systems that facilitate the expected behavior of the foundation, namely drainage, water- and dampproofing, and insulation systems are equally important to consider. Drain rock, geotextiles of various designs, piping, sumps, and pumps are all, in various combinations, parts of drainage systems. The long-term performance of a substructure is dependent on a seal that can span minor cracks in buried walls and remain serviceable for a very long time. Plastic, fiber-reinforced paper, asphaltic and synthetic rubber sheets, and chemical, asphaltic, cementitious, or bentonite clay coatings are all used for waterproofing in subterranean structures. Synthetic rubber water stops are commonly installed at cold joints (the joint between adjacent concrete pours, one of which has set) in retaining walls and slabs-on-grade. Asphaltic and cementitious plaster coatings are used for dampproofing. Mineral batt, plastic foam, polystyrene, and glass fiber insulation may be called for in various locations underground or in basements to prevent heat loss in structures. Anyone reviewing a set of foundation drawings should expect to find some combination of these systems.

Graphic Expression in Foundation Drawings

Projection Types

Foundation plans, drawn in $1/16$", $1/8$", and $1/4$" = 1'-0" scales (1:200, 1:100, 1:50 in SI) and sections, drawn in $3/4$", $1 1/2$", and 3" = 1'-0" scales (1:20, 1:10, 1:5) are the most common projection types. Axonometric drawings and oblique drawings are occasionally used, as the situation requires.

Line Types

The line types commonly used in foundation drawings are found in Figure 7.2.

Symbols

In plan views, material symbols are not commonly used, with the exception of site and landscape plans; however, design professionals who are trying to identify special conditions will employ a variety of symbols of their own to set those conditions apart from others. Hash marks or poché showing a raised curb on a concrete slab-on-grade, or alternating dots and dashes within a special firewall on a floor plan, are examples. Normally, these customized symbols are explained in a note or legend

The drawing at right shows: 1) The superstructure, that portion of a structure that is above the ground; 2) the substructure, the habitable portion of the structure below ground; and 3) the foundation, the lowest part of the structure, which supports both the substructure and superstructure. Shallow foundations—spread footings, continuous footings, mat or raft foundations, and floating foundations—transfer loads to the earth at the base of the substructure. Deep foundations—timber, concrete, and steel piles, and caissons, installed under pile caps, transfer loads to the earth at some point below the substructure.

a. Spread footing most often support individual columns.

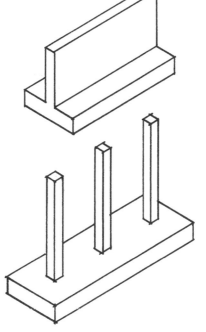

b, c. Continuous footings support walls or a series of columns.

d. Mat or raft foundations are used when the size of spread footings gets to the point that it is cheaper to cast a thickened slab.

e. Floating foundations exchange the weight of the soil removed from an excavation for the weight of the structure.

Figure 7.1 Common foundation systems.

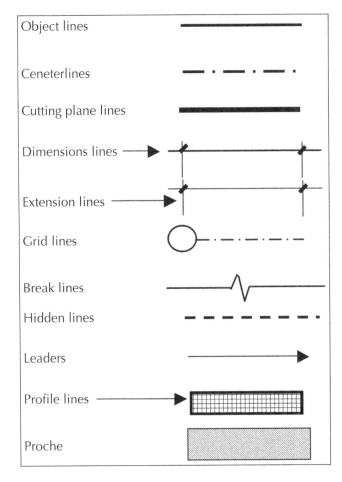

Figure 7.2 Common line types in foundation drawings.

on the same sheet as the conditions they describe. The material symbols commonly used in section views and other details are shown in Figure 7.3.

Foundation Plans

Foundation plans are views of the foundation looking down from above—taken at the slab elevation (for a slab-on-grade) or at some point close to the finished floor level. Gridlines identifying the configuration of structural system are commonly used and are effective reference tools. The National CAD Standard (NCS) suggests that letters be used to identify the vertical gridlines on a drawing sheet, and numbers be used for horizontal gridlines (north is commonly the top of the drawing sheet). Figure 7.4 shows a foundation plan that uses a grid system, though not the one that the NCS embraces.

Elements below the horizontal plane represented on the plan—footings, for example—are hidden from view in reality, and therefore are designated by hidden (dashed) lines. Some architectural offices will lightly superimpose walls depicted on the floor plan onto the foundation plan (whether there are footings under them or not) to help the user of the drawings correlate the location of foundations elements with the floor plan.

*Symbols in the NCS are somewhat different from these, which were selected from and are common to several different drawing sets spanning 20 years.

Figure 7.3 Material symbols commonly used in foundation drawings.

Figure 7.4 This foundation plan uses a grid referencing system, though not the one promoted by the National CAD Standard. Note the idiosyncrasies in this drawing: north is normally the top of the page. (From *The Professional Practice of Architectural Working Drawings*, 2nd edition, by Osamu Wakita and Linde, Richard, John Wiley & Sons, Inc., 1995. Used with permission of John Wiley & Sons, Inc.)

Vertical sections are taken at various locations in the foundation to provide critical additional views. While the sections themselves are typically shown on foundation detail pages, their location and orientation are identified on the plan (see Figure 7.5).

Foundation Details

As noted in Chapter 4, builders need to quickly and efficiently navigate detail pages, and the detail referencing system that the design professional chooses is critical. The architectural office that produced the project that is partially described in Figure 7.5 used a very helpful referencing system commonly known as the split-bubble system, which lists the detail number in a half-circle above the horizontal line, the page on which the detail can be examined on the lower right, and, on the lower left the page on which the detail originated. This system, not yet adopted by the NCS, is extremely help-ful to drawing users in that a detail can be reviewed and then promptly put into its proper context in the project. This referencing technique is analogous to an index in a book; a subject of interest can be located in the index, and the page (or pages) on which the subject is discussed is listed for the user's benefit.

Seeing the foundation in the mind's eye is impossible without contiguous views of its components. The height of the foundation in detail 2 S-2 in Figure 7.6, for example, is indeterminable in the plan view because the third dimension—depth—is not given. This is the major drawback to all multiview drawings: only two dimensions can be shown simultaneously. Depth, as well as other details such as the insulation on the inside of the foundation stem wall, the capillary break under the slab-on-grade, and the slab itself, must be shown in transverse sections of the foundation or described in plan notes which, as previously mentioned, should be described using generic terms to reduce the probability of conflicts with the specifications.

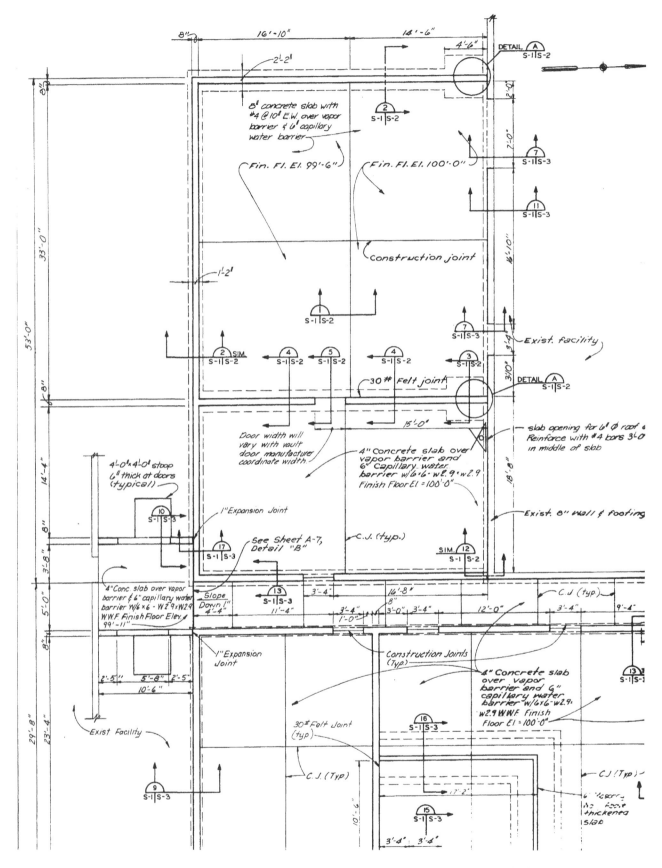

Figure 7.5 This foundation plan uses the **split-bubble referencing** system, easily the most useful detail referencing systems in construction drawing.

#4 Bars @ 10" O.C.E.W.

30 # Felt Joint

#4 @ 10" E.W.

El. Varies
See Plans

8"

6"

#4 Dowels @ 10" O.C.

1" Perimeter
insulation

Vapor Barrier (typ.)

Capillary Water Barrier (typ.)

3"

3-#4 Bars

Elev.= 96'-0"

2'-2"

SECTION ② S-1 | S-2

SCALE: $\frac{3}{4}$ INCH = 1 FOOT

Figure 7.6 Plan views of foundations do not show footing depths. That remains to details such as this one. Refer back to Figure 7.5 to see where this detail applies.

▣ Translating the Drawings into Work

Constructability

The desires of the design professional and certain drafting conventions are sometimes at odds with what is practical and realistic to produce in the field. A concrete crew would not produce the sharp 30-degree plane breaks depicted in Figure 7.7, for example, because of the difficulty of doing it and the negligible benefit that would result.

The undersides of slabs are typically shown by designers as straight lines; and elevation changes, when they occur, are drawn at an angle that is easy to produce with drafting instruments, in this case, a 30-60-90 drafting triangle. Although computers are currently the drafting instrument of choice, it is reasonable to assume that architects and engi-

neers will continue to follow such conventions. The underside of the slab-on-grade is not something anyone would see, and more gradual plane breaks and a gentler slope in the capillary break would not affect the structural integrity or the performance of the slab-on-grade, assuming that the minimum requirements described in the detail were observed (mainly slab thickness and depth of the capillary break). The constructor is required to achieve what is described in the plans, if possible, but does so by balancing plan requirements with the capabilities of the industry and the realities of doing business in a very competitive environment. Although the circumstance just described is a very simple example and would not be an issue on most jobs, dilemmas such as this occur, and the more problematic ones should be broached with the designer and resolved to the satisfaction of the interested parties. Failing to do so could result in the contractor removing and replacing work.

Elev. 99'-6"

8"

6"

6"

El. 100'-0"

8"

6"

#4 @ 10" E.W.

8"
Min.

Vapor Barrier (typ.)

Capillary Water Barrier (typ.)

SECTION ① S-1 | S-2

SCALE: $\frac{3}{4}$ INCH = 1 FOOT

Figure 7.7 Certain features of a building component are frequently not constructed exactly as drawn: the underside of slabs-on-grade is an example. Refer to Figure 7.5 to see where this detail applies.

4" Concrete w/ 6x6
W2.9 x W2.9 W WF
Elev = 100'-0"

Control joint (typ)

#4@10" E.W.(typ.)
30" felt joint (typ.)
Elev. varies
see plan

Vapor barrier (typ.)

Capillary water barrier (typ.)

#4 Dowels @ 10"

3 #4 Cont

Elev = 96'-0"

2'-4"

SECTION 4
S-1|S-2

SCALE: ¾ INCH = 1 FOOT

Figure 7.8 Detail 4 S-2 is a detail of a cast-in-place concrete building wall.

Recognizing the Extent of the Work

Constructing any part of a building can be more complex than the details would at first lead one to believe, in part because the static quality of all drawings makes the numerous steps required much harder to see—they exist in one's imagination. Drawings, after all, describe a finished product, not the processes required to construct them. Consider what detail 4 S-2 in Figure 7.8 requires:

- The distance from the bottom of the concrete footing to the finished floor (FF) is 4' (FF elevation is 100'-0", bottom of footing is 96'-0").
- The footing is 2'-4" wide and 12" high.
- The stem wall that sits on the footing is 8" wide and 3' high.
- There are cold joints depicted in two places: on top of the footing under the stem wall and at the finished floor level.
- There are three #4 longitudinal reinforcing bars with #4 dowels 10" on center (center to center) in the footing.
- The stem wall and wall above the floor line have #4 reinforcing bars 10" on center both ways.
- The slabs-on-grade vary in thickness: one is 8" thick, with #4 bars 10" on center both ways, and the other is 4" thick, with .193"-thick welded wire fabric 6" on center both ways.

The plan and details suggest the following construction sequence (assume that the soil under the building has been properly engineered and the pad constructed):

1. Lay out the wall.
2. Excavate for the footing; dispose of or stockpile spoils; clean trench.
3. Form footing (or use sides of excavation).
4. Install footing steel; clean trench.
5. Inspect forms and steel.
6. Place, consolidate, and finish (float tops of) concrete footings.
7. Install stem wall steel.
8. Form stem wall to elevation 100'-0" (except at doorway; see plan in Figure 7.5); align and brace.
9. Inspect forms and steel.
10. Place and consolidate concrete, cure.
11. Strip forms; patch tie holes below the finished floor elevation (if required by the specifications).
12. Backfill and compact both sides of the stem wall to subgrade level (99'-2" on the left and 98'-10" on the right, 98'-4" at the door).
13. Form slab edges; install capillary break and vapor retarder.
14. Install slab-on-grade steel and control joints between slab and wall.
15. Inspect steel.
16. Place and finish slab-on-grade; cure.
17. Form wall above slab, one side.
18. Install door buck, wall reinforcing steel, and embedments.
19. Inspect steel.
20. Form wall above slab, second side; align and brace; clean forms.
21. Inspect forms.
22. Place concrete in wall forms; consolidate, cure.
23. Strip forms.
24. Patch and sack wall (if required by the specifications).

Consideration of Other Tasks

There are several important points to make regarding the preceding sequence. As detailed a list as this one might seem, it does not reveal all of the work that has to be done or the impact of that work on the planning and execution of the project.

Excavating for the footing described in 4 S2 requires the builder to dig a trench varying roughly from 9' to nearly 18' wide (about 3 to 6 meters) at the subgrade level (top of the trench) to 3' or 4' deep (1 to 1.2 meters), depending mostly on the soil type. The width of the trench is a function of the footing depth, the type and stability of the soil, and the task that will take place in it (see Figure 7.9).

The geotechnical engineer will make recommendations as to the appropriate layback (slope), for a trench based on soil tests and visual inspections at the site, but a 1-to-1 slope is frequently the minimum. For deeper excavations (5' or more), benching may be the only practical method of exposing the work area, since trench shoring, which would be required in a trench over 5' in depth, severely restricts the effective work area, in addition to interfering with the component being constructed.

In addition to determining the size of the trench, the contractor has to select the equipment to be used and determine how the excavation will be performed. In an open area, a pull shovel (backhoe or excavator) can easily and quickly cut a long wide trench, but in a confined area, such as the one in which this project was constructed, a small backhoe (rubber-tired pull shovel) or even handwork might have to be employed. Imagine the shock (not to mention the cost) of estimating a project believing that one can use specialized power equipment, discovering that a site is too confined to use it, and having to excavate it by hand in clayey soil with 4" to 6" river cobble!

The reality is that detail 4 S2 looks the same on the drawings whether it is constructed in highly cohesive or cohesionless soil, but the constructor who fails to recognize the differences in the two might, in the case of cohesionless soil, have to excavate, backfill, and compact 60 percent more material than was estimated (the difference in the cross-sectional areas of the 1-to-2 and 2-to-1 cuts shown in the lower half of Figure 7.9). Complicating matters further, excavation generates spoils (excess earth or rock), which have to be disposed of or stockpiled on or off site. Although the issue is not discussed here, the presence of groundwater within an excavation can totally change the character (and especially the cost) of work. In most cases, the contractor pays for oversights and lapses of judgment regarding the work entailed in an excavation out of his or her profit.

Figure 7.9 Excavating plans must consider safety and adequate workspace.

Inspections

The type and frequency of inspections are driven by the agencies—federal, state, or local—having jurisdiction over various aspects of the work. Requirements vary widely, and the constructor is well advised to investigate them prior to bidding work. Many agencies require project developers to hire independent consulting engineers to inspect and monitor earthwork, the quality and placement of reinforcing steel, structural concrete, and structural welding for building construction projects. Building departments may accept inspections by independent firms as substitutes for their own, or require them as additional inspections. Some agencies—highway departments, city utility departments, and departments of public works, for example—employ full-time quality control consultants who bear the title Resident Engineer (RE) or Clerks of the Works. Regardless of the specific requirements, inspections (which can add one day per inspection to a schedule) must be factored into the construction sequence.

Installation of Reinforcing Steel

The phrase "install reinforcing steel" sounds reasonably straightforward, and it is; however, a number of steps have to occur before the steel is actually installed. Once the type and quantity of required material has been determined, or perhaps while the quantities are being determined, a plan for installing the steel must be developed. Shop drawings are the detailed instructions produced by the contractor, subcontractor or supplier as to how an assembly, component, or system will be installed.[1] They are the subcontractor's opportunity to inform the general contractor and design professional of his or her plan to install or construct an assembly and have it approved prior to the commencement of work. Once approved by the design professional for compliance with the intent of the drawings, shop drawings are used by the skilled tradespeople to properly assemble the elements of a structure. For a simple footing like the one described in Detail 111 (refer ahead to Figure 7.11) and photographed in Figure 7.12, the installation of the reinforcing steel consists of only a few steps:

1. Tie steel for column bases (this can be done on site or in a shop).
2. Set column base mats in excavation.
3. Install longitudinal bars.
4. Tie column base to longitudinal bars (if required) and bar laps.
5. Set steel on dobies (small precast cubes of concrete used to raise reinforcing steel off the earth) or chairs.

Tying steel in other situations—for example, in a wall form on a tilt-up concrete project—a thick slab-on-grade

composed of two mats of steel, or in complicated members in a cast-in-place concrete frame, can be far more complicated, especially when post-tensioning tendons are also used in the member.

It is worth pointing out that the sequence of steps just enumerated completely ignores other important considerations, for example that the steel needs to be transported to the site, unloaded from the delivery truck, sorted and stored (often under specific conditions set forth in the specifications), and delivered to the ironworkers at either a prefabricating area on site or in proximity to the locale where the steel will be installed.

Placing, Consolidating, and Curing Concrete

The project location, access to components that will be cast, the quantity of concrete to be placed, the size and shape of the forms or excavation, and the number, size, and grouping of reinforcing steel are important considerations when placing concrete. Although the A/E may mandate or preclude a specific placement method, it is generally the contractor who decides how the concrete will be placed. Placement methods are often considered during plan review, and are settled upon during the job-costing period.

The success of reinforced concrete as a system depends on the quality and proportions of the ingredients in the concrete, the quality and location of the reinforcing steel (where it is placed within the structural component), and the quality of the bond between the concrete and steel, so the consolidation effort is a critical one. External or internal vibration methods (outside the forms or within the concrete itself) are commonly used to consolidate concrete. For vertical elements such as walls, a handheld electric cable vibrator immersed in the concrete is widely used. Horizontal elements such as slabs-on-grade and wall panels for a tilt-up project are frequently consolidated externally by motorized screeds that have vibratory shafts mounted in them. Architectural precast concrete panels often require that the form tables (panel casting beds) be vibrated.

The concrete curing process can have a dramatic effect on the production rate of a structure. For critical parts of the structure (beams, elevated slabs, and walls, for example), the specifications will describe the acceptable curing method(s) and the time period—usually, 7, 14, or 28 days—before subsequent work can occur (for instance, between placing and backfilling a retaining wall, or placing and loading a concrete beam). The builder may be required to abandon work on parts of the project until the concrete reaches its minimum design strength, or may choose to obtain permission from the designer to change the concrete mix or use methods that will accelerate the curing period. Figure 7.10 shows a shear wall (a wall designed to control lateral forces applied parallel to the wall) that was cast 56 days after the adjoining

---

[1] Shop drawings are discussed in Chapter 14.

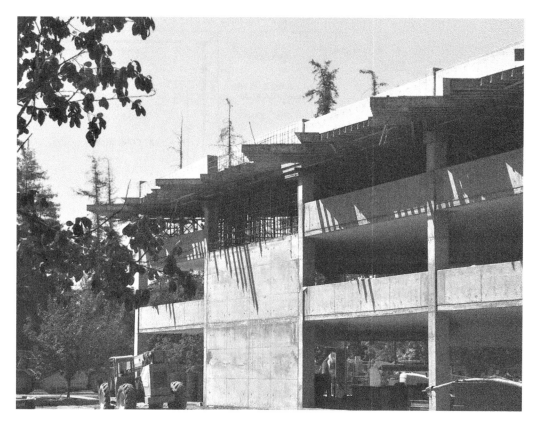

Figure 7.10a Concrete curing specifications may require that portions of the work be delayed. In this project, the shear panels were cast 56 days after the deck and columns. (Photo by the author.)

Figure 7.10b Photograph of completed building. (Photo by author.)

(2)#5 CONT

MC12x10.6 W/(2) ROWS
3/4"ø x6" HSA @ 24"

CONST JOINT

#4 $\frac{36"}{2}$ @ 18"

EXT SLAB
SEE CIVIL

24" MIN

1'-0"

4'-2"

1/8"

SOG - SEE PLAN
FOR REINF

BACKFILL PER SOILS REPORT
AFTER CONC WALL REACHES
3000 PSI BREAKING STRENGTH

8" CONC STEM

#4 HORIZ @ 18"

#5 VERT, CENTERED @ 18", ALT
STD HOOK

8"   8"   1'-2"

(3)#5 CONT

2'-6"

DETAIL
SCALE: 3/4" = 1'-0"

111

Figure 7.11 Detail for a dock at a truck terminal. (Drawing courtesy of Robert S. Blair, Architect.)

concrete. Large concrete slabs-on-grade and walls are frequently cast, finished, and cured in a checkerboard or alternating pattern to allow sections of the concrete to shrink prior to casting adjacent sections. This also allows the constructor to use less form material overall.

It is worth mentioning that the concrete curing process by itself can constitute a significant amount of work. Forms help to protect concrete while it cures; however, exposed surfaces (bridge decks, slabs-on-grade, and site concrete, for example) may require large sheets of plastic, insulated curing blankets, carpet, frequent watering, or chemical curing compound to cure properly.

Backfilling Operations
The horizontal pressure—surcharge—applied to a wall or curb being backfilled and compacted on just one side can be considerable, especially when heavy equipment is used in the compaction effort. Newly cast concrete is particularly vulnerable to such forces, since it does not reach its design strength immediately. Consequently, constructors have to schedule and perform backfilling operations carefully.

In Detail 4 S-2 (Figure 7.8), the stem wall is fairly thick (8") in relationship to its height (3'), and it contains steel that will resist horizontal bending when the backfilling operation is performed. Nevertheless, the backfilling operation would most likely be done on both sides of the wall simultaneously (probably with a small motorized compactor). This minimizes the bending stresses imposed on the wall, and the drawings call for backfill on both sides anyway. As noted previously in comments

on soil types and their effects on excavations, construction drawings are frequently moot on such subtleties, although the specifications may not be. The architect who designed the truck dock detail (Detail 111 in Figure 7.11) considered this issue; he notes in the detail that the backfilling operation cannot occur until the concrete has reached 3,000 pounds per square inch (psi) strength.

Errors, Omissions, Oversights, and Artistic License
As pointed out in Chapter 2, the amount and variety of information required to produce even a simple building is considerable. Mistakes and oversights occur, and professionals differ as to the means and methods of construction and design. While it is not necessarily an oversight, the details in Figures 7.5, 7.6, 7.7, and 7.8 show the slab-on-grade lying directly on the vapor retarder. The American Concrete Institute, an authority on concrete products, recommends the installation of a 1"- to 3"-thick layer of sand under slabs-on-grade.

There are some who believe that the sand facilitates the curing process by allowing water to escape from the underside of the slab, thereby avoiding the warping and cracking that can occur when one side of a slab cures more rapidly than the other. Concrete, which cures by hydration, needs water to cure properly, and it could be that while the sand allows water to escape from the concrete, significant moisture is trapped for a period of time between the concrete and vapor retarder in the sand layer. A sand layer may also help prevent the vapor retarder from being punctured during the concrete placement process by spreading a worker's body weight over a larger area. However, when a structural slab is cast over an impervious layer such as metal decking, the sand layer is not recommended or used.

What is the difference in the two situations? Building officials and the industry are divided on this issue. There are concrete subcontractors who will refuse to warrant slabs-on-grade that are installed directly over vapor retarders, not only because of the warping and cracking that may occur, but also because it is unrealistic to expect that the vapor retarder—commonly, plastic of some sort that is ten-thousandths of an inch thick—will survive a concrete installation without being punctured, which defeats its purpose.

Close scrutiny of Detail 111 reveals that the steel in the concrete slab-on-grade has two bends near the end of the wall dowels. It is probably not the intention of the design professional that every slab bar be bent as shown and installed under the wall dowel. When reinforcing steel has to be modified from its standard shipping condition (straight bars, usually in 40' lengths), additional costs are incurred; and in this case, the bend is not necessary because the slab steel can be attached just as effectively alongside the wall dowel. This is an example of artistic license at work: There is no other reasonable way for the designer to graphically show that the slab bars should both line up with and overlap the wall dowels. Overlapping bars are frequently shown this way.

Both parties (the A/E and constructor) want the same thing—a product that will perform its intended function at a reasonable cost. That said, adjustments to the original construction plan are frequently required to realize that function.

Subtleties in the Drawings

Referring back to Figure 7.8, note that Detail 4 S-2 shows two *cold joints* (the interface between consecutive concrete pours, one of which has set), one underneath the stem wall (on top of the footing) and the other at the finished floor level. They are shown as straight horizontal lines. Having two joints results in the reinforcing steel having to be installed in at least three trips: (1) prior to casting the footing, (2) after one side of the stem wall has been constructed, and (3) after one side of the forms for the wall steel has been installed (or after the footing has been cast and prior to the installation of forms, if prefabricated two-sided wood and metal forms are used). It is entirely possible that the engineer would require this sequence to be followed; however, it is worth pointing out that the stem wall and wall above the floor (which is 14' high) could easily be formed as a single unit, thus eliminating unnecessary fabrication (cutting and packaging the stem wall steel), one trip by the reinforcing steel subcontractor (to install the stem wall steel), and at least one inspection. Here is the adjusted sequence:

1. Lay out the wall.
2. Excavate for the footing; dispose of or stockpile spoils.
3. Form footing (or use sides of excavation).
4. Install footing steel; clean trench.
5. Inspect forms and steel.
6. Place, consolidate, and finish (float tops of) footings.
7. Form one side of stem wall and wall above finished floor.
8. Install door buck, wall reinforcing steel and imbedments.
9. Inspect steel.
10. Form remaining side of stem wall and wall above finished floor; align and brace.
11. Place concrete in wall forms; consolidate, cure.
12. Remove forms, brace wall; patch wall below and above finished floor (if required by the specifications).
13. Backfill and compact both sides of the stem wall to subgrade level (98'-4", 98'-10", or 99'-2").
14. Install capillary break and vapor retarder.
15. Install slab-on-grade steel.
16. Inspect slab steel and edge forms.
17. Place and finish slab-on-grade; cure.

This installation results in a finished product that is, for all intents and purposes, as good as that prescribed by the A/E, perhaps better, but with fewer steps and less overall time to the construction schedule. The contractor may improve the profitability of the project by compressing the schedule in this manner. When opportunities such as the one presented here are discovered and are clearly and carefully presented well in advance of the scheduled installation, the A/E will frequently grant permission to change the construction method, especially if it results in a savings to the owner.

Reconciling the Drawings and Required Work

Every construction project is unique, and foundation work—particularly in an urban setting or when groundwater is present—can be very involved. As noted in Appendix A, understanding how construction work is performed, knowing the strengths and weaknesses of a wide variety of materials, and being able to visualize a project under construction are essential to an effective takeoff.

The foundation detail shown in Figure 7.11 is relatively innocuous in terms of the demands placed on the builder, but nevertheless requires a number of steps. The photographs in Figure 7.12 and 7.13 summarize how it was constructed.

The contractor who built the retaining wall described in Detail 111 had the benefit of a clear workspace and a paved parking lot in which to build, which he used to

Figure 7.12 The builder has chosen to use the sides of the excavation for forms, and is installing the reinforcing steel in the footing. (Photo courtesy of Blueline Construction.)

advantage. The photograph in Figure 7.12 clearly shows that the asphaltic concrete paving was first saw-cut and removed, after which the 12" -deep footing was excavated. The builder elected to use the excavation as formwork for the footing concrete, thus saving considerable labor and materials, installing and removing forms, and additional excavation costs, with a slight increase in concrete quantity cost. Spoils were generated by the footing excavation—the footing concrete occupies the entire space excavated— however, the detail shows nearly 3' of fill on the right side of the wall, so the spoils were no doubt temporarily stock-piled. Figure 7.13 captures the steps required to construct the wall to the point where it can be left to cure.

It is worth pointing out that work remains to be done before the slab that forms the dock surface can be cast, namely:

- After the concrete has attained 3000 psi strength, the wall forms have to be disposed of or, more commonly, stripped, cleaned, sorted, stacked, and hauled away.
- Voids caused by snap ties on the inside face (right side) of the wall may have to be patched, to prevent the intrusion of water into the wall and the subsequent corrosion of the reinforcing steel that could occur.

- This wall does not show dampproofing and a drainage system; however, they are frequently required behind walls, and are commonly installed after the forms have been removed.
- The wall has to be backfilled and compacted, which may involve importing material.
- Miscellaneous channel (C-shaped rolled steel sections designed to protect the slab edge from damage) with headed stud anchors welded to the inside corners of the channel has to be attached to the outside face of the wall and adequately braced against movement during the casting of the slab.
- The slab steel remains to be installed, and the slab-on-grade placed, finished and cured.

The Takeoff

A material quantity takeoff for the wall just described would have to include all of the materials described in the drawings and specifications, as well as many that are not (the formwork for example). Determining material quantities in building construction and identifying the required equipment are not usually the challenge—after all, the required elements are described in the drawings

The footing has been cast.

The form sill has been installed on one side.

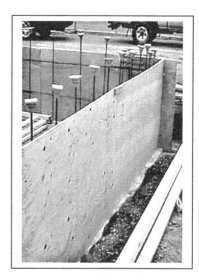

One side of the forms has been set.

The ties and slab dowels have been installed and the second side of the forms is being installed.

Forms installation is complete and the wall has been aligned and braced.

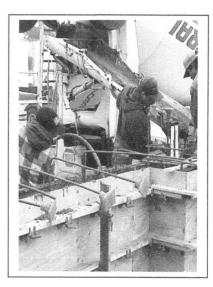

Concrete is being placed, and consolidated with an internal vibrator.

Figure 7.13 Evolution of a CIP wall. This sequence of photographs shows common steps in the construction of a cast-in-place concrete wall, in this case the dock wall described in Figure 7.11. (Photos by the author.)

TABLE 7.1 Shopping List for the Concrete Wall in Figure 7.11

SHOPPING LIST Project Name and Number_____		By:_____ Page_____
Components	Operations	Materials
02200 — 02300*Site prep & structural earthwork		
Dock wall (det 111, S3) Prep site	Lay out wall; cut and remove existing AC paving (LF, LS?).	Easy access; sufficient room to work.
Footings	Excavate footings (CY; compact to required density (p. 70 specs).	Stockpile onsite for backfill, ample room for backhoe;
Stem wall	Backfill dock behind wall (CY).	Use footing spoils, import balance in common earth (CY).
Slab-on-grade (det 111, S3, foundation plan S2)	Place rock under slab (CY).	Drain rock (tons). (specs, p. 78)
03100* –Concrete Forms and Accessories		
Dock wall (det 111, S3) continuous footing	Form footings 12" high.	Use excavation.
Dock wall	Form dock walls (SFCA).	Plywood, lumber, nails, ties, hardware — Reuse forms.
	Form corners and column supports (SFCA).	As above, but column anchor bolts by others.
Slab-on-grade (det 111, S3, foundation plan S2)	Place and brace MC edge protection (LF?).	Labor only, MC channel by others.
03300* – CIP concrete		
Dock wall (det 111, S3)	Place and consolidate concrete (CY).	3000 PSI concrete, (specs p. 80), internal vibration.
	Patch tie holes behind wall (EA).	Sand, cement, (specs. p. 81).
	Sack and patch exposed concrete (SF).	As above.
Slab-on-grade (det 111)	Place concrete (CY).	3000 PSI concrete (specs p. 80).
S3, foundation plan S2)	Finish concrete (SF).	Rough broom (specs p. 81).
	Cure concrete (SF).	Chemical cure (gals) (specs. p. 81).

*From the 1995 edition of CSI's MasterFormat.

and specifications, and the contractor is normally required to construct only what is described by the A/E. The challenge (and risk) more often lies in determining the work—the labor, equipment and time—involved in constructing the component or assembly.

As noted in Appendix A, contractors evaluate construction costs by identifying a variety of components and the operations and materials associated with them, then translating the operations into units that can be related to costs ("buy" or purchase units and units of productivity), and summarizing the resulting figures in a useful matrix (the estimate). The operations into which a component is separated are distinguishable by the steps associated with the material itself, or associated with a specific work category. For example, the operation called "place and finish slab-on-grade" requires that the concrete be placed, consolidated, struck off, floated,

finished, and cured, steps that occur when concrete is used, no matter the application. While the attention to detail and involvement of a crew and machinery in each of these steps varies according to the individual project requirements, these steps nevertheless occur when concrete is used as a material. For example, in a laser screed operation, the machine known as the laser screed consolidates, strikes off, and floats the concrete after it has been deposited out of the chute of a truck or placed by a pump. It remains for the cement finishers to finish and cure the concrete.

Operations are commonly listed in what are known as "shopping lists," which are developed when the estimator initially reviews the project for categories of work. For the contractor planning to construct the component depicted in Figure 7.11 with his or her own forces, the shopping list might look like that shown in Table 7.1.

Figure 7.14 Various ancillary systems are shown in this schematic drawing of a basement wall. (From *Fundamentals of Building Construction Materials and Methods*, 3rd edition, by Edward Allen, John Wiley & Sons, Inc, 1999. Used with permission of John Wiley & Sons, Inc.)

Ancillary Systems

The basement wall in Figure 7.14 describes the essential components of the ancillary systems often incorporated into the substructure, including the waterproofing (or dampproofing) membranes on wall and slab, the drainage mat and pipe, and protective boards. The only other elements not shown in this drawing that often play a role are insulation (rigid or batt insulation, depending on where it is installed) and parging, which in at least one application

FIGURE A

Figure 7.15a (Drawing by Tom Krejci, after Keith Bisharat.)

FIGURE B

Figure 7.15b (Drawing by Tom Krejci, after Keith Bisharat.)

is a thin layer of plaster applied to CMU walls as a substrate for a damp- or waterproofing membranes.

▨ Summary

Interpreting foundation drawings from the viewpoint of the constructor requires an understanding of the environment in which foundations perform and the goals and responsibilities of the design professionals who have been retained to design them. This understanding leads to more rapid absorption of the numerous details that describe the work and the efficient integration of the details and processes required to construct foundations. It also empowers the constructor to contribute effectively to the resolution of constructability and performance issues, since for a variety of reasons what is finally done in the field is not always reflected in the foundation drawings.

What appear to be relatively simple elements in a foundation can result in complex undertakings, depending on the nature of the soil, the level of the groundwater, and the project being constructed. The prudent builder will recognize subtleties in the drawings and account for them in the project review stages. Although foundation drawings normally direct the builder to construct an element or system in a specific way, opportuni-

ties to produce the desired result in more efficient ways exist. Awareness of the potential to increase profits by focusing on efficiency in the field is one hallmark of the successful constructor. Both parties (the A/E and constructor) want the same thing: a product that will perform its intended function at a reasonable cost. Adjustments to the original construction plan, though, are frequently required to realize that function.

CHAPTER 7 EXERCISES

1. Analyze the two footings shown in Figures 7.15a and 7.15b and identify the differences in how they would be constructed.

2. Using Figure 7.15a of question 1, determine the amount of concrete required for 200 LF of the footing only. Ignore the slab-on-grade and assume the wall is a 2 × 4 framed wall.

3. Develop the installation plan for the two foundations described in Exercise 1 using words and drawings. Assume that the soil is sandy silt.

4. Identify the principal foundation types and the proper applications of each.

5. List the common forces resisted by foundations.

6. What other systems are often incorporated into foundations?

8 Framing Systems

Key Terms
Beam
Cast-in-place concrete
Column
Compressive stresses
Falsework
Fenestration
Formwork
Girder
Glued laminated beam
Precast prestressed concrete
Shoring
Sitecast concrete
Tendons
Tensile stresses

Key Concepts
- The corollaries made between buildings and the human body are instructive. Skeleton, tendons, joints, sheathing, footings—vascular, plumbing, nervous, and control systems—are all terms that can (and do) effectively describe both.
- Construction is a business of details.

Objectives
- Identify the purpose of framing system drawings.
- List and describe the common framing systems.
- Identify common graphic conventions in framing system drawings.
- Identify the roles of the architect and design consultants in construction drawing production.
- Reproduce drawings of simple framing systems.
- Correlate construction processes with construction drawings.

Construction Graphics: A Practical Guide to Interpreting Working Drawings,
Third Edition. Keith Bisharat.
© 2025 John Wiley & Sons Inc. Published 2025 by John Wiley & Sons Inc.

Purpose of Framing Systems

Framing systems represent the bones and connective tissue of superstructures—it is through the framing system that the loads imposed on a structure are transferred to the foundation. The variety of structural systems is impressive—numerous combinations of the basic systems in nearly any conceivable configuration can perform the duties expected of structural framing systems. The basic systems are:

- Cast-in-place (CIP) reinforced concrete
- Cast-in-place post-tensioned concrete
- Precast prestressed concrete
- Structural steel
- Reinforced masonry
- Heavy timber
- Wood light framing

Cast-in-place reinforced concrete wall panels tilted into place, with tube steel columns, glued-laminated roof beams supporting manufactured wood trusses, with dimensioned lumber roof joists and plywood or oriented strand board sheathing between trusses is one example of a common structural design for warehouses.

CIP Concrete

Every system has its strengths and weaknesses. The most successful of them are synergistic in nature (1 + 1 = 3). Concrete is extremely durable in a variety of environmental conditions and is widely available throughout the world. Indeed, there are numerous examples of concrete structures that not only have survived for hundreds of years, but remain functional today. Cast-in place reinforced concrete is economical, fire-resistant, very durable, and extremely flexible from the standpoint of design. The Achilles' heel of concrete—its inability to resist tensile forces—is made manageable by the use of reinforcing steel in areas of high tensile stress. Additionally, the elements in this system are normally chemically compatible and respond similarly when the temperature fluctuates. Cast-in-place post-tensioned concrete is also very economical, and raises simple cast-in-place reinforced concrete to the level that it can support long bridge spans.

Precast Prestressed Concrete

All of the principal structural elements—columns, beams, slabs, and wall panels—can be made of precast prestressed concrete, which offers speed and flexibility in the construction process, and greater strength and higher-quality finishes than site-cast concrete does. The members are manufactured in a controlled environment, transported to the site, and erected as rigid elements, which allows the constructor to assemble the structure

in inclement weather, particularly cold weather, and to avoid forming, concrete cure time, and form stripping. In some situations, precast concrete is selected because it enables a project to be constructed with minimal disruption to sensitive environments. Slender, somewhat more graceful proportions in structural elements are possible with precast prestressed concrete primarily because the tendons are more easily oriented for structural efficiency in a plant, and each element is steam-cured, which translates to lower weight and higher strength. Precast prestressed concrete is more limited from the design standpoint than site-cast concrete is, and transporting the elements can present logistical problems for a builder.

Structural Steel

Structural steel is in many ways the ideal construction material in that it has nearly limitless applications, if its shortcomings—susceptibility to corrosion and intense heat—are properly accounted for in the design process. Light in comparison to its strength, steel has for years been the structural material used for record-setting bridges and tall buildings. It lends itself well to rapid construction (the Empire State Building, for years the tallest building in the world, was completed in just over a year), particularly when the design of the structure bears redundancies, and it is one of few isotropic construction materials.

Reinforced Masonry

Reinforced masonry—brick and concrete masonry units combined with reinforcing steel and concrete grout—is an economical, flexible structural system that shares many of the advantages of precast concrete. Reinforced masonry systems are rapidly constructed, they require minimal *falsework*, (e.g., bucks for openings), and there is a nearly endless variety in the shapes, sizes, colors, and textures of the units.

Heavy Timber

Heavy timber construction, consisting of large new and recycled natural timbers and manufactured wood columns and beams, is commonly used in smaller buildings such as churches, schools, warehouses, and, occasionally, residences, although larger structures like small stadiums have been constructed with them as well. Imbued with great flexibility in design and the potential for visual interest and a warmth that other structural systems simply do not have, heavy timber construction shares many of the strengths that precast concrete possesses: It has high strength in proportion to its weight, lends itself to rapid installation, and has remarkable resistance to fire for a combustible material. It is, however, susceptible to damage, particularly by water and pests; and unlike welded steel, masonry, and concrete connections (including precast concrete), heavy timber connections are not integrated connections—they remain forever mechanical connections.

Wood Light Framing

Wood light framing is the predominant construction material for single-family residences, apartments, condominiums, and small commercial building projects. It is the least expensive form of durable construction and it offers as many variations in design as there are designers to conceive them. Wood light framing systems are rapidly assembled, frequently without hoisting equipment, using a minimal number of small hand and power tools. Natural wood used for framing systems has a very high strength-to-weight ratio. Surprisingly, one species of wood is second to the high-strength steel used as tendons in post-tensioned and pretensioned concrete structures. Spruce that is free of knots and loaded parallel to the grain has a *breaking strength* in tension of 12.5 miles (20 km), significantly higher than aluminum—just over 6 miles (10 km)—and mild steel—just over 3.1 miles (5 km)![1] The ability of a properly designed and constructed wood light frame structure to resist high intermittent loading (caused primarily by wind, snow loads, and seismic events) for short periods of time makes it a good choice in virtually any region. Chemically treated plywood sheathing and studs can be used to frame basement walls in residences. Since it is a naturally occurring material, lumber, like soil, is subject to some performance uncertainty; however, the uncertainty is minimized by quality-control inspections performed by independent groups such as the Western Wood Products Association (WWPA). The weaknesses of wood light framed structures are that they are susceptible to damage by fire, water, and pests, and connections between members have to be properly detailed and constructed; however, such obstacles to a durable building are easily overcome.

■ What to Expect in the Drawings

The programming requirements of a project strongly influence the selection of a structural frame. In building construction projects, the choice of one or another type is the architect's. A retail center, for example, depends heavily on attracting patrons by eye-catching storefront displays. Large expanses of window and open space close to customer entrances tend to typify this sort of project. Adjacent to the truck docks of a warehouse, room to maneuver forklifts that are loading and unloading trucks is critical. Elsewhere in the warehouse, the space requirements could be quite different. Where longer spans are required, different methods of bracing the structure laterally are also required. These requirements supersede the additional costs that are normally associated with longer spans—larger columns, footings, deeper beams, and moment-resisting bracing. The selection of a structural frame for engineering and industrial construction projects is the engineer's. Architects and engineers routinely conduct cost-benefit analyses during the design phase to help them balance their priorities until such time as the structural system is decided upon.

Framing Systems in Architectural Drawings

In building construction projects, architectural floor plans are the initial source of information regarding the structural system to be built. These drawings identify the locations and proportions (primarily length and thickness) of exterior and interior walls, column locations, and other important information. The third dimension—height—is left to building elevations and sections, which commonly succeed floor plans in the order of drawing sets. It should be pointed out that graphic depictions of framing systems in architectural drawings are to some extent schematic—the specific characteristics of the structure are normally left to the structural engineer to describe in detail. In highway construction projects, bridge and retaining wall structures often follow highway plan and layout drawings, profiles, details (of everything but structures), quantities lists, planting plans, and the like. Other engineering projects often reflect a different order.

■ Common Graphic Conventions

Projection Types

The drawings that describe the structural system of a project make use of all of the different projection types, including plan views, elevations, sections, and pictorial drawings. Figures 8.1a–g comprise a sampling of the projections used to depict structural systems in projects.

Applicable Line Types

There are frequently thousands of different elements to describe in structural drawings. Consequently, all of the line types available to the design professional are used in the course of describing the project. The most common line types are shown in Figure 8.2.

Symbols

The symbols used in structural drawings describe the materials common to the superstructures of projects

---

[1]*Shaping Structures—Statics*, by Waclaw Zalewski and Edward Allen, John Wiley & Sons, Inc., 1998.

Figure 8.1a Structural systems drawings employ every type of projection used in technical drawing of structures. This plan view shows the required nailing on the roof of an office building. (Drawing courtesy of Buehler and Buehler Structural Engineers.)

Figure 8.1b In addition to showing elevations of one side of a structure, structural drawings may show partial building elevations and elevations of individual components. (Drawing courtesy of Buehler and Buehler Structural Engineers).

Figure 8.1c This braced frame for a mid-rise building is shown in elevation, that is, as if one were at a distance examining the surface features.

Figure 8.1d In this drawing, the plan view of a column base plate (upper drawing) is juxtaposed against a section of it (lower drawing). This orientation not only explains the component, it makes it easy for the design professional to draw. An explanation of the welding requirements indicated in this drawing appears in Appendix E.

FOR DETAILS, SEE S5.3.

BRACED FRAME ELEVATION
GRID 4/E-F

E
S3.1 **SCALE: N.T.S.**

Figure 8.1e As in other structural drawings (floor framing plans, for example), this one is somewhat schematic in nature; that is, it does not duplicate or even come close to what the actual frame will look like.

Figure 8.1f In graphic depictions of structural details, plan views, elevations, and sections (cuts) are all used, often side by side. In this drawing, an elevation (left) accompanies a section (right). (Drawing courtesy of Buehler and Buehler Structural Engineers.)

Figure 8.1g In this detail, a plan view of the connection is shown on the left, with an elevation view shown on the right. A photograph of this condition appears in Figure 10.4p. An explanation of the welding symbols in this drawing is given in Appendix E. (Drawing courtesy of Buehler and Buehler Structural Engineers.)

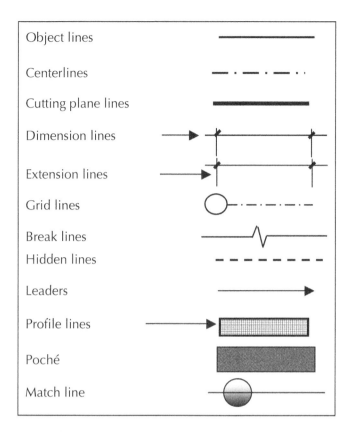

Figure 8.2 Line types commonly found in framing drawings.

(the portion of the project above ground). The short list includes those shown in Figure 8.3a.

Scale

The scale of framing system drawings is the same as that of the architectural drawings:

- Floor plans: $1/16$", $1/8$", $1/4$" = 1'-0" (1:200, 1:100, 1:50)
- Exterior elevations: same as the floor plans
- Building sections: same as the exterior elevations, or larger if possible ($1/2$" = 1'-0", or 1:20)
- Wall or partial sections: $1/2$" − 1'-0", $3/4$" − 1'-0" (1:20, 1:10)
- Details: depending on the complexity and size of the object, $1/2$" = 1'-0", 1" = 1'-0", $1 1/2$" = 1'-0", and 3" = 1'-0" (1:20, 1:10, 1:5)

■ Framing System Drawings: Subject Matter and Orientation

As noted in Chapter 2, the information within disciplines in construction drawings goes from general to specific. The task of designing and describing the framing system in commercial building projects is relegated to structural engineers under contract with the architect. For engineering projects such as highways and bridges, the

Figure 8.3a Common symbols in framing system drawings.

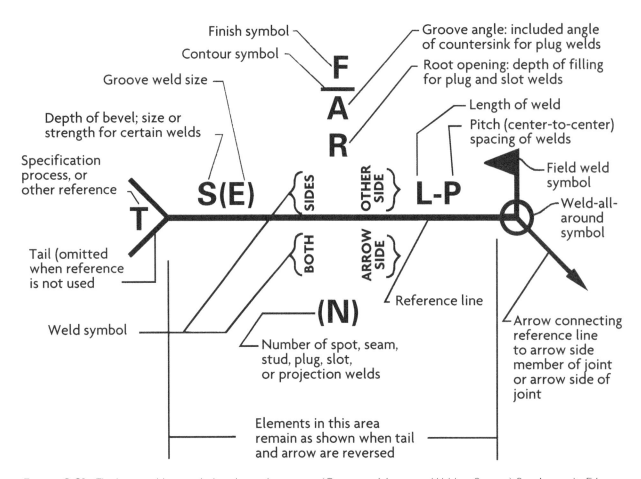

Figure 8.3b The basic welding symbol is a line with an arrow. (Courtesy of American Welding Society.) See Appendix E for more on welding.

entire project is normally the responsibility of civil engineers who are frequently employed by highway departments. Specific aspects of a highway project—bridge structures, for example—are the responsibility of structural engineers in the state highway departments or private-sector firms retained by those departments.

Structural drawings, which generally follow architectural drawings in a drawing set, reveal the structure in the same sequence the architectural drawings do. In their character, structural drawings amount to X rays and CAT scans of a building: They reveal its "bones," their size and shape, and the manner in which they are connected to other bones and "tissues" in the system. Other characteristics of the structure—the cosmetic treatment of the exterior face of a tilt-up panel, for example—are often ignored completely by the structural engineer, who most frequently refers the reader of the drawings to the architectural sheets for that information.

The first structural sheets describe the project as a whole. If the structure is a multistory building, the ground floor is described first, followed by successive floors (second, third, fourth, etc.) and the roof plan. If there are ancillary structures, such as mechanical buildings or penthouses, or portions of the project that are of particular interest, the plans and sometimes the details of them

are shown next, followed by sections and elevations of the structure or portions of it. The balance of pertinent framing system information is shown in increasing detail. The final sheets in a structural set describe tens or hundreds of individual connections within the structure.

Connections between foundations and walls, walls and floors, walls and roof, and detailing around openings, within and through walls and structural components such as beams are all of particular interest to builders, since it is the details wherein the devil resides.

Figures 8.4 through 8.7 are excerpts from construction drawings that illustrate how various framing systems are depicted graphically. Worth noting in a review of these excerpts is the amount of work required to adequately describe (and interpret) what amounts to small parts of the buildings that were the subject of the entire drawing set (e.g., Figure 8.6).

■ Translating the Drawings into Work

As noted in an earlier chapter, the documents created and assembled by design professionals are the graphic and textual descriptions of a structure in its com-

(text continues on page 151)

Figure 8.4a This plan view of a cast-in-place concrete highway bridge shows an existing structure as well as its replacement. Notice the 1:200 SI scale.

Figure 8.4b This elevation was drawn in line with and above the plan view in 8.4a on the same sheet. It shows four rows of existing piers to be removed, and the new piers to be constructed (pier 2), as well as old and new profiles of the earth.

Figure 8.4c The bridge is symmetrical in the two principal directions. Consequently, a single section suffices for the entire structure. The dimensions are given in meters, except at the crown of the road where the thickness of the box is given in millimeters.

Figure 8.4d This foundation plan shows existing piles and grade beams to be removed, as well as the new south abutment and the piers in the center of the span.

ABUTMENT 1 PLAN & PILE LAYOUT
1:50

ABUTMENT 1 ELEVATION
1:50

Figure 8.4e These two drawings identify the number and spacing of piles supporting the south abutment, as well as the layout of the wing walls, using two views.

Figure 8.4f The piles designed to support the abutments are detailed in these two views.

Figure 8.4g This drawing is a transverse section of the bridge abutment. Notice several detail references, in this case to standards developed and required by the highway department.

Figure 8.4h The girders supported on the two main piers are to be constructed according to these drawings.

Limits of ⊘BO-5/5-10⊘ or ⊘BO-5/5-11⊘ & distribution reinf, typ.

915 | 915
₵ Pier

75

#16 @ 300

110 mm clr to main reinf

#19 @ 300 within column area

100 mm fillet

#25 tot 9

#19 ▭ evenly spaced vertically

75

#36 tot 14
#19 tot 8
#32 tot 14

100 mm clr to main reinf

#16 1450 @ 300 each way
Bar shall hook #36 reinf
Hooks to be field bent to 135°

Omit upper #16 bars

Note: Column reinf. not shown for clarity.

SECTION A-A
1:20

₵ Col

₵ Pier

Approximate location of bent cap bottom reinf at column location

(Typical at all columns)
SECTION B-B
1:20

NOTES:

1. For bar spiral splice and spiral anchor and hoop detail **see** "MISCELLANEOUS DETAILS" sheet.

Figure 8.4i In Figure 8.4h, the reader is introduced to the reinforcement required at and between the piers at center span. Here the details around each pier are shown in larger scale.

Figure 8.4j In this drawing, attention is focused on the reinforcement in the piers. CISS pile stands for cast-in-steel-shell pile.

Notes:

1. Splices in main reinforcement shall be mechanically butt spliced. Splices must be staggered a minimum of 1500 mm. Only one splice permitted per bar in each allowable splice zone.

2. #16 spiral or hoop shall be butt welded or mechanically spliced.

3. Splicing does not apply to casing.

4. For specified tip and cut-off elevation, see pile data on "General Plan" sheet.

Figure 8.4k The piers at the abutments (top) and in the center of the span are detailed in these section views. Section A-A pertains to piles that support the bridge abutments; section B-B pertains to center span support.

Figure 8.4l This is a partial view of the typical bridge section. It focuses on the dimensions of key elements of the box girder. Little other information is offered in this drawing, but see the next figure.

Figure 8.4m This drawing is devoted to describing the longitudinal steel in the box girder (shown as dots and circles). The steel that is parallel to the plane of the drawing (the stirrups, for example) are lines.

Figure 8.4n This and the next drawing show the girder steel as well; however, this view focuses on the spacing of the transverse steel.

Figure 8.4o This very simple drawing shows the spacing of the transverse girder steel. It appears directly under the drawing in Figure 8.4n in the drawing set.

Figure 8.4p This drawing depicts the upper deck steel. Notice the concentration of steel over the pier 2 girder, where tensile stresses in the structure are high.

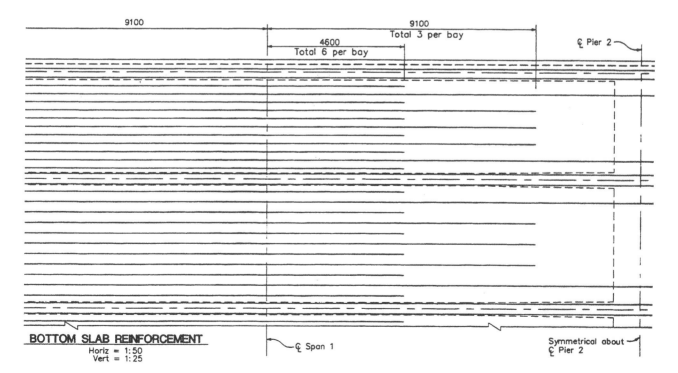

Figure 8.4q The steel in the bottom of the deck is most closely spaced at midspan, where tensile stresses are highest in the bridge.

Figure 8.5a Drawings of structural steel framing systems begin with the foundation plan, which is where the columns and footings that carry the frame are described. (Drawing courtesy of Buehler and Buehler Structural Engineers.)

Figure 8.5b Footings are often depicted in wall sections on subsequent sheets, but in this instance the engineer is showing just a footing section, denoted C S3.1 on the plan in 8.5a. (Drawing courtesy of Buehler and Buehler Structural Engineers.)

Figure 8.5c The first-floor framing plan commonly shows column locations and lists girders and beams by size. The floor deck is also described on the plan. The girder designation W21 × 50 C = + 3/4" (above gridline F) and 30-5-30 (below gridline F) is, respectively, the girder size and camber and number of headed stud anchors required in each third of the beam (left, center, right). The beam designation is slightly different (see lines perpendicular to girder lines): Above the beam line following the beam size is the number of headed stud anchors to be uniformly distributed between columns on the top of the beam, with the camber listed below the beam line. (Drawing courtesy of Buehler and Buehler Structural Engineers.)

Figure 8.5d On the plan shown in 8.5c, the reader is referred to several sections and details, 16 S5.2 (this one) among them. This detail refers the reader to yet another detail, 14 S5.1. (Drawing courtesy of Buehler and Buehler Structural Engineers.)

Figure 8.5e The information trail on one connection in this structure (16 S5.2) ends with detail 14 S5.1, shown here. (Drawing courtesy of Buehler and Buehler Structural Engineers.)

Figure 8.5f Detail 13 S5.1, introduced in the plan shown in Figure 8.5c, makes reference to two other details, A S5.1 and 1 S5.1. (Drawing courtesy of Buehler and Buehler Structural Engineers.)

510S100–1

Connection Schedule

Beam Size	No. & Dia. A-325-N Bolts	Shr ℞ 't'	◁w◁	Max. load capacity
C8 & C10	2 - ⅞"φ	¼"	³⁄₁₆"	11.1ᵏ
W8 & W10	2 - ⅞"φ	¼"	³⁄₁₆"	11.1ᵏ
W12 & W14	3 - ⅞"φ	¼"	³⁄₁₆"	22.4ᵏ
W16	4 - ⅞"φ	³⁄₈"	¼"	35.7ᵏ
W18	5 - ⅞"φ	³⁄₈"	¼"	49.4ᵏ
W21	5 - ⅞"φ	³⁄₈"	¼"	49.4ᵏ
W24	6 - ⅞"φ	³⁄₈"	¼"	63.0ᵏ
W27	7 - ⅞"φ	³⁄₈"	¼"	76.7ᵏ
W30	8 - 1"φ	³⁄₈"	³⁄₁₆"	101ᵏ
W33	9 - 1"φ	½"	³⁄₈"	119ᵏ
W36	10 - 1"φ	⅝"	³⁄₈"	158ᵏ

1. This schedule applies to _non-frame_ connections, typical.

Figure 8.5g Detail A S5.1, first referenced in 13S5.1 (see Figure 8.5f), turns out to be a connection schedule that lists beam sizes; number, diameter, and type of bolt; shear plate requirements; welding size and type; and the load capacity of the connection. (Table courtesy of Buehler and Buehler Structural Engineers.)

Figure 8.5h Detail 1 S5.1 (see detail 13S5.1 in Figure 8.5f), is a detail of the bolting pattern and welding information for all beam-to-girder connections. Notice how this detail refers the user to AS5.1, the connection schedule in Figure 8.5g. (Drawing courtesy of Buehler and Buehler Structural Engineers.)

Figure 8.6a In this partial floor plan for a reinforced masonry structure, the wall descriptions are very simple. Note the conservative use of the masonry symbol and the consequent uncluttered appearance of the drawing. The split-bubble referencing system used throughout these drawings directs the reader's attention to several details, depicted on other pages as well as the page on which they originate. Details 1 A-4/A-6 and 3 A-4/A-6 are building sections; details A and B A-4/A-4 are details of the connection to existing concrete columns; and detail E A-4/A-11 is a roof connection detail. In the upper right part of the drawing is the reference to an exterior elevation (A A-4/A-5).

Figure 8.6b Details A A-4/A-4 and B A-4/A-4, shown here, were drawn on the same page as the floor plan in 8.6a. These two details describe how the new masonry construction is supposed to connect to the existing concrete columns shown on the floor plan in 8.6a.

NG
SPOUT

ROOF MOUNTED
EXHAUST FAN

EXIST CONC
BEAM

SHT. METAL
SHPAN

NEW SHT METAL
DOWNSPOUT

3/4" H.M. INSUL.
ITY DOOR.

NEW WEATHER-
PROOF LOUVER
W/SECURITY BARS
(TYP)

3'-4"

4'-0"

ELEVATION

A

A-4 A-5

FOOT

Figure 8.6c This partial elevation depicts the view from outside the building, looking at the hydraulic room in 8.6a. It is first referenced on the plan depicted in Figure 8.6a.

CONC. BEAMS & COLUMNS
OND

NEW MECH. EXHA

TEEL JOISTS

OPEN
ICE BAY

HYDRAULIC ROOM
118

CONC. COLUMNS BEYOND

ONC FLOOR SLAB
PILLARY WATER BARRIER

5
A-6 A-6

6
A-6 A-6

1
A-4 A-6

BUILDING SECTION

EXHAUST

SCALE: 1/4 INCH = 1 FOOT
12" 0   5

Figure 8.6d This drawing is part of building section 1 A-4/A-6, which was taken through the hydraulic room (see Figure 8.6a for location and orientation).

Figure 8.6e This wall section is one of two taken in the hydraulic room. Notice how the building is revealed, piece by piece, in increasing detail.

Figure 8.6f This is the second wall section pertinent to the hydraulic room.

JAMB DETAIL
B
A-3 | A-8

SCALE: 3 INCHES = 1 FOOT

1¾" INSUL. H.M. SECURITY DOOR

SWEEP STRIP

THRESHOLD - BED IN SEALANT - (TYP.)

CONC. FLOOR SLAB

4" x 10" PRECAST CONC SILL W/ 2 #5 BARS AND NOSING

PAVEMENT

2"

1"

SILL DETAIL
C
A-6 | A-8

TYPICAL DOOR DETAILS

SCALE: 3 INCHES = 1 FOOT

0   3"   6"   12"

Figure 8.6g This door detail is one of three that the reader is referred to by wall section 6 A-6/A-6 (see Figure 8.6f).

½" GWB ON 3⅝" METAL STUDS @ 16" O.C. W BATT INSUL. W/ VAPOR BARRIER BETWEEN THE STUDS

8" CMU WALL

8" x 16" H. C.M.U. BOND BEAM LINTEL W/ 3 #5'S

DAMPPROOFING - (TYP.)

8" C.M.U. BOND BEAM LINTEL W/ 2 - #5'S

SEALANT - JOINT TYPE MJB BOTH SIDES (TYP.)

5¾" P.S. DOOR FRAME - FILL W/ GROUT

1¾" INSUL. H.M. SECURITY DOOR

SEALANT - JOINT TYPE MJB (TYP.)

HEAD DETAIL
L
A-6 | A-8

SCALE: 3 INCHES = 1 FOOT

HEAD DETAIL
M
A-8

SCALE: 3 INCHES = 1 FOOT

5¾"   1⅞"

⅜"   2"   2"

Figure 8.6h The detail on the left belongs to wall section 6 A-6/A-6. The detail on the right occurs elsewhere, but the similarity in information in both details compelled the architect to save time writing notes twice.

BUILT-UP ROOFING
OVER RIGID INSULATION

STRIP FLASHING

SHT. METAL FASCIA

WOOD NAILERS
W/ 3/8" Ø EXP. BOLTS @ 4'-0"
O.C. KERF AND TREATED

TOP OF CMU BOND BEAM.
ELEV. VARIES

CONT. EDGE STRIP (TYP.)

8"X16" HIGH CMU BOND BEAM
W/2-#5 BARS, CUT TO
MATCH ROOF SLOPE

8" CMU W/JOINT REINF. @ 8" O.C.

BATT. INSULATION W/VAPOR BARRIER
1/2" GWB ON 3 5/8" METAL STUDS
@ 16" O.C.

METAL ROOF DECK
OVER STL. JOISTS

PREMOULDED
CLOSURE STRIPS

DAMPPROOFING
(TYP.) @ INT. FACE
OF ALL WALLS FURRED
OUT FOR INSULATION

ROOF DETAIL

SCALE: 3 INCHES = 1 FOOT

Figure 8.6i This is the wall-to-roof connection for wall section 6 A-6/A-6 (see Figure 8.6f).

Figure 8.7a This partial roof framing plan shows the glued-laminated girder and beam system. Note the weight of AC unit 1 and how the structural engineer has addressed the additional loading where mechanical equipment is supported by the roof. (Drawing courtesy of Buehler and Buehler Structural Engineers.)

Section $\frac{L}{\overline{a}\overline{a}}$ — ½"=1'-0"

Figure 8.7b This view, a cut through the girder, shows a section of column above the roof line that is designed to support the mechanical equipment screen. (Drawing courtesy of Buehler and Buehler Structural Engineers.)

Detail $\frac{10}{\overline{a}\overline{a}}$ — 1"=1'-0"

Figure 8.7c This is a detail of the connection between a tube steel column and glued-laminated girders along gridline F.

Detail $\frac{9}{\overline{a}\overline{a}}$ — 1"=1'-0"

Figure 8.7d This detail addresses the condition that occurs when a purlin beam does not connect to the column. Blocking installed between roof joists is connected to a steel strap (CMST16), which is welded to a steel angle welded to the column. See also Figures 10.4w, x, and y. (Drawing courtesy of Buehler and Buehler Structural Engineers.)

Figure 8.7e Here the engineer is describing a connection to a glued-laminated wood beam abutting a column that penetrates the roof: a beam pocket is supposed to be welded to the column. (Drawing courtesy of Buehler and Buehler Structural Engineers.)

Figure 8.7f These are the instructions as to how to construct the mechanical equipment screen shown in 8.7b. Notice the use of plan views (top and second from bottom) and elevations (bottom and second from top). (Drawing courtesy of Buehler and Buehler Structural Engineers.)

Figure 8.7g Although this detail does not pertain to any part of the roof structure outlined in 8.7a, it is a good example of how the engineer develops structural continuity between glued-laminated beams. (Drawing courtesy of Buehler and Buehler Structural Engineers.)

pleted condition. Determining how to get there is the responsibility of the builder, and that effort always begins with a review of the drawings and specifications.

As they familiarize themselves with the projects they are estimating or planning to manage, usually according to some sort of work breakdown structure or deconstruction methodology, many estimators, project managers, and other key personnel take notes. Within a short review period, the quality of the construction drawings, particularly the effectiveness of the detail referencing system, becomes apparent. The characteristics of a drawing set are taken into consideration during the job costing effort. After all, it takes time to hunt for information—*chasing information* it is sometimes called—and the longer it takes, the more costly it is to administer the project, for everyone involved.

There are many things to learn in a new project, and time is always short, so it is frequently the tasks of significant magnitude and the unusual characteristics of projects that tend to get the most attention, since much of the work in projects is routine. In a cast-in-place reinforced concrete tilt-up project, for example, the major items of work are sitework, concrete, floor framing and decking (if the project is two or more floors), and the roof structure and roof covering. Performing a cursory review of any major item of work, or worse yet, overlooking it completely, can have a very adverse effect on a construction company's profitability. Within these categories of work the reviewer is on the lookout for unusual details, for instance out-of-plane wall panel thickening (pilasters integrated into the wall panel), unusual exterior details

or finishes, the building footprint as it relates to panel size and number, or specifications calling for a pristine slab-on-grade (such specifications preclude crane traffic on the slab and have very stringent quality requirements, among other things). Access to the work is always a consideration; in fact, in urban environments, it is one of the most challenging aspects of large building projects.

Subtleties in the Drawings

The effects on the construction process of seemingly innocuous details could easily fill a book, Some examples of those effects are worth exploring in this chapter.

Figure 8.8 is a rudimentary site plan for a two-story 72,300 square foot (6717 m^2) commercial office building. The exterior walls are precast on-site reinforced concrete, commonly known as concrete tilt-up, with 12" (305 mm)- and 8¾"- (222 mm-) thick panels (see plan in Figure 8.8 for wall thickness locations). The floor system is structural steel with corrugated metal decking and concrete, and the roof structure consists of glued-laminated girders and purlin beams, with dimensioned lumber roof joists and ½" (13 mm) Structural 1 plywood.[2] The following bulleted list describes several pertinent architectural features:

---

[2]Structural 1 plywood consists of more laminations of higher-quality wood from specific species of trees than do other plywoods of the same thickness. It is commonly specified where shear and cross-panel strength are critical.

Figure 8.8 This simplified sketch shows the project site for a two-story tilt-up office with structural steel floor framing and heavy timber roof. Column spacing is 30' (9144 mm) on center. (Drawing by Tom Krejci, after Keith Bisharat.)

- The building is roughly 210' × 190' (64 000 mm × 57 910 mm). Column spacing (and therefore panel width) is 30' (9144 mm) both ways.
- Second floor elevation is 15' (4570 mm) a.f.f. (above finished floor). Roof elevation varies from 23' to 32' (7010 mm–9750 mm) a.f.f. The top of the parapet is 34' (10 360 mm) on the north façade, the north quarter of the east façade, the west half of the south façade, and the south half of the west façade. Elsewhere the parapet height is 42' (12 800 mm).
- Local planning ordinances require that rooftop machinery be concealed behind a screen or wall. The 42' (12 800 mm) high panels fulfill this requirement; and where the parapet is lower, either a concrete screen panel or metal screen wall set back from the face of the building is planned. The elevation of screen panels and the screen wall is 42' (12 800 mm). Figure 8.9a is a plan view of the concrete screen panel that was installed over the roof in two locations. Figures 8.9b and 8.9c show how the panel was formed and how it appeared after the concrete was cast, cured, and the forms stripped.

Figure 8.9a This is the architectural detail in plan view of the spandrel panel connection above the roofline. This panel screens the mechanical equipment from view. At least 13 details in the structural drawings were generated to address this component alone. (Drawing courtesy of Comstock Johnson Architects.)

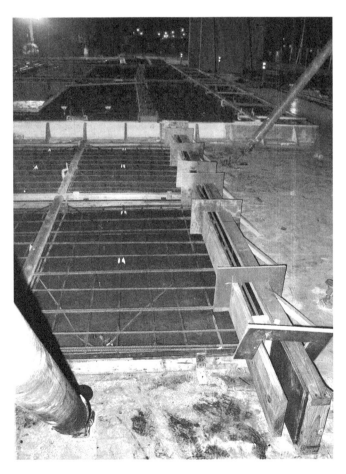

Figure 8.9b The photograph, taken at night just before the casting of the panels, shows how the contractor formed the panel depicted in Figure 8.9a. (Photo by the author.)

Figure 8.9c This is a photograph of the screen panel after it was cast and the forms were stripped. (Photo by the author.)

Figure 8.9d This is a photograph of the complete screen assembly described in Figure 8.9a. Bear in mind that Figure 8.9a is a horizontal section, and the photo is the equivalent of an exterior elevation. (Photo by the author, with the permission of the Sacramento County Office of Education.)

- The vision glass and spandrel glass are designed as an uninterrupted band across several panels, with the face of the glazing inset 3/4" (19 mm) from the face of the panel Figure 8.10a shows a partial building elevation, and Figures 8.10b and 8.10c show the pertinent details.

Cast-in-place Tilt-up Concrete: The Process

For purposes of this chapter, all of the other work that would occur simultaneously with most building concrete work—underground plumbing and electrical within the building for instance—is ignored. The sequence below assumes the building pad is in place.

1. Continuous and spread (column) footings are laid out, excavated, and cleaned. Column footings are formed such that the top edge of the form is lower than the finished floor elevation.

2. Reinforcing steel and imbedments in footings are installed and inspected.

3. The footings are placed and finished.

4. The edge forms of the slab are set several feet inside the building to create the closure strip (some contractors use a laser screed with or without slab edge forms).

5. The materials under the slab—2" of sand over vapor retarder over coarse porous fill over subgrade, or aggregate base over vapor retarder over subgrade or some other prescription, depending on specification—are installed and leveled.

6. Slab steel, if required, is installed and inspected. (Mixing glass fibers into the concrete is a popular substitute for labor-intensive steel slab reinforcement.)

7. The slab is cast, finished, saw-cut to control shrinkage cracking (if some other method to control cracking is not used) and cured (see Figure 8.11).

8. Where panels will be cast, the slab is treated with a chemical release, the panels are laid out, edge forms are installed, and reveals and form liner are attached to the slab (see Figures 8.12a–b).

9. The reinforcing steel, imbedments, lifting hardware and lifting steel, and floor and roof ledgers are installed (see Figure 8.13).

10. The panel steel is inspected.

Figure 8.10a This partial elevation shows the fenestration on this project. The glazing between gridlines 5 and 6, where the hatched and clear glass are shown, is where detail F5 in Figure 8.10c applies. (Drawing courtesy of Comstock Johnson Architects.)

Figure 8.10b The vision glass is inset 3/4" (19 mm) from the face of the panel. (Drawing courtesy of Comstock Johnson Architects.)

Figure 8.10c The faces of the spandrel glass and vision glass are flush in this section through the jamb at the edge of the window opening. This detail was a key generator of some unusual requirements in the construction of the building. (Drawing courtesy of Comstock Johnson Architects.)

Figure 8.10d This photograph captures part of the west façade shown in Figure 8.10a. (Photo by the author, with the permission of the Sacramento County Office of Education.)

Figure 8.11 While the slab was still "green," the contractor created the control joint using a special concrete saw. (Photo by the author.)

11. The panels are cast, finished, and cured.

12. The perimeter footing is prepared for the panels. This may involve installing grout pads to grade (grout pads are used to temporarily support the wall panel during erection).

13. The pick-points and brace points are exposed, strongbacks are installed, and the top end of the pipe braces is connected to the panels (see Figure 8.14).

14. The panels are erected, plumbed, aligned, and braced (see Figures 8.15a–f).

Figure 8.12a The reveals (the light strips forming a grid inside the forms), which are inexpensive milled lumber or medium-density fiberboard, leave grooves in the face of the concrete. (Photo by the author.)

Figure 8.12b Form liners come in a variety of styles and are commonly installed between reveals. (Photo by the author.)

Figure 8.13 The structural drawings called for two mats of steel. The chord bars (not shown in this photo) are 30' (9144 mm) long and, depending on the panel, either two #7, three #7 (22 mm), or four #10 (32 mm) bars with large plates welded to the bars at both ends. They had to be threaded through the two layers of steel shown here, which explains why the mats are sitting up above the edge forms—the chord bars had not yet been installed. The lifting and bracing hardware is also not yet installed. (Photo by the author.)

Figure 8.14 Tall narrow panel sections occasionally must be temporarily reinforced to resist bending during the erection process. Precautions such as this are not addressed in the structural drawings at all; they are the responsibility of the contractor, as is the lifting steel cast into the panel, and the picking and bracing hardware. (Photo by the author.)

Figure 8.15a This panel weighed 208,000 pounds (94 349 kg), and along with three other panels in the building had to be erected with two cranes—a 200-ton (181 440 kg) and a 300-ton crane (272 160 kg; rated capacity). Between the two cranes, their counterweights, the rigging and the panel, over a million pounds (453 600 kg) of weight were exerted on the slab. (Photo by the author.)

Figure 8.15b Once the panel has been raised, the cranes move forward to place the panel on the footing. (Photo by the author.)

Figure 8.15c With the panel hovering above the slab, the cranes drive toward the perimeter of the building to place the panels. This operation required two crane operators, two drivers, and a spotter. (Photo by the author.)

Figure 8.15d Another 10 feet and the cranes will place the panel on the footing. The grout pads and footing are clearly visible in the lower eighth of the photograph. (Photo by the author.)

Figure 8.15e Once the panel is close to its final resting position, it is muscled into place by workers with large steel pry bars. (Photo by the author.)

Figure 8.15f Here, several panels have been placed and temporarily braced. (Photo by the author.)

15. The panel chord bars are welded together, as are the panel and footing imbedments (see Figures 8.16a–b).

16. The exterior surface of the panels is patched, sacked, and otherwise made acceptable for finishes.

17. The slab over the column footings (usually a couple inches thick) is removed, and the column anchorage exposed.

18. The space under panels between grout pads is grouted.

19. The under-slab materials are installed in the closure strip and leveled (see Figure 8.17a).

20. The closure strip steel is installed and inspected (see Figure 8.17b).

Figure 8.17a Plumbing, in this case the drain pipe that will ultimately be connected to the roof drain, as well as electrical conduit and other underfloor piping, is installed in the closure strip prior to the installation of reinforcing steel. (Photo by the author.)

Figure 8.16a The panel and footing embedments are welded together prior to grouting the panel and casting the closure strip. Notice the grout pads to the right and left of the panel connections. (Photo by the author.)

Figure 8.16b A steel plate welded to the chord bars imbedded in each panel connect to the tops of the panels. The picking hardware is visible in both panels, above and below the 4 × 12 Douglas fir roof ledgers. The uniformity of the ³⁄₄" (19 mm) joint between panels is evident in this photograph. (Photo by the author.)

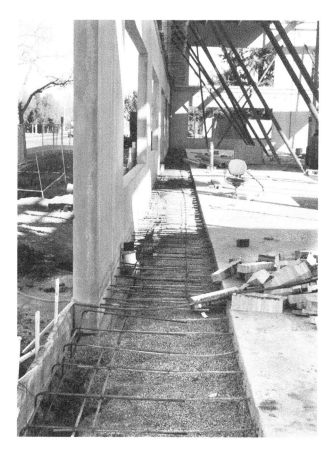

Figure 8.17b The reinforcing steel is installed in the closure strip, and edge forms are installed in openings, just prior to casting the closure strip. Independent consultants as well as the local building inspector commonly inspect reinforcing steel before the concrete is cast. (Photo by the author.)

21. The closure strip is placed and finished.

22. Once the columns have been set, they are grouted and the slab around them cast and finished.

23. Exposed connections, brace and pick points, and other voids are grouted or patched and sacked as required.

24. Following the completion of the roof structure, the braces are removed from the panels.

Design Decisions Affecting the Construction Process

In the example project, the architect's choice to use some tall walls to hide mechanical equipment and 30' × 30' (9144 mm × 9144 mm) column spacing resulted in rather large panels (one panel was 35' wide on the bottom, 40' (10 668 mm) wide at the top (12 192 mm), and 42' (12 802 mm) tall. Panel configurations vary, but 24'-wide × 30'-high × 7¼"-thick [7315 mm × 9144 mm × 184 mm] panels are common). Ceiling and floor elevations and, more significantly, the decision to make the spandrel glass and vision glass flush on the exterior resulted in very thick panels, which necessitated the installation of two curtains of reinforcing steel in each panel, as well as diagonal bracing above the roof to stabilize the parapet (Figures 8.18a–c).

Consider the effect that the spandrel glass detail alone had on the construction process. In order for the spandrel

C4 **TYPICAL ROOF PENETRATION**
1 1/2"=1'-0"

Figure 8.18a This drawing is from the architectural detail sheet. Notice how rudimentary the graphic information regarding the connection is. The architect refers the reader to the structural drawings for details on the connection. This is legal and professional courtesy at work–the architect is focused on flashing the roof penetration and defers to the structural engineer for specific information about the connection. (Drawing courtesy of Comstock Johnson Architects.)

glass to finish flush with the vision glass, the spandrel frame had to be recessed into the face of the concrete. Stretching the panel to 42' (12 800 mm) to create a machinery screen, and having the floor and roof at approximately the

Figure 8.18b Although the solution the engineer is requiring is similar in this and the next detail (8.18c), the underlying conditions are not. This detail shows the bracing connection to a wide flange beam in the roof. Notice the instruction to the contractor to provide temporary bracing until the structural connections are complete. (Drawing courtesy of Buehler and Buehler Structural Engineers.)

7'-0"

4'-0"

+42'-0"
TOC

6"ø Schedule 40 pipe
(2 pipe at notch)

4'-3"

varies

1" clr
typ

12¾"

ø panel
notch

12" conc. tilt
up panel

SP

varies
roof tof

5½" glb purlin

as in
uno

Section **½"=1'-0"**

Figure 8.18c In this detail, the source of support for the brace is a glued-laminated purlin beam in the roof structure. (Drawing courtesy of Buehler and Buehler Structural Engineers.)

one-third and two-thirds points in the wall, caused bending in the panel to become an even bigger issue than it normally is. The engineer determined that the minimum panel thickness could be 8¾" (222 mm), with two curtains of reinforcing steel. The spandrel panel framing added 3¼" (83 mm) to the panel thickness (above, between, and below the windows) resulting in a 12" (305 mm-) thick panel overall (the panel thickness, in other words, was due in large measure to the spandrel panel framing detail). The resulting very tall and thick parapet wall added significant mass above the roofline. To stabilize the mass, the engineer specified that steel braces be attached to the roof structure columns and beams. The engineer also called for the bracing imbedments in the wall to be at the same elevation, which meant that the braces themselves (due to the roof slope) were all different. The consensus of the decision makers involved was to cast 30' × 42' × 12"-thick (9144 mm × 12 800 mm × 305 mm) panels, create voids where the spandrel panels occur, and use two cranes to lift the four heaviest panels. The forming for the spandrel framing void is visible in Figure 8.15c.

Using two cranes to erect panels in tilt-up construction is not a common occurrence; however, in this circumstance it was deemed necessary. This decision

had its ramifications, too. The cranes used were large rubber-tired mobile cranes—one 200-ton (181 440 kg) and one 300-ton (272 160 kg) capacity crane. The contractor had to carefully calculate the slab-to-panel ratio to allow enough room to maneuver the cranes on the slab (which meant that the location of the temporary panel bracing had to be taken into consideration); in fact, it was necessary to remove counterweights and a rolling dolly just to get the cranes into the proper position on one panel. Additionally, the weight of the panels meant that the cranes had to get very close to the panel base to make the pick (cranes lose capacity rapidly as the boom tip moves away from the base pin), which forced the concrete subcontractor to leave the steel pipe braces off during the erection process—something that is not normally done. Workers in a personnel lift and on the slab attached the braces while the cranes held the panels in place (Figure 8.19).

The concrete subcontractor determined that the most efficient way to manage this challenge was to cast and erect panels in two stages. It is worth pointing out that the average production rate for panel erection for the first phase on this project was seven panels per day. On other projects, with more slab space on which to work

Figure 8.19 Because the cranes had to be close to the panel at the start of the pick, the steel pipe braces were installed after the panel was erected. (Photo by the author.)

and smaller panels, the production rate for erection approaches 50 panels per day. Figures 8.20a–b are copies of the subcontractor's panel layout, casting plan, and erection program for the example project.

The Takeoff

Takeoffs for framing systems are as varied as the systems, themselves. As with other systems, the contractor analyzes the drawings, sorts the various kinds of work required by them into prioritized categories, and sets about determining the quantities of the materials and the required operations. Table 8.1 is a matrix of common units used in takeoffs of concrete, masonry, steel, and wood framing. More than one unit is listed under some

materials in this table, reflecting the difference between purchase quantities and installation quantities (concrete is purchased and placed by the cubic yard, for instance, but it is finished by the square foot), and the individual preferences of estimators. In addition to the materials described in the drawings and specifications, all of the temporary structures have to be quantified.

Once the quantities have been determined, the estimator determines crew makeup and costs, production rates and equipment requirements, task duration, and so on, then converts the units to time and money values. When a constructor looks at a set of drawings, visions of one or more methods for constructing what is shown begin to crop up, and decisions as to how the work will be done begin to be made, before or while the quantities

Figure 8.20a This drawing is a copy of phase 1 of the subcontractor's casting plan for the example project. The four panels requiring two cranes are labeled P1, P2, P3, P4, and were picked in that order. (Drawing courtesy of Blueline Construction.)

Figure 8.20b Phase 2 of the casting plan. (Drawing courtesy of Blueline Construction.)

are taken off. Examples of very simple, innovative methods for casting the windowsill in the exterior panels of the example project and for creating the snap-tie pattern that the architect desired are shown in Figures 8.21a–c. Such creative problem solving characterizes successful builders.

■ Summary

The amount of information related to the superstructure of a project is voluminous and detailed, but by virtue of

various work breakdown approaches is made manageable. There appears to be no painless way of determining what work must be done, what materials procured, and what labor is required; it is a matter of visualizing the project under construction and seeing people performing work in the mind's eye. It helps to be able to recognize what the design professionals are trying to accomplish and to ask good questions: How am I going to form this panel? How will I get workers 40′ in the air to perform the work? What piece of equipment am I going to need to perform this task? How will I protect the adjacent structures during construction? None of the answers to these questions will be found in the drawings.

TABLE 8.1 Sample Unit Designations for Materials and Work—Framing Systems

	Worker hours	EA each or Lump Sum LS	PR pair	LF lineal foot	SF square foot	Square (CSF, or 100 sf)	SFCA square foot of contact area	CF cubic foot	CY cubic yard	FBM foot board measure*	MFBM thousand board feet	LB pound	T ton
Concrete													
Shoring		X											
Formwork		X		X			X						
Edge forms				X									
Place concrete									X				
Finish, cure concrete					X								
Post-tension tendons				X								X	
Precast concrete structural members		X		X									
Masonry													
Concrete masonry units (CMU)		X			X			X					
Reinforced brick		X		X	X			X					
Hardware, ties		X		X									
Flashing		X		X	X								
Mortar								X					
Grout		X							X				
Steel													
Reinforcing steel (material)												X CWT, 50 KG***	X
Install reinforcing steel													X
Structural steel (material)		X		X									X
Erect steel	X												X
Plumb & line	X												
Decking					X	X							
Headed stud anchors		X											
Wood light framing													
Layout	X	X											
Plate and Detail	X												
Frame walls		X		X							X		
Erect Walls	X			X									
Plumb & line	X												
Install joists		X		X						X			
Install sheathing		X			X	X							
Install trusses		X											
Install beams		X		X						X			
Install blocks		X								X			
Install bridging			X							X			
Install backing		X								X			

*One board foot = a piece of lumber 1" × 12" × 12" or ˝44 cubic inches using the nominal dimensions of the boards.

**Metric drawings require different units: liters (liquids), kilograms (weight), lineal meters (length), cubic meters (volume), square meters (area).

***Hundredweight (CWT) or 50 kilogram.

STOREFRONT SILL @ 2nd FLOOR
SCALE: 3" = 1'-0"

Figure 8.21a This drawing describes how the architect wants the sill at the storefront glazing system to be constructed. (Drawing courtesy of Comstock Johnson Architects.)

Figure 8.21b This photograph shows how the contractor addressed the formwork requirement in this location. Note how the contractor formed the slope in the window sill: a saw kerf halfway into the form board (center of photo) allowed the contractor to bend the form at an angle. The top of the panel is to the right. (Photo by the author.)

Figure 8.21c The concrete subcontractor who constructed the wall panels in the foregoing photographs devised a simple, clever method of duplicating the pattern left by snap ties in walls; it called for attaching plyform to the slab, drilling holes into the slab on the appropriate layout, installing wooden dowels through snap-tie cones, and securing them with a drywall screw.

CHAPTER 8 EXERCISES

1. List the common framing system types.
2. Contact a local contractor to request a tour of a construction site in your area. If you are equipped with the proper clothing—hard hat, safety vest and glasses, long pants, and thick rubber soled boots—you stand a good chance of being allowed on the site to observe construction work being done. For many projects, particularly in urban areas, access to the site is not necessary; a great deal can be safely observed from a distance (remember the title "sidewalk superintendent"?).
3. Who selects the structural framing system for a project?
4. What are the areas of interest to builders in framing systems?

9 Cladding, Glazing, and Exterior Doors

Key Terms

Backer rod
Break metal
Capillary action
Cladding
Curtain wall
Differential pressure
Dry sealant
Exterior insulation and finish
    system (EIFS)
Elastomeric materials
Gaskets
Glazing
Lockstrip
Mullion
Rain screen
Safing
Sealant joint
Solid
Surface tension
Vapor barrier
Vapor pressure
Vapor retarder
Wet sealant

Key Concepts

- Cladding, like an interstitial ceiling space, represents a distinct design zone and should be treated as such in its design and construction.

Objectives

- Distinguish between cladding and glazing systems.
- Describe the forces that cladding and glazing systems face.
- Identify the design challenges that cladding and glazing systems represent.
- Identify fundamentally good details involving cladding and glazing.

Construction Graphics: A Practical Guide to Interpreting Working Drawings,
Third Edition. Keith Bisharat.
© 2025 John Wiley & Sons Inc. Published 2025 by John Wiley & Sons Inc.

Purpose of Cladding, Glazing, and Exterior Doors

While it has other definitions, cladding commonly refers to the nonload-bearing exterior enclosure of buildings (not including the roof). Its primary role is to protect the building frame and interior spaces from the intrusion of sound, water, wind, heat and cold, and sunlight. In fulfilling this role, the cladding system must mitigate secondary design issues such as thermal expansion (particularly significant with aluminum and glass cladding and glazing systems), moisture expansion, structural movement (short- and long-term), fire, and esthetics. Much like our own skin and physical features, cladding is a significant part of a building's identity.

Glazing refers to the design, fabrication, and installation of glass in various framing systems, and it performs many of the functions cladding does; in fact, metal and glass systems frequently serve as the cladding system for a whole building.

Exterior doors provide access to and egress from buildings by people and machines and, when thoughtfully and creatively integrated into an entrance, offer a warm welcome to a building's users.

Cladding may consist of unit masonry or stone panels, precast concrete panels, glass fiber-reinforced concrete (GFRC), metal and glass, or a system of plastic foam boards attached to a backup wall (called an *exterior insulation and finish system*, or EIFS), or combinations of these systems. A variety of glazing and exterior door systems exist as well.

Effectively resisting the elements just described and articulating the solutions to other design problems characteristic of a building's exterior results in a variety of challenging and interesting details. Table 9.1 is a matrix of common design issues involving cladding, glazing, and exterior door systems. The solutions that are developed to resolve them are reflected in the drawings set and specifications. They manifest two design philosophies for cladding and glazing systems and exterior doors: (1) create a barrier to the elements, and (2) develop a system that resists the elements but acknowledges and mitigates problems caused by imperfections in the barrier. Of the two approaches, the second represents the most prudent philosophy to the design of cladding and glazing systems and exterior doors.

TABLE 9.1 Common Design Issues of Cladding, Glazing, and Exterior Door Systems

Issue	Problem	Design Solution
The Elements		
Water	Water intrusion by surface tension and capillary action	Create a barrier wall; create drip edges and capillary breaks.
Wind	Air infiltration causing loss of energy, water vapor condensation within the building's cavities, transmission of sound	Use sealants, air membranes, and gaskets effectively; neutralize the force of the air by creating pressure-equalization chambers.
Differential Vapor Pressure	Condensation of water vapor in interstitial spaces	Identify whether a vapor retarder is required (evaluate wall materials carefully); if one is, strategically locate it, and ventilate internally.
Light	Excessive light, glare, solar heat	Use sun shades, thermal insulation; select glass thoughtfully.
Heat and Cold	Uncomfortable interior environment	Use sun shades, thermal insulation, thermal breaks; select glass carefully; create heat sinks (mass).
Sound	Sound transmission	Sound attenuation measures, including mass.
Fire	Injury to occupants	Install firestops between cladding and floor; select materials carefully.
Constructability	**Problem**	**Solution**
Worker Access	System installation	Devise systems that take advantage of easily and safely handled parts, installed as close to waist height as possible, in ample light and workspace.
	Location of connections	Locate connections where workers can access them easily.
Specified Quality	Limits of skilled trades	Develop and specify realistic quality requirements.
Extent of Assembly	The more work left to be done in the field, the lower the quality and the higher the cost in terms of materials	Specify that components rather than elements be assembled into systems; reduce to a minimum number of parts.
Degree of Standardization	Many unique pieces results in lower worker productivity	Make as many components as possible the same.

What to Expect in the Drawings

The exterior elevations and architectural detail sheets contain most of the graphic information pertinent to cladding systems. The elevations provide the initial view of the cladding system, along with reference detail notes directing the builder to partial elevations and various details that describe the cladding system further. Building and wall sections will also offer information pertinent to cladding. However, like the elevations, they primarily offer reference information; that is, they direct the constructor to the appropriate detail. Structural sheets frequently address cladding issues, as do the cladding subcontractor's shop drawings (which are the most specific of all the drawings). The structural engineer may design and describe the connections to the building frame, or may simply identify where the connections should be and leave the details to the fabricator of the cladding system. Qualitative issues pertinent to the cladding are, or should be, contained in the project specifications.

The cladding system itself acts as the initial barrier to the elements, though the architect frequently requires a secondary barrier. Back-up walls consisting of concrete masonry units or steel studs are common choices; the wall serves the important function of blocking air and (depending on the wall type) providing thermal insulation to interior spaces. Figure 9.1 illustrates one assembly, a brick masonry spandrel panel over an insulated steel-stud back-up wall.

With an air space between walls and weep holes in the masonry, a pressure equalization chamber is formed, thus reducing the capacity of the air to carry moisture with it into the recesses of the wall.

Close inspection of this drawing will reveal numerous elements: glazing, interior trim materials, a vapor

Sill of aluminum window

Rowlock brick sill

Flashing and weep holes

Brick veneer

Cavity

Wire masonry tie screwed to steel stud

Steel studs screwed and welded to slab edge

Asphalt-saturated felt air barrier

Gypsum sheathing panels

Steel studs and glass fiber insulation

Flashing and weep holes

Steel shelf angle

Head of aluminum window

Wood stool

Treated wood subsill

Gypsum board

Vapor barrier

Vinyl base and flooring

Bent plate slab edge with rod anchors, cast into concrete topping

Steel angle supporting frame for shelf angle

Suspended gypsum board ceiling

Figure 9.1 This drawing illustrates a variety of materials used in cladding systems. Pictured here is a masonry spandrel panel system. (Drawing from *Fundamentals of Building Construction Materials and Methods,* 3rd edition, by Edward Allen, John Wiley & Sons, Inc., 1999. Used by permission of John Wiley & Sons, Inc.)

retarder, thermal insulation, steel stud frame, an air barrier, exterior gypsum wallboard sheathing, steel angle bracing, a steel shelf angle, masonry ties, and brick masonry. Depending on the system, these and other materials will be represented on the drawings.

Common Graphic Conventions in Cladding, Glazing, and Exterior Doors

Projection Types

As was pointed out in Chapter 4, architects describe very large objects, as well as the minutest details, within them. The exterior of a high-rise building, for example, may consist entirely of cladding. Elements of that cladding system may be fractions of an inch in size. Consequently, every graphic tool available to the architect is used to describe these systems. Plan views, elevations, and sections of every scale are used to describe cladding systems.

Applicable Line Types

The lines used to describe cladding systems run the gamut of line types and are found in Figure 9.2.

Symbols

The variety of materials that can be used in cladding systems is wide, but the common symbols include

concrete, unit masonry, reinforcing steel and wire, steel, aluminum, tempered and laminated glass, fire and thermal insulation of various types, sealant and backer rod, stone, and plaster. Reference symbols—major and minor sections and detail symbols—are also used. Figure 9.3 is a collection of common symbols.

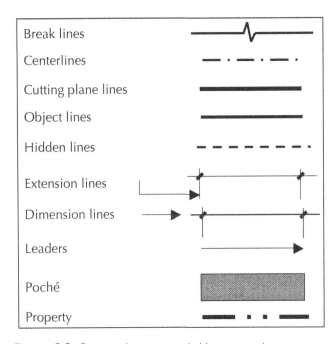

Figure 9.2 Common line types in cladding system drawings.

Figure 9.3 Symbols commonly used in cladding and glazing systems.

Scale

Cladding and glazing systems may be shown on floor plans, elevations (and partial elevations), and in details. Consequently, the guidelines used to select scale for cladding are the same as in other systems described by the architect.

- Floor plans: 1/16", 1/8", 1/4" = 1'-0" (1:200, 1:100, 1:50)
- Exterior elevations: same as the floor plans
- Partial elevations: 1/4" = 1'-0", or larger if possible (1:50 or larger)
- Building sections: same as the exterior elevations, or larger if possible (1/2" = 1'-0", or 1:20)

- Wall or partial sections: 1/2" = 1'-0", 3/4" = 1'-0" (1:20, 1:10)
- Details: depending on the complexity and size of the object, 1/2" = 1'-0", 1" = 1'-0", 1 1/2" = 1'-0", and 3" = 1'-0" (1:20, 1:10, 1:5)

Figures 9.4a–p reveal one sequence followed by architects to produce drawings of cladding and glazing systems. The subject of these drawings is a metal and glass cladding system at the main entrance to an office building and a storefront glazing system used in the tilt-up concrete wall panels and under concrete spandrel panels.

Figure 9.4a Although information as to the cladding and glazing system can be derived from floor plans like this one, it is generally introduced on building sections and elevations, in the manner shown in 9.4b, 9.4c, and 9.4d, which were drawn at 1/8" = 1'-0" scale. The numbers in boxes correspond to keynotes on the same page of the drawing. The numbers and letters in oval boxes designate finish schemes. In this drawing, the reader is directed to two building sections (G6C and G6D on page A3.2) and a partial elevation (E4 A4.3) for more detailed information. The floor plans and roof plan also refer the reader to the same two building sections. (Drawing courtesy of Comstock Johnson Architects.)

Figure 9.4b This exterior elevation is the first drawing to bear real fruit in the search for cladding information: Two sections, G6C and G6D, are referenced, between gridlines C and D.

Figure 9.4c Shown here is the west end of building section G6C on A3.2 (see Figure 9.4a), a cut through the entry lobby and stair landing. (Drawing courtesy of Comstock Johnson Architects.)

SECTIONS

Figure 9.4d This section, G6D, adjacent to the section shown in 9.4c (see 9.4a for the correlation between the two), combines a cut of the main building entry lobby and an exterior elevation of the glazing system. Behind the glass in the elevation is the stair landing through which G6D cuts. (Drawing courtesy of Comstock Johnson Architects.)

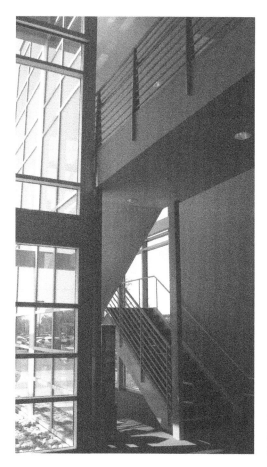

Figure 9.4e This picture shows most of the lobby staircase described in the drawings shown in Figures 9.4c and 9.4d. (Photo by the author, with the permission of the Sacramento County Office of Education.)

Figure 9.4f This drawing (E4 on A 4.3), introduced in 9.4b, is a partial elevation, drawn at twice the scale as the elevation in 9.4b. The shaded portion of the building represents spandrel glass, and the solid dots on the doors and adjacent windows (visible also on the drawing in 9.4b) represent tempered glass, required by building code for glass doors, for windows next to doors, and for windows within 18" of the floor. The three lites of glass to the left of the entryway above the door (shown with dark dots) are tempered because of their proximity to the stair landing (see next drawing). There are two different glazing systems in this project: this one, which functions as a cladding system for the entryway, and the glazing system used in the tilt-up panels. The pertinent details for this second system (G2 and G3 on A9.1) are referenced in the second-floor window to the right of the entryway. (Drawing courtesy of Comstock Johnson Architects.)

Figure 9.4g The north exterior elevations described in the drawing in 9.4b and 9.4f are shown here. (Photo by the author, with the permission of the Sacramento County Office of Education.)

Figure 9.4h This plan view appears directly below the drawing in 9.4f. Up to this point, we know little about the glazing systems other than their general type, their locations within the building, the size of the glass lites, and the orientation and number of the mullions. Some of the character of the metal and glass cladding system is coming to light however; note how the mullions stand outside the stair landing and how the three structural steel columns are integrated into the system (identified by keynote #27). (Drawing courtesy of Comstock Johnson Architects.)

Figure 9.4i The role of one of the steel columns in this part of the building is made clearer in this elevation of part of the glazing system (first referenced in the drawing in 9.4g). The column supports a concrete panel above the glazing system at gridline 1.6, as well as the tube steel beam supporting the entry canopy. Several details pertaining to the connection between the roof frame and the glazing system are identified on this drawing: (E2, E3, and E4 of page A 9.1 and detail G3 on A1.2). We may find what we need to know about the glazing system by examining these details. (Drawing courtesy of Comstock Johnson Architects.)

G2 BEAM @ CURTAINWALL SECTION
SCALE: 1 1/2"=1'-0"

Figure 9.4j These three details help us to understand how the entry canopy attaches to the metal and glass curtain wall (cladding system). G2 shows how the steel beam supporting the tube steel canopy joists connects to the column adjacent to the stair landing (referenced in the drawing in 9.4g). G3 shows the tube steel beam that carries the joists over the entryway, and G4 describes the connection between the tube steel joists supporting the laminated glass roof of the canopy and the roof glass itself. There are three other details on the same sheet as these three that show other views of the canopy. (Drawing courtesy of Comstock Johnson Architects.)

G3 ENTRY CANOPY @ CURTAINWALL
SCALE: 1 1/2"=1'-0"

E2 CURTAINWALL @ ROOF PARAPET
SCALE: 1 1/2"=1'-0"

E3 CURTAINWALL @ ROOF FRAME
SCALE: 1 1/2"=1'-0"

TYPICAL GLASS PANEL JOINT
G4
SCALE: 3" = 1'-0"

Figure 9.4k These details show the connection between the curtain wall and the structural steel beams in the roof. Note how the architect defers to the structural engineer regarding roof structure information. (Drawing courtesy of Comstock Johnson Architects.)

Figure 9.4l The architect proceeds down the wall with these three details until the glazing system is described from top to bottom. The safing between floors is evident in E5. (Drawing courtesy of Comstock Johnson Architects.)

Figure 9.4m The progression of drawings for the second glazing system—known here as the "storefront" system—is the same as for the cladding system: The architect starts at the second-floor window head (F2) and describes a mullion 2' below the head (F3) then . . . (see Figure 9.4n). (Drawing courtesy of Comstock Johnson Architects.)

F4 STOREFRONT SILL
SCALE: 3" = 1'-0"

- LINE OF CONCRETE BEYOND
- TYP. STOREFRONT & GLAZING
- CONT. BEAD SEALANT O/ BACKER ROD TO MATCH STOREFRONT COLOR
- CONC. PANEL

F5 STOREFRONT DOUBLE MULLION
SCALE: 3" = 1'-0"

- CONC. PANEL
- LINE OF SINGLE SPANDREL PANEL
- 2"X3" STOREFRONT AT SPANDREL PANEL, TYP.
- CONT. BEAD SEALANT O/ BACKER ROD TO MATCH STOREFRONT COLOR
- 4 1/2" STOREFRONT AT GLAZING AREAS, TYP.
- DUAL GLAZING AT VISION GLASS AREAS
- LINE OF CONCRETE PANEL ABOVE

3" 8" 3 1/2" 3/4"

F6 CURTAINWALL HEAD @ PANEL
SCALE: 1 1/2"=1'-0"

- CONCRETE WALL PANEL
- CONT. BEAD SEALANT O/ BACKER ROD TO MATCH CURTAINWALL COLOR
- +28'-0" T.O. CURTAINWALL
- CURTAINWALL SECTION
- SPANDREL UNIT. SEE ELEVATIONS.
- +24'-0" T.O. CURTAINWALL
- HORIZONTAL MULLION

Figure 9.4n ... the sill for first- and second-story windows (F4) and the jamb where the spandrel glass and vision glass connect (F5). F6 describes the connection between the two-story storefront system and concrete spandrel panels in two locations on the project. (Drawing courtesy of Comstock Johnson Architects.)

G2 STOREFRONT HEAD/ JAMB, SIM.
SCALE: 3" = 1'-0"

- CONC. PANEL
- CONT. BEAD SEALANT O/ BACKER ROD TO MATCH STOREFRONT COLOR
- 4 1/2" STOREFRONT & GLAZING
- LINE OF CONCRETE JAMB, SILL BEYOND

G3 STOREFRONT SILL (RAISED)
SCALE: 3" = 1'-0"

- LINE OF CONCRETE JAMB BEYOND
- 4 1/2" STOREFRONT & GLAZING
- CONT. BEAD SEALANT O/ BACKER ROD TO MATCH STOREFRONT COLOR
- LINE OF CONCRETE SILL
- CONC. PANEL

Figure 9.4o These two details address the connection between the window frame and concrete wall panel at the head and raised sill. Can you think of a way to improve either of these details from the performance standpoint? (Drawing courtesy of Comstock Johnson Architects.)

Figure 9.4p There are two sill conditions with this window system: a raised sill (G3 in Figure 9.4o) and a sill that is flush with the floor (G4). The architect has kept the wall panel in a separate zone from the floor to avoid conflicts; where gaps might occur G4 instructs the builder to install safing between floors, contained by metal deck angle. G5 is the detail for the mullion that, in certain conditions, is 2' above the floor. (Drawing courtesy of Comstock Johnson Architects.)

■ Translating the Drawings into Work

Building construction contractors frequently delegate the design, production, and installation of cladding and glazing systems to experienced, highly competent contractors who specialize in a particular cladding system type. These contractors spend considerable time and energy developing the shop drawings that are normally required by the design professional as a part of the submittal process. Shop drawings, particularly for cladding and glazing systems, can consist of tens of sheets that describe in great detail how the specialty contractor intends to execute the architect's design for the cladding or glazing system. They are organized in much the same way as the architectural drawings; for example, shop drawings for

glazing systems often begin with cover sheets that include the project identity; drawing sheet index; various schedules addressing the finish, sealant, and glass required; and frequently identify the abbreviations and symbols used. Subsequent sheets include floor plans (to orient the user), elevations corresponding to the elevations on the architectural sheets, large-scale elevations, sections and myriad details of mullions, sills, break metal, doors, and glass, among other things. The drawings in Figures 9.5a–g are excerpted from the shop drawings created for the glazing system described in Figures 9.4a–p. A handful of drawings were selected to demonstrate the similarity in the shop drawings to the architectural set and to indicate the level of detail involved in such drawings. The project is the office building first introduced in Chapter 8, and the entire shop drawing set consisted of 38 pages.

Figure 9.5a Floor plans are used on shop drawings for reference; here, six elevations are listed, four of which are drawn in greater detail on page 309 of the shop drawing set. (Drawing courtesy of Entelechy.)

Figure 9.5b Page 309 of the shop drawings contains a reiteration of the floor plan showing where wall sections are taken, as well as the detailed elevations that were referenced in the partial plan shown in 9.5a. (Drawing courtesy of Entelechy.)

ELEVATION 'C7'

SERIES 3250 (2 1/2" X 7") PRESSURE PLATE SYSTEM
(1) REQUIRED THUS
REF. A3.1/A4.3

Figure 9.5c The detailer drew a partial floor plan identifying where in the structure four elevations, including this one, occur in the building. Count the number of details listed just on this elevation! Notice how the detailer has referenced the architectural drawing sheet pertaining to this elevation. (Drawing courtesy of Entelechy.)

SECTION 'C7'

Figure 9.5d The partial floor plan on page 309 refers the reader to four wall sections; here is the section pertinent to elevation C7. (Drawing courtesy of Entelechy.)

Figure 9.5e This detail shows the connection between the window frames and the horizontal structural steel member over the doorway; it is a vertical section through the beam and frames. (Drawing courtesy of Entelechy.)

Figure 9.5f This horizontal section through the tube steel column supporting the beam described in 9.5e shows how the vertical window frame members connect to the column. Note the absence of any connection to the column other than sealant; the window system connects at the sill and the head only. (Drawing courtesy of Entelechy.)

Figure 9.5g This is a section taken through the mullion adjacent to the concrete wall panel to the right of the entryway (as you face it from outside). Note the angle attached to the concrete panel that supports the backer rod on the inside (top center of drawing). The closure metal (bottom right in the drawing) is set back from the face of the mullion, in line with the glass, to allow the mullion to retain its character in the façade. (Drawing courtesy of Entelechy.)

Subtleties in the Drawings

An evaluation of the effects of details on the construction process is basic to every review of a construction drawing set performed by a builder. Architects who approach detailing with the contractor in mind develop "user-friendly" details—details that reflect an understanding of the construction process and project costing as well as consideration for the people who do the work. Those who do not produce such details raise the cost of construction for their clients, since details that are difficult to understand or construct inevitably cost more.

Where cladding systems are concerned, of primary importance to the constructor is the location from which the cladding will be installed. Cladding and glazing systems that can be installed from inside rather than outside the building are likely to result in better installation productivity rates, lower equipment costs, and greater worker safety. The floor of a building provides a stable, large, level work surface that is not significantly affected by the elements, especially wind; and since floor construction precedes cladding installation, the work platform does not have to be moved as work progresses.

Unit masonry cladding systems are more costly if the mason installs each piece while standing outside the structure. If in these situations panels can be prefabricated in a manufacturing setting and installed on the structure with hoisting equipment, the cost is likely to be lower. Cladding designs that take standard material length and shapes into consideration (extruded aluminum comes in 24' (7315 mm) lengths, for example, and is either rectangular or square in section) generally result in lower overall costs.

If the architect has established a design zone just for the cladding, there will be fewer conflicts with other systems in the building. Effective details—that is, details that acknowledge normal tolerances in the various trades, parts that fit easily and cannot easily be improperly installed, that have adjustable connections, that recognize the succession of construction work from rough to finish, and that conceal minor imperfections without altering the esthetic effect—will result in the best value for the owner and a much less contentious construction experience. Figures 9.6a–j comprise a sampling of cladding system details that illustrate some of these ideas.

Bad

Connections and
sealant joints are
hard to reach

Good

All connections
and sealant joints
are accessible

5. Cladding Panel Connections

Figure 9.6a This sketch compares a well-thought-out connection detail on the right with an ill-conceived one on the left. In both cases, the panel joint is a good one; it considers gravity, momentum, air pressure differentials, surface tension, and capillary action. Both connections are accessible in the detail at right. (Drawing from *Architectural Detailing–Function, Constructability, Aesthetics,* by Edward Allen, John Wiley & Sons, Inc., 1993. Used by permission of John Wiley & Sons, Inc.)

Wedge insert

Askew-head bolt

Shims

Shelf angle

3. Adjustable Support for Shelf Angle

Figure 9.6b The connection in Detail 3 accounts for adjustments in all three dimensions. It also shows foresight where long-term changes are concerned; the weight of the material that the angle supports–probably masonry units–will have the effect of tightening the connection, since the bolt head slides on a ramp. (Drawing from *Architectural Detailing–Function, Constructability, Aesthetics,* by Edward Allen, John Wiley & Sons, Inc., 1993. Used by permission of John Wiley & Sons, Inc.)

Shims

4. Adjustable Panel Anchor

Slotted hole pairs

5. Support Frame for Brick Curtain Wall

Figure 9.6c This detail allows for considerable adjustment as well. Once the panel has been properly located, the connection can be welded. (Drawing from *Architectural Detailing–Function, Constructability, Aesthetics,* by Edward Allen, John Wiley & Sons, Inc., 1993. Used by permission of John Wiley & Sons, Inc.)

Figure 9.6d This drawing clearly requires that the soffit work follow the installation of the glazing system (note how the aluminum flashing at the top of the glazing system on the left-hand side is installed well below the sealant between the panel and the glazing frame, thus precluding the installation of sealant between the window frame and soffit framing after the flashing is installed). This soffit construction involves rough carpentry, drywall, finished sheet metal, and sealant, which, depending on the project, may mean the involvement of four different subcontractors. Can you design a solution that would allow the builder to schedule the work differently without changing the character of this detail? (Drawing courtesy of Comstock Johnson Architects.)

(b)

Aluminum window sill

Cut stone wall panel

The support plate is shimmed and bolted, then welded to an anchor plate cast into the edge of the floor topping

Mortar setting bed

Epoxy

Support plate with rod welded on

Pocket milled into edges of panel (see detail)

Safing

DETAIL OF EDGE POCKET IN STONE PANEL

Steel angle struts brace the panels against tilting

Bolt in expansion anchor

Drip slot

Aluminum window head

Figure 9.6e This and the next drawing show two methods for attaching cut-stone panel cladding to a structure. (Drawing from *Fundamentals of Building Construction Materials and Methods,* 3rd edition, by Edward Allen, John Wiley & Sons, Inc., 1999. Used by permission of John Wiley & Sons, Inc.)

Stone panels

Steel grid strut

Sealant

Aluminum support

Bearing tape

(b)

Figure 9.6f An alternate method for that shown in Figure 9.6e. (Drawing from *Fundamentals of Building Construction Materials and Methods*, 3rd edition, by Edward Allen, John Wiley & Sons, Inc., 1999. Used by permission of John Wiley & Sons, Inc.)

Figure 9.6g Details that successfully neutralize the forces acting on cladding systems are fundamentally the same. Here a simple wall with a masonry exterior exemplifies the principles of an effective cladding approach, where an exterior wall is created, a pressure equalization chamber is developed, and an air barrier exists in the form of a concrete masonry unit wall. Any required reinforcing steel has been eliminated for clarity.

BARRIER WALL

PRESSURE EQUALI-
ZATION CHAMBER

AIR BARRIER (MASTIC
OVER CMU)

FLASHING

WEEP
HOLES

Figure 9.6h In this photograph, architectural precast concrete panels have been installed on the left; what appears to be the framing for an EIFS is under construction on the corner of the building. (Photo by the author.)

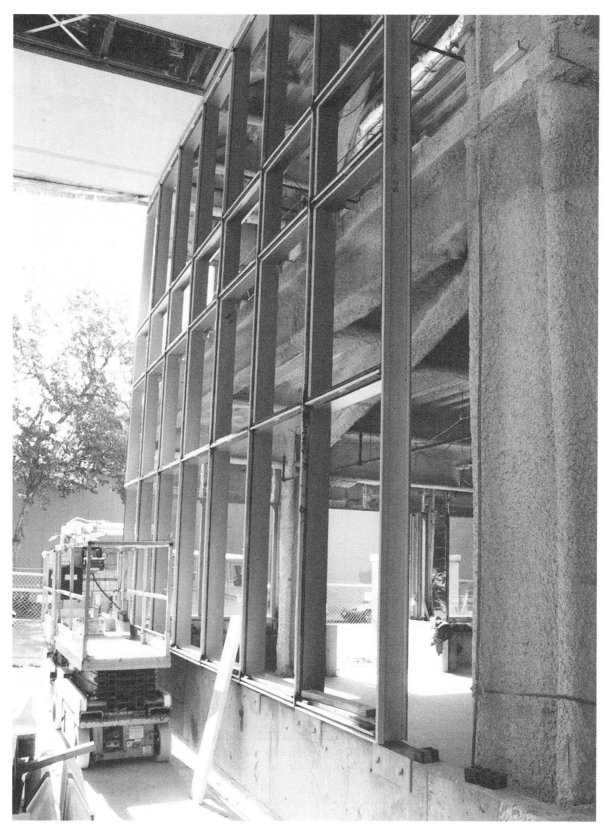

Figure 9.6i The curtain wall system designed for this project is under construction; notice how the system exists independently from the structural steel frame. (Photo by the author.)

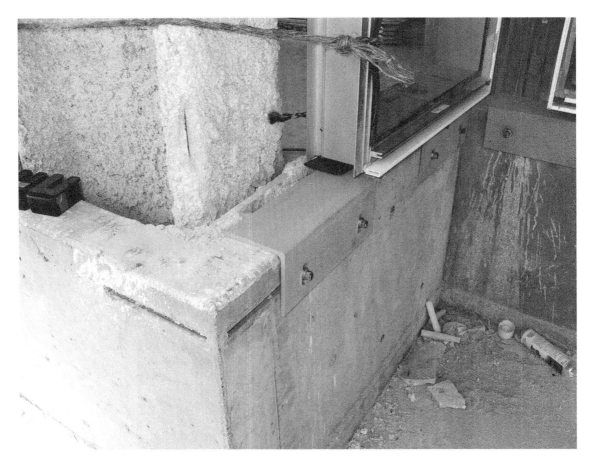

Figure 9.6j The manner in which the curtain wall system connects to the structure is clear in this photograph. A metal channel has been cast into the concrete, to which bolts holding an angle are connected. A stack of shims sits on top of the concrete wall to the left in the photograph. (Photo by the author.)

The Takeoff

Installing cladding systems, glazing, and exterior doors are, for the most part, all *labor-paced operations* (work performed by the physical effort of people). There are two common approaches to determining labor costs in labor-paced operations: (1) worker hour factors and (2) labor unit costs. In the former, the total number of units determined in the takeoff is multiplied by a productivity factor, and the resulting figure (time) is multiplied by the cost per unit of time (usually a composite [weighted average] hourly rate). In the latter, the total units determined in the takeoff are multiplied by a unit cost to arrive at the total labor cost. With both approaches, the challenge and risk lie in identifying how much a worker or crew will accomplish under varying circumstances.

It stands to reason then that the builder should closely evaluate details affecting labor productivity. If workers have to install a cladding system from outside the building while supported on scaffolding hung from upper floors in a tall building, their productivity is likely to be lower than the crew that installs the cladding from within the building. Likewise, panel connections that require workers to reach, that exist in poor light, or are made up in limited space take longer to complete. If just five minutes more per panel are required to make a connection, a building with 500 cladding panels—a modest number—will take an additional week of the erector's time. Masonry cladding assembled as panels in a factory take advantage of a manufacturing environment—a protected environment and ideal ergonomic circumstances for the skilled trades. Architectural precast concrete panels boasting different colors of concrete within the same panel are more expensive to produce than those that are not, and a building in which there are many different panel styles and locations will result in a costly project. The forms for architectural precast panels are fabricated for each project out of steel, so the more identical panels there are, the lower the form costs and the more economical the project. Worth considering as well, particularly when it comes to metal and glass cladding systems and doors, are the size, shape, and weight of the elements of the systems, and the degree of assembly of components. A large door, for example, can weigh close to a hundred pounds and may have to be transported and installed using a hand truck.

TABLE 9.2 Sample Unit Designations for Materials and Work—Cladding, Glazing, and Exterior Doors

	LS lump sum	EA each	Opng Opening	M 1000	PR pair	LF lineal foot	SF square foot	Square (CSF, or 100 sf)	SFCA square foot of contact area	CF cubic foot	CY cubic yard	LB pound	T ton	
Cladding system complete	X													Cladding
Architectural precast concrete panels		X					X							
Metal and glass						X	X							
Unit masonry panel				X			X				X			
Stone panel		X					X							
Glass fiber-reinforced concrete panels		X				X	X							
EIFS	X	X					X							
Panel supports		X				X								
Glazing system complete	X													Glazing
Vision glass		X	X				X							
Spandrel panel		X					X							
Metal frame						X								
Exterior doors complete	X													Exterior doors
Storefront door		X			X									
Hardware		X			X									

*Metric drawings require different units: liters (liquids), kilograms (weight), lineal meters (length), cubic meters (volume), square meters (area).

The takeoff units commonly used to estimate cladding and glazing systems and exterior doors are shown in Table 9.2.

■ Summary

Describing cladding systems as simply protective envelopes is certainly accurate, but that generalized description opens the possibility of overlooking the critical details of the system, which, when they act together, serve to protect the cavities and interior spaces of a structure from invasion by the elements. Gravity, momentum, air pressure, surface tension, and capillary action are the forces that drive water into the recesses of a building; and washes, labyrinths and pressure equalization chambers, drips, and capillary breaks all are effective at neutralizing these forces. Sealants and gaskets, pressure equalization chambers, and air barriers neutralize the force of the wind. Thermal breaks, thermal insulation, and dual-glazed windows help to neutralize the effects of heat and cold, and vapor retarders help minimize the effects of water vapor diffusion. Within the cladding system or secondary barrier, mass and sound-absorbing materials control sound. Cladding details must also take into consideration the movement of the building frame caused by wind, settlement, and seismic events, and the effects of age on the structure. When all these details are developed with the builder in mind, a productive, relatively conflict-free construction experience is likely.

CHAPTER 9 EXERCISES

1. Identify the forces that work against the success of cladding systems in keeping the elements out.

2. Identify several techniques that will neutralize the forces acting on a cladding system.

3. Redesign the detail in Figure 9.6d so that the builder can schedule the work differently and still prevent the penetration of water by surface tension and capillary action.

10 Roof Systems

Key Terms

Built-up roof
Cant strip
Crickets
Curb
Flashing
Hydrostatic head
Low-sloped roof
Multiple-layer roof
Parapet
Pitched roof
Reglet
Roof jack
Shake
Shingle
Single-layer roof
Vapor diffusion

Key Concepts

- The expectations for roof systems and coverings are different from those of cladding systems because of the roof system's alignment to the elements.
- More than half of the lawsuits filed against contractors involve problems with the roof.
- Most energy, particularly in one-story buildings is lost through the roof.
- Low-pitch roofs are susceptible to ponding, which can result in a catastrophic roof failure.

Objectives

- Summarize the critical weaknesses of roof coverings.
- Identify effective design remedies for roof maladies.
- Identify cost contributors in roof design.

Construction Graphics: A Practical Guide to Interpreting Working Drawings, Third Edition. Keith Bisharat.
© 2025 John Wiley & Sons Inc. Published 2025 by John Wiley & Sons Inc.

■ Purpose of Roof Systems

Roofs epitomize humankind's concept of shelter, as evidenced by the expression "a roof over my head" used to describe a complete shelter. In warm climates, roofs may be the only shelter required. Roof systems constitute the sloping or horizontal plane that protects people gathered underneath it or, in a building, its interior spaces and occupants from intrusion by water, wind, sunlight, heat and cold, and sound. Roof systems include the water conveyance systems that are critical to its proper functions. The roof structure provides the required support of the roof covering and other "residents" of the rooftop such as antennae, flagpoles, mechanical equipment, equipment screens, maintenance walkways, roof hatches, crickets, diverters and jacks, and exhaust ports, to name a few, as well as the necessary geometry to divert water to collection points. The roof covering, or membrane, performs the task of protecting the roof structure, and the drainage system conducts water to the appropriate facilities located downstream.

Roof systems can be divided into two principal groups: pitched roofs and low-sloped roofs (also known generically, although inaccurately, as "flat" roofs—roofs with slopes less than 3:12). From a system standpoint, roofs can be classified as single-layer, multiple-layer, and composite systems that share characteristics of both. The ubiquitous built-up roof, wood or concrete shingles, and the composition shingle roof, respectively, are examples of the three system types. The decision to use a steep or low-sloped roof often precedes the selection of the functional type, although it is certainly possible, perhaps necessary in certain circumstances, to select a material first and design the roof system around it. Each system has its advantages and disadvantages and appropriate application.

Low-sloped roofs offer the designer an economical way to cover a very large building, but they are less forgiving of carelessly performed work and complex geometry. Hydrostatic pressure exerted by water that gathers in puddles is sufficient to force the water through very small openings in the membrane. That ponding can occur means that these roofs are also susceptible to catastrophic failure when the size and depth of the pond exceeds the structure's capacity to support it.

Steep roofs perform particularly well and are far more tolerant of careless work in the installation of the membrane, and the materials used to cover them are easily installed with simple tools, but the structures themselves are more complicated geometrically, they require more materials than a low-sloped roof of the same span, and they create greater work accessibility problems, to name a few considerations.

The roof structure, in addition to supporting the vertical loads imposed on it by the deck, membrane, mechanical equipment, and other loads also transfers lateral loads to the walls, which carry them to the foundation, where they are dissipated into the earth. As such, roof coverings are diaphragms, like walls and floor systems, and can be just as intriguing structurally; in fact, it is appropriate to think of roofs as sloped walls, although the performance expectations are different because of a roof's orientation to the elements.

Principal among the forces acting on a roof system is the gravitational (vertical) force vector, which plays an increasingly significant role as the roof pitch diminishes, to the extent that a truly flat roof with a parapet is, for all practical purposes, an accident waiting to happen. The hydrostatic pressure exerted by just a cubic inch of water is more than half an ounce; enough pressure to force water through small openings in the membrane.

The roof structure, parapet, deck, roof membrane, drainage systems, vents and flashing components, and elements such as the vapor retarder, thermal insulation, and ballast, though not always found together in the same roof, are critical elements and sources of failure in the roof system. The materials used in these components and elements are numerous and varied, ranging from wood, stones, and reeds to titanium and plastics. All can be found in drawings of the roof system.

■ What to Expect in the Drawings

The components and elements just mentioned act in concert to carry out their roles. As with any true system, the interdependence of components is typical. Ballast (stone, aggregate, interlocking pavers) protects the roof membrane from the destructive effects of light and the erosive effect of water droplets falling onto a dusty surface, and is supported by the roof structure. Ballast also prevents the membrane from being lifted up in a wind. The roof membrane protects the structure from intrusion by water. The insulation, if it is placed above the membrane, protects the membrane as well as the structure from temperature extremes. When it is installed below the membrane, insulation protects the structure and interior spaces, but not the membrane. Components such as ridges, valleys, crickets, and sloped planes divert water to internal or external drainage systems, which are frequently connected to the site storm drain system, relieving the structure from bearing an excessive load and the membrane from being penetrated by water. The vapor retarder, ventilation system, and membrane (depending on where it is placed) protect the insulation that protects the structure and its occupants. The user of the drawing should expect to see a variety of materials and components in various combinations in drawings of roof structures and systems.

Common Graphic Conventions in Roof Structures and Coverings

Information regarding the roof system is generally first offered early in the architectural sheets, starting with the roof plan, or perhaps even earlier in the set; for example, in a multistory project with a façade that steps back on higher floors, floor plans will show portions of the roof that happen to occur on the same level (see Figures 10.1a–b).

Elevations, which frequently follow floor and roof plans, may also show parts of the roof structure, particularly in projects in which a steep roof is used. Building and wall sections add still more information, and detail sheets

Figure 10.1a Information on the roof may be introduced as early as in the architectural floor plans, especially on projects having a receding façade on upper floors.

Figure 10.1b In this partial roof plan, the slope of the roofs is apparent, as is the location of some of the rooftop equipment.

complete the explanation, at least for as much as the architect is responsible.

As noted in an earlier chapter, there may be a number of professionals involved in one system, each with a different responsibility, so a review of sheets in other disciplines is essential. Structural engineers dispatch their contractual duties regarding the roof with the structural performance in mind; they do not address other performance issues—flashing for example—in the course of their work. Some of the equipment required to heat and cool a building is perched on the roof, and the mechanical engineers who design the HVAC and plumbing systems often detail the curbs and flashing around the equipment pads and penetrations of all kinds in the roof. The order of these drawings is similar to the architectural sheets—the general precedes the specific.

Projection Types

Drawings for roof systems make use of the range of projections, including multiview, axonometric, and oblique projections. Plan views, elevations, sections, and details are all required to adequately describe the roof system.

Applicable Line Types

The lines used to describe roof systems run the gamut of those available to an architect. Some common examples appear in Figure 10.2.

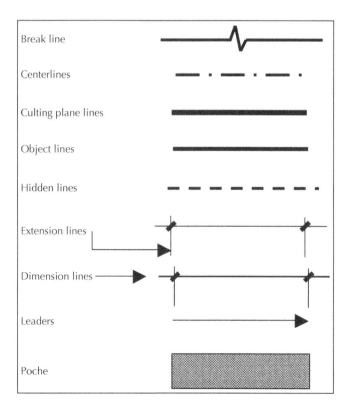

Figure 10.2 Common line types in roof system drawings.

Symbols

Figure 10.3 is a sampling of symbols commonly found in roof systems, including concrete, unit masonry, reinforcing steel and wire, steel, aluminum, tempered and laminated glass, fire and thermal insulation of various types, sealant and backer rod, stone, and plaster.

Figure 10.3 Symbols commonly found in roof system drawings.

Reference symbols—major and minor sections, and detail symbols—are also used.

Scale

Roof systems may be shown on floor plans, elevations (and partial elevations) building and wall sections, and in details. Consequently, the guidelines used to select scale for roofs are the same as in other systems described by the architect:

- Floor plans: $\frac{1}{16}$", $\frac{1}{8}$", $\frac{1}{4}$" = 1"-0' (1:200, 1:100, 1:50)
- Exterior elevations: same as the floor plans

- Partial elevations: $\frac{1}{4}$" = 1'-0", or larger if possible (1:50 or larger)
- Building sections: same as the exterior elevations, or larger if possible ($\frac{1}{2}$" = 1'-0", or 1:20)
- Wall or partial sections: $\frac{1}{2}$" = 1'-0", $\frac{3}{4}$" = 1'-0" (1:20, 1:10)
- Details: depending on the complexity and size of the object, $\frac{1}{2}$" = 1'-0", 1" = 1'-0", $1\frac{1}{2}$" = 1'-0", and 3" = 1'-0" (1:20, 1:10, 1:5)

Figures 10.4a–gg consists of drawings of one roof system that serves as an example of how drawings introduce critical information to the user.

Figure 10.4a This is part of an architectural roof plan showing the mechanical equipment and referencing two sections and three roof details. (Drawing courtesy of Comstock Johnson Architects.)

Figure 10.4b Building sections offer information about the roof plans, too. (Drawing courtesy of Comstock Johnson Architects.)

Figure 10.4c Wall sections offer information about critical connections, like the wall-to-roof connection. (Drawing courtesy of Comstock Johnson Architects.)

Figure 10.4d Detail C3 is a section through the parapet wall at the roofline, first referenced in the wall section in 10.4c. D3 is an isometric drawing of the same section of parapet wall, and it includes information about the transition from roof to wall. (Drawing courtesy of Comstock Johnson Architects.)

D5 PARAPET WALL FLASHING
1 1/2"=1'-0"

D6 SADDLE FLASHING @ "SLOT"
SCALE: 1 1/2"=1'-0"

Figure 10.4e Section D5 is a cut through the exterior wall panel where the parapet wall is very tall. The flashing shown in D6 is designed to protect the panel joints over which the saddle sits. Notice that the responsibility to accommodate the flashing resides with the concrete subcontractor. This detail imposes an unusual, subtle requirement on the concrete subcontractor, who could overlook this detail in the panel construction stage. There is no practical way to cut a groove into the concrete to accept the flashing after the concrete has been cast, other than cutting the panel after the edge forms have been stripped and before the panels are placed side by side during the erection process. The practical way to carry out this instruction is to insert a T-shaped strip against the form and remove it after the concrete has set. (Drawing courtesy of Comstock Johnson Architects.)

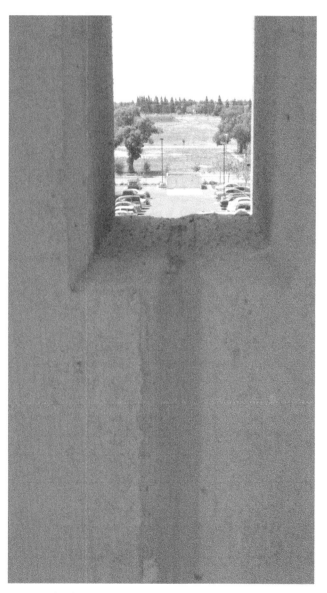

Figure 10.4f The saddle flashing described in detail D6 was apparently changed in favor of this approach. (Photo by the author, with the permission of the Sacramento County Office of Education.)

F5 REGLET FLASHING @ PARAPET
SCALE: 3" = 1'-0"

20 GA. CAST-IN PANEL SHEET METAL SNAP-LOCK TYPE REGLET

SPRINGLOK FLASHING

BUILT-UP ROOFING

Figure 10.4g Rather than bringing the roof membrane all the way up the parapet wall, which was not possible under the circumstances (the parapet was over 6' [1829 mm] high), this architect chose to direct the contractor to cast a sheet metal reglet into the wall, which acts as counterflashing for the roof membrane. (Drawing courtesy of Comstock Johnson Architects.)

STEEL PIPE SUPPORT WELDED TO ROOF COLUMN. SEE STRUCTURAL PLANS.

WELD STEEL CAP ONTO COLUMN.

SEALANT ALL AROUND

SHT. MTL. "UMBRELLA" FLASHING

SHEET METAL ROOF JACK

BUILT-UP ROOFING O/ PLWD. DECK. SEE ROOF PLAN.

TUBE STL. ROOF COLUMN EXTENDED THRU ROOF DECK. SEE STRL. PLANS.

C4 TYPICAL ROOF PENETRATION
1 1/2"=1'-0"

Figure 10.4h The architect proposed this solution to stabilize the screen hiding the mechanical equipment. While an extended column is an excellent way to stabilize the screen, it results in numerous roof penetrations. (Drawing courtesy of Comstock Johnson Architects.)

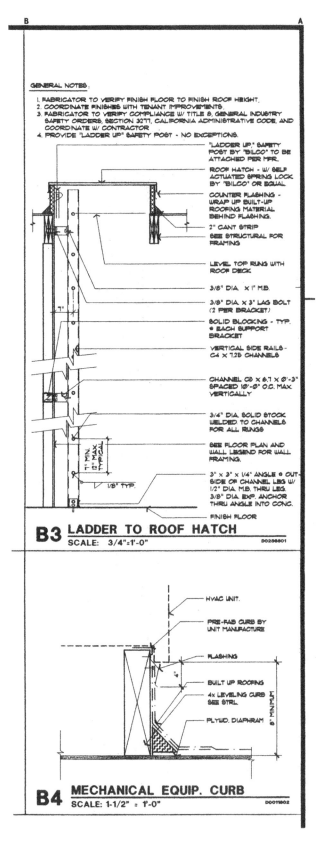

GENERAL NOTES:

1. FABRICATOR TO VERIFY FINISH FLOOR TO FINISH ROOF HEIGHT.
2. COORDINATE FINISHES WITH TENANT IMPROVEMENTS.
3. FABRICATOR TO VERIFY COMPLIANCE W/ TITLE 8, GENERAL INDUSTRY SAFETY ORDERS, SECTION 3277, CALIFORNIA ADMINISTRATIVE CODE, AND COORDINATE W/ CONTRACTOR
4. PROVIDE "LADDER UP" SAFETY POST - NO EXCEPTIONS.

"LADDER UP," SAFETY POST BY "BILCO" TO BE ATTACHED PER MFR.

ROOF HATCH - W/ SELF ACTUATED SPRING LOCK BY "BILCO" OR EQUAL

COUNTER FLASHING - WRAP UP BUILT-UP ROOFING MATERIAL BEHIND FLASHING.

2" CANT STRIP

SEE STRUCTURAL FOR FRAMING

LEVEL TOP RUNG WITH ROOF DECK

3/8" DIA. X 1" M.B.

3/8" DIA. X 3" LAG BOLT (2 PER BRACKET)

SOLID BLOCKING - TYP. @ EACH SUPPORT BRACKET

VERTICAL SIDE RAILS - C4 X 7.25 CHANNELS

CHANNEL C3 X 6.1 X 0'-3" SPACED 10'-0" O.C. MAX. VERTICALLY

3/4" DIA. SOLID STOCK WELDED TO CHANNELS FOR ALL RUNGS

SEE FLOOR PLAN AND WALL LEGEND FOR WALL FRAMING.

3" X 3" X 1/4" ANGLE @ OUTSIDE OF CHANNEL LEG W/ 1/2" DIA. M.B. THRU LEG. 3/8" DIA. EXP. ANCHOR THRU ANGLE INTO CONC.

FINISH FLOOR

B3 LADDER TO ROOF HATCH
SCALE: 3/4"=1'-0"

HVAC UNIT.

PRE-FAB CURB BY UNIT MANUFACTURE

FLASHING

BUILT UP ROOFING

4x LEVELING CURB SEE STRL

PLYWD. DIAPHRAM

B4 MECHANICAL EQUIP. CURB
SCALE: 1-1/2" = 1'-0"

Figure 10.4i These details describe two other features of the roof system, the access ladder and the curbs for the mechanical equipment pads. (Drawing courtesy of Comstock Johnson Architects.)

Figure 10.4j These three details describe the roof drains and overflows and the condition where the plumbing penetrates the exterior wall. (Drawing courtesy of Comstock Johnson Architects.)

Figure 10.4k Structural drawings include extensive information on the roof system. This is an excerpt from the roof plan, showing the same portion of the roof as in 10.4a. (Drawing courtesy of Buehler and Buehler Structural Engineers.)

Figure 10.4l The size of the air handlers and other rooftop HVAC equipment is apparent in this photograph. Compare the weights of the units shown in Figure 10.4k with the size of these units. (Photo by the author, with the permission of the Sacramento County Office of Education.)

Figure 10.4m This drawing is a section through the roof structure and mechanical equipment screen, referenced in Figure 10.4k (called out as "concrete spandrel panel" in the lower right corner of the plan). (Drawing courtesy of Buehler and Buehler Structural Engineers.)

Figure 10.4n Where a wall panel of one height intersects with a panel of a greater height, the architect called for this flashing detail. (Drawing courtesy of Comstock Johnson Architects.)

Figure 10.4o One connection between the roof structure and wall panel, which is referenced in the roof plan in Figure 10.4k, is detailed here. The condition that this detail addresses appears in the photograph in Figure 10.4p, upper right corner. An explanation of the welding symbols shown in this drawing is given in Appendix E. (Drawing courtesy of Buehler and Buehler Structural Engineers.)

Figure 10.4p This is the condition for which the detail in Figure 10.4n was created. The photograph also describes the structural elements of the connection shown in Figure 10.4o. (Photo by the author.)

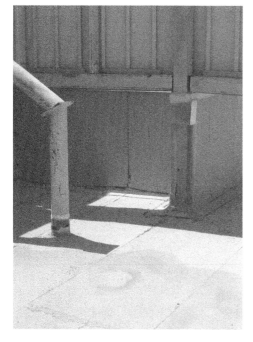

Figure 10.4q Although the picture taken in Figure 10.4p looks up at the structure from the ground floor, the connection to this photograph, taken from the roof level, becomes apparent after a few moments' review. (Photo by the author, with the permission of the Sacramento County Office of Education.)

Figure 10.4r This photograph, taken from other side of the metal screen shown in Figure 10.4q, completes the image of the part of the building described in several details, including Detail 11 S5.4 (see Figures 10.4n and 10.4o). (Photo by the author, with the permission of the Sacramento County Office of Education.)

Figure 10.4s The architect describes the mechanical equipment screen in this detail, including how the roof penetration is supposed to be flashed. At least two notes refer the reader to the structural drawings for certain details. (Drawing courtesy of Comstock Johnson Architects.)

Figure 10.4t This is the structural engineer's rendition of the condition shown in Figure 10.4s. Notice how the drawing focuses entirely on structural issues; flashing and finishes are completely ignored. (Drawing courtesy of Buehler and Buehler Structural Engineers.)

Figure 10.4u The connection described in Figures 10.4t and v is shown in this photograph. (Photo by the author.)

Figure 10.4v The structural engineer prescribed this connection for columns that penetrate the roof. (Drawing courtesy of Buehler and Buehler Structural Engineers.)

Figure 10.4w This photograph shows the blocking for the 8'- (2438 mm)- long metal strap mentioned in Section L S3.3 (Figure 10.4t). (Photo by the author.)

Figure 10.4x Section L S3.3 (shown in Figure 10.4t) calls for Simpson CMSTC16 8'-long metal strap (see note in drawing). There is no beam to tie the strap into, so a row of blocks between the beams, visible in this photograph, was installed to provide nailing. Another drawing of this condition is shown in Figure 8.7d. (Photo by the author.)

Figure 10.4y The strap is welded to a steel tab on the column. See Figure 8.7d for the drawing prescribing this connection. (Photo by the author.)

Figure 10.4z This photograph depicts the final solution to the problem posed by the penetration of the roof covering by structural elements. Can you identify any evidence of a problem associated with a design decision here? (Photo by the author, with the permission of the Sacramento County Office of Education.)

Figure 10.4aa The columns in the preceding figures support the mechanical equipment screen on this project. (Photo by the author.)

Figure 10.4bb This picture shows the final outcome of the screen wall shown in Figure 10.4aa. (Photo by the author, with the permission of the Sacramento County Office of Education.)

Figure 10.4cc The structural engineer required steel pipe bracing on the parapet walls on 42'- (12 802 mm)-high panels, shown in this detail. The actual braces are shown in Figure 10.4ee and 10.4ff. (Drawing courtesy of Buehler and Buehler Structural Engineers.)

Figure 10.4dd The hold-down in Section L in the previous figure is shown in this photograph. (Photo by the author.)

Figure 10.4ee Here and in the panel details, the engineer called for bracing embedments in the wall panels 4' (1219 mm) below the top of the parapet. The length of the vertical brace is adjusted to accommodate the slope of the roof. The advantage to this approach is that conventional flashing systems, like those used for plumbing penetrations, can be adapted here. This detail affects the miscellaneous metals subcontractor; however, virtually every brace is different. In the foreground is a dual-legged brace required where the roof beam aligns with the panel joint. (Photo by the author.)

Figure 10.4ff This picture is taken from a very similar location to the one in Figure 10.4ee. (Photo by the author, with the permission of the Sacramento County Office of Education.)

Figure 10.4gg While the idea of using round pipe for structural elements that penetrate the roof covering is a good one, close inspection will reveal the shortcomings of this detail. (Photo by the author, with the permission of the Sacramento County Office of Education.)

Correlating the Drawings with Work Concerns

Breaking down a roof system into functional parts helps to absorb the graphic information related to it more readily. One effective approach is to first review how any water falling on the roof's surface is diverted to collection points and then to examine increasingly smaller portions of the roof for critical information. In the analysis of the details, it is important to first understand what the drawings call for and then evaluate them in light of the forces acting on the system, the common challenges faced by the skilled trades, and subcontract work scope.

The roof structure is a critical component of the roof system, and it certainly warrants a review in any discussion of a roof system, its incorporation in the framing drawings notwithstanding. The reality on the construction site is that on a given project, various subcontractors may cooperate to build the roof structure, especially in composite systems like structural steel columns that support a heavy timber roof, or structural steel deck with cast-in-place concrete over corrugated decking. In steel/heavy timber systems, the ironworkers install the columns, and carpenters install the heavy timbers, purlins, joists, and roof sheathing. Subcontractors who specialize in roof structures may refuse to perform carpentry work below the roof line and above the sheathing (skylight curbs, for example), in which case the prime contractor must find yet another subcontractor to perform that work or do it with his or her own forces. Other subcontractors are likely to be involved in the completion of the roof system; the roofing contractor typically installs the flashing furnished by the mechanical contractor, and furnishes and installs the roof covering; the plumbing contractor may supply roof jacks for the plumbing, and the electrical contractor frequently must do the same for electrical piping that penetrates the roof.

It is worth reiterating that complexity and higher construction costs generally go hand in hand; the more complex the geometry of a roof system, the more costly the roof in the principal cost categories—labor, equipment, and materials. Although steeper roofs function more reliably, with greater tolerance for poorly executed work, they require more material and may create work access challenges; imagine installing a roof membrane on a steeply pitched roof several hundred or a thousand feet in the air. In terms of membrane installation alone, the differences between low-sloped and steeply pitched roofs are significant. The typical covering for pitched roofs is small units that are easily transported and held in place by workers, whereas the typical low-sloped roof uses much larger pieces that, during installation, are held in position by gravity. Combining larger units, say a standing seam metal roof and a steep pitch (10:12 or more), results in very different challenges for the installers of

the roof covering. Whether composition shingles can be installed on a steeply pitched roof depends on the weather; work is limited to the early morning work hours on hot days because of the damage that can be done to the surface of the shingles by the installers.

The more turns, twists, and elevation changes there are, the more custom fitting of metal flashing is required and the greater the potential for leaks. Movement due to temperature fluctuations on the roof—the most extreme of any system in a building—can result in a well-executed roofing job turning into a liability simply because of a design that fails to account for the movement of the membrane (expansion joints for the membrane are required on larger roof surfaces). Surface-mounted reglets installed on the inside face of concrete or masonry parapet walls are notoriously unreliable, due in part to the shearing effect on sealants resulting from the differential thermal expansion of the concrete or masonry parapet and the galvanized metal flashing, and because the caulking duct traps water (see Figure 10.5a).

Parapet walls are frequently treated as extensions of the wall out of which they grow, but the environment that surrounds them is different from the wall below the roofline. The reasoning behind the pressure-equalized cladding system applies to a parapet wall cap as well, although details of them do not always reflect that reasoning. In general, roofing on the low side of a roof is much more likely to fail than on the high side, yet a single detail applying to both conditions is frequently called for.[1]

The builder is called upon first to determine the cost of constructing a project, and the law requires that it be built in accordance with the drawings and specifications. That said, there are occasions where the drawings impose an unrealistic expectation on the industry. A prudent constructor will be alert for those occasions and either price the risk accordingly, include the additional time and money that may be required, or suggest alternatives that are more realistic.

The Takeoff

Roof systems are estimated in the same way as other systems, in that they are "deconstructed," the constituent parts are listed and quantified, and the labor and equipment required to install the work is combined with the material costs to arrive at a figure. Calculating the costs of labor, not surprisingly, is the challenge, since determining the material quantities and identifying the installation requirements are both relatively simple. The units commonly used in estimating roof coverings are listed in Table 10.1.

Much of the work installing roof coverings is not difficult to do; however, it is self-regulated—oversight by an

[1]From *Building Pathology*, Samuel Harris, New York: John Wiley & Sons, Inc., 2001.

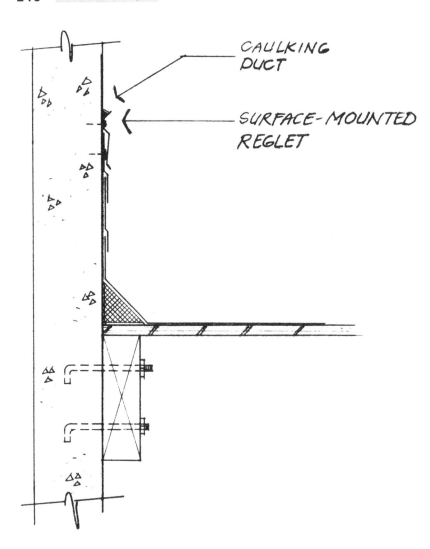

CAULKING
DUCT

SURFACE-MOUNTED
REGLET

Figure 10.5a Surface-mounted reglets are problematic details; the shearing effect caused by the differential expansion between the panel and sheet metal makes the joint totally dependent on the quality of the sealant and its connection to the panel and sheet metal.

Figure 10.5b At some point during the construction of this project, the decision to deviate from the original plan for flashing was made. The original design employed a cast-in reglet (see Figure 10.4e, detail D5), however at some point the decision to use a plant-on reglet was made, and this photograph shows the final outcome. (Photo by the author, with the permission of the Sacramento County Office of Education.)

TABLE 10.1 Sample Unit Designations for Materials and Work*—Roof Coverings

Category	Item	LS lump sum	EA each	Opng Opening	M 1000	PR pair	LF lineal foot	SF square foot	Square (CSF, or 100 sf)	Box	G Gallon	Roll	LB pound	T ton
Single Layer	EPDM membrane							X						
	slip sheet							X						
	adhesive										X			
	insulation							X					X	
	ballast													X
	Built-up roof								X					
	cant						X							
	plys							X				X		
	asphalt												X	X
	edge metal						X							
	walkway							X						
Multiple Layer	Wood shingle								X					
	step shingles		X					X						
	valley flashing						X							
	wall/roof flashing						X							
	fasteners									X			X	
	ridge shingles						X							
Composites	Composition shingle								X					
	underlayment											X		
	valley flashing						X							
	wall/roof flashing						X							
	edge molding						X							
	ridge shingles													
Accessories	Roof hatch		X											
	Skylights		X											
	Jacks		X											
	Drains and overflows		X											
	Scuppers		X											
	Gutters and leaders						X							

*Metric drawings require different units: liters (liquids), kilograms (weight), lineal meters (length), cubic meters (volume), square meters (area).

independent inspector is not often required, as it is with reinforcing steel placement, welding, and earthwork, for example. The quality of the roof covering, therefore, is dependent upon the integrity of the installers and the pride they take in their work. For the person charged with the task of managing the construction work, recognizing this reality is critical.

Summary

The horizontal orientation of roof structures is the principal distinguishing characteristic of them as enclosures, and it is the principal source of roof system problems. This orientation introduces a host of problems that vertical enclosure systems simply do not have. Simple details, uncomplicated rooftop geometry, adequate slopes and drainage systems, and designs that seek to balance differential forces are likely to succeed in the challenging environment of the roof top.

CHAPTER 10 EXERCISES

1. What two key words summarize roof systems?
2. List the principal force acting on roof systems that has a much lesser effect on wall enclosures.
3. List the advantages and disadvantages of low-sloped and pitched roofs.
4. What is the purpose of ballast on low-sloped roofs?
5. List several causes of problems in roof systems. What would you do, as a designer, to mitigate the potential for failure in these areas?

11 Interior Construction

Key Terms
Backing
Base
Casework
Chair rail
Door and hardware schedules
Finish schedules
Fire rating
Handing
Panic hardware
Register
Tenant improvements
Terrazzo
Wainscot

Key Concepts
- Interior construction may involve as many or more subcontractors than the building shell itself.
- Tenant improvement work is generally performed by a wide variety of skilled workers under tight schedules in restricted space, creating coordination and access-to-work challenges.

Objectives
- Define interior construction and identify some of its special characteristics.
- Sketch components and assemblies that are typical of interior construction.
- Identify some of the areas where work that is not called for in the drawings is nevertheless required.

Construction Graphics: A Practical Guide to Interpreting Working Drawings,
Third Edition. Keith Bisharat.
© 2025 John Wiley & Sons Inc. Published 2025 by John Wiley & Sons Inc.

Purpose of Interior Construction

Interior construction refers generally to the nonstructural fixed and demountable partitions that permanently or temporarily subdivide the typical office or commercial building, as well as walls mounted on the inside face of cladding and exterior walls. Interior windows; doors, stairs and handrails; toilet, bath, and laundry accessories and bathroom partitions; floor coverings; ceilings and finishes are part of this type of construction as well. Mechanical and electrical components, interior lighting, and the fire suppression systems, though classified under the service assemblies in UniFormat, are installed at approximately the same time as the interior construction components.

Interior partitions can be constructed of cast-in-place concrete, masonry, glass, wood, and metal stud framing with a variety of surfaces; and finishes can run the gamut from stone flooring to gold-foil ceilings. Although interior walls may perform a structural role—elevator cores are common structural elements in a building—most interior partitions are distinguished from others by the absence of a support role in the structural performance of the building. It is not as though they completely lack structural integrity or fail to make a contribution in some way to structural performance in, say, a seismic event; in fact, in some areas, interior wall construction must meet certain seismic design requirements; it is just that many interior walls are simply not included in the structural engineer's calculations for the performance of the building overall.

Interior construction is referred to in the vernacular as "TIs" (the acronym for tenant improvements). The managers and workers who perform TI work frequently constitute separate divisions of construction companies because the work is often contractually distinct from the construction of the building shell, and because of the unique requirements of the work—fast-paced, labor-intensive, sequentially complex work involving many different steps, materials, and subcontractors. Individuals and investment groups frequently construct building shells on a speculative basis, which they then lease to business entities as office space. TIs are therefore often made for leaseholders, who contract with architectural or interior design firms and contractors for TI design and construction services and bear the expense of the improvements (depending on the lease agreement).

What to Expect in the Drawings

As noted, interior construction involves numerous trades, and TI construction may involve as many or more subcontractors than the building shell itself, particularly for highly technical businesses, and the designs are often more unusual than that of the shell. How is the information presented? Since tenant improvements exist within the building shell, drawings of them typically begin with plan views of the floor or portion of the floor that the space under consideration occupies. These drawings show the constraints of the building shell, for example, the exterior walls, glazing in the exterior walls, the location of the columns, stairs, any fixed corridor spaces, elevator lobbies, and the like. The location and width of new partitions are described on TI floor plans, which show the general layout of the improvements. Beyond that, sections are used to describe the vertical spaces between the floors of the shell, which for entrance lobbies, for example, can be quite elaborate. Elevations are used to describe the surfaces of partitions, and reflected ceiling plans show the pertinent details of lighting, supply and return air grills, and finishes in ceilings. Details and schedules provide the balance of the required graphic information, and of course, the specifications play a critical role in describing the work to be done.

Schedules

The word "schedule" has three common definitions in construction: (1) A project schedule is the proposed sequence and duration of the principal tasks involved in the construction itself, usually presented in graphic formats known as a *precedence diagram*; or "bar chart"; (2) a door and hardware, equipment, or room finish schedule is a matrix identifying types, numbers, and locations of these components in the project; (3) a schedule of values refers to the line-item list of costs established at the beginning of the project that the owner and contractor agree will be used to allocate monthly progress payments. The summation of all the line items results in the total contract value (other than unforeseen changes that occur during the course of the work).

Much of the information related to interior construction—including framing information—can be derived from door, window, and room finish schedules contained in the architectural sheets of a drawing set. Ceiling heights, for example, are not always provided on the drawings in sections. In such instances, a resourceful drawing user will explore the room finish and perhaps the door and window schedules for helpful information. Common schedules are shown in Table 11.1 and Figure 11.1.

Common Graphic Conventions in Interior Construction

Not unexpectedly, tenant improvement drawings take advantage of all the projections types available to the drafter of such drawings, although by far the most common

ROOM FINISH SCHEDULE

RM NO	ROOM LOCATION	FLR	BASE	WALLS NORTH	WALLS SOUTH	WALLS EAST	WALLS WEST	CEILING MAT	CEILING TYPE	CEILING HEIGHT	REMARKS (SEE NOTES BELOW)
105	CORRIDOR	B	B	E E	E	B	E	E	E	8'-0"	1, 2
108	T. O. LIBRARY	E	B	E	B	E	N	N	E	8'-0"	
109	WOMEN'S RESTROOM	N	B	N	E	N	B	N	N	8'-0"	
114	EQUIPMENT ROOM	N	N	N E	N	N	E	N	N	VARIES	
117	SINGLE MISSILE MAINT. BAY	N	N	N E	N	N	N	N	VARIES	
118	HYDRAULIC ROOM	N	N	N	E	N	N	N	N	VARIES	

LEGEND: N – NEW
E – EXISTING
B – BOTH NEW AND EXISTING

Figure 11.1 Room finish schedules contain a variety of information that might be useful when interior construction drawings fall short of fully describing a room. Here is a very simple one.

TABLE 11.1 Door and Hardware Schedule

Symbol	Size	Fire Rating	Door Type	Door Material	Door Finish	Frame Type	Frame Material	Frame Finish	Details Sheet #	Details Threshold	Details Jamb	Details Head	Hardware Group	Keing Pattern	Remarks
424	3°–7°	1 HR	F*	HM**	PT***	F1	HM[1]	PT	A 12	4	3	2	A	XX-2	

*Elevations of door types either accompany the schedule or are found elsewhere in the documents. In this case, the architect is directing the user's attention to door type F.

**HM is a common abbreviation for a hollow metal door, a door type in which the door frame is sandwiched between two light-gauge metal panels.

***PT indicates that the door should be painted.

[1] A hollow metal door frame consists of a light gauge hollow metal frame that is attached to the structure or a substrate by clips or tabs.

type is the multiview projection. The number, type, and location of partitions to be included in the work are commonly introduced early in the drawing set in floor plans. Once the partitions have been located, it is up to the designer to reveal the other notable characteristics of the partitions, namely their characteristics in the third dimension, height. The tool used to describe partitions in the third dimension is of course vertical sections, taken at appropriate locations throughout the floor plans and on elevations, which focus on the surface of the partitions. Axonometric and oblique projections are used, for example, in details describing how nonload-bearing partitions are braced or connect to the structure. Figure 11.2

is an example of how axonometric projections are used to describe details in interior construction.

Applicable Line Types

The line types commonly used to describe interior construction are listed in Figure 11.3. The symbols used in interior construction drawings are as varied as the materials, which is to say that virtually all of the symbols available in construction drawing are used. Examples of common symbols are included in Figure 11.4. Figures 11.5a–gg are a collection of interior construction drawings.

Figure 11.2 This drawing shows how the architect expects the typical metal stud frame wall to be constructed and braced. (Drawing courtesy of Comstock Johnson Architects.)

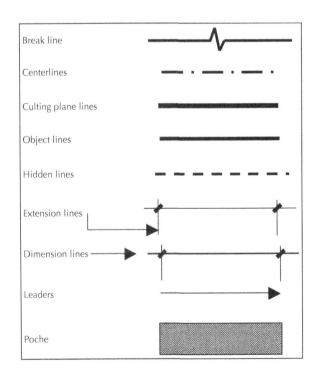

Figure 11.3 Common line types in interior construction drawings.

Material	Symbols*
Concrete	
Concrete masonry unit	
Brick	
Steel–plate	
rebar	
Aluminum	
Glass – regular	
laminated	
Safing	
Insulation – rigid	
Insulation – batt	
Sealant and backer rod (in section)	
Wood – continous blocking	
Plywood – large-scale	
small-scale	
Plaster	
Gypsum wallboard	

*Symbols in the NCS are somewhat different from these, which were selected from and are common to several different drawing sets spanning 20 years.

Figure 11.4 The material symbols used in construction drawings run the gamut of those available; virtually every material is represented in interior construction. This is merely a sampling.

Figure 11.5a A portion of the floor plan for a military facility is shown in this drawing; a medium-width line referring the reader to page A-11 for more information circumscribes the restroom.

Figure 11.5b A partial building section is taken to reveal the structure and orientation of the restroom; the location of the section is given on sheet A3 (Figure 11.5a).

Figure 11.5c The partial building section in 11.5b directs the reader to detail A on the same sheet for details pertinent to the tops of the walls, the acoustical ceiling (left side) and gypsum wallboard ceiling (right side).

Figure 11.5d Building sections give the reader generalized information about the structure, as well as specific information such as the elevations of certain elements (here, the height of the parapet on the left, and the joist-bearing elevations at both ends of the roof joists). Reference to critical connections are made, frequently by encircling the subject of the detail and assigning a number to it (in this case details B-A6A6 and 2-A6A7).

PARAPET COVER
FLASHING
I" EXPANSION JOINT
CANT
BUILT-UP ROOFING
RIGID INSULATION
STEEL ROOF DECK

S
A-7 A-11

TOP OF EXISTING
WOOD PARAPET
ELEV. 114'-1½"

EXISTING 2×4 STUD
WALL W/ NEW ½" PLYWOOD
SHEATHING @ WEST SIDE.
CUT OUT OPENINGS @
JOIST BEARING POINTS

JOIST BEARING
ELEV. 111'-4½"

STUFF GAPS W/ MINERAL
FIBER INSUL.

BOTTOM OF EXIST.
CONC BEAM
ELEVATION VARIES

STEEL BAR JOISTS
EXTENDER

SUSPENDED ACST
TILE CEILING

EXISTING WOOD STUD WALL
W/ NEW ⅝" FIRE-RATED
GWB BOTH SIDES TO
CONCRETE STRUCTURE.

CORRIDOR
104

EXISTING BUILDING
7030

PAINT EXISTING INTERIOR WALL
EXIST 2" CHAMFER @ FOUNDATION
WALL
NEW RESILIENT BASE
I" EXPANSION JOINT

V.C.T. ON
CONCRETE FLOOR SLAB
OVER VAPOR BARRIER
(TYPICAL)

EXISTING 2"×4" WOOD STUD WALL
NEW RESILIENT BASE
EXISTING CONCRETE FLOOR

FINISH FLOOR
ELEV. 100'-0"

EXISTING GRADE BEAM

RRIER

CAPILLARY WATER BARRIER

WALL SECTION
SCALE: ¾ INCH = 1 FOOT

2
A-6 A-7

Figure 11.5e In increasing detail, the design professional explains the structure. This wall section is referenced in the building section shown in Figure 11.5d. Note that the encircling technique is used here, too.

SHEET MET. PARAPET COVER
INTERLOCKS W/ PARAPET
FLASHING

CONT. EDGE STRIP

NEW ½" EXT. GRADE PLYWD. THIS
SIDE OF PARAPET ONLY - NOTCH
PLYWD. AROUND JOIST CHORDS &
SHIM PL. TO CREATE JOIST POCKETS
& MAINTAIN SEISMIC JT.

1" SEISMIC JOINT

CONT. FLASHING

BASE FLASHING

2x6 WOOD NAILERS W/
KERFING. BOLT TO METAL
DECK.

B.U.R.

RIGID
INSULATION

SLOPE

2x12 WD.
NAILER BOLT
TO BENT
PLATE

BENT STL. PL.

STEEL JOIST EXTENDER

1" COMPRESSIBLE FILLER

SLOPE

EXIST. 2x4 WD. STUDS W/ ½" PLYWD. @ WEST SIDE
(REMOVE & SCRAP STUCCO & ½" PLYWD. @ EAST SIDE
OF EXIST. PARAPET.)
TOP OF PARAPET
ELEV. 114'-1½"

SHEET MET. CAP FLASHING W/ INTERLOCKING EDGE
STRIP - BEND EXIST. COUNTER FLASHING TO INTERLOCK
AS INDICATED & SEAL. REPAIR OR REPLACE ANY
EXIST. FLASHING OR ROOFING SYSTEM COMPONENTS
DAMAGED DURING CONSTRUCTION TO
MAINTAIN INTEGRITY OF
EXIST. ROOFING.

EXIST. COUNTER
FLASHING, BASE
FLASHING &
CANT

FRAME AROUND
JOIST POCKETS AS REQ'D.
W/ DOUBLE 2x4 HEADER

EXIST. ANGLE IN
CONC. BEAM

CANT

JOIST BEARING
ELEV. 111'-4½"
FACE OF STUD
(BEYOND)

NEW ½" PLYWD.
(BEYOND)

EXIST. CONC. "T" BEAM
ROOF SYSTEM

EXIST. CONC. BEAM

SEE STRUCTURAL STEEL SHIM
@ JOIST BEARING ONLY

EXIST. ANGLE IN
CONC. BEAM

DETAIL AT JOIST BEARING S
A-7 | A-11
A-12

Figure 11.5f The connection between an addition and an existing structure is critical, and is therefore shown in great detail by the designer.

Figure 11.5g Within this portion of the floor plan of a five-story research and office facility, there exist no fewer than 17 different wall frame and surface treatments (the description follows in the next two figures).

FULL HEIGHT STUD WALL:
(C.B.C. Table 7-B, Item 16-1.1)
3 5/8" x 20 ga. full height metal studs @ 24"o.c. to structure or deck above w/ 5/8" type "X" gyp. bd. @ both sides. run gyp. bd. to 6" above ceiling. Walls will be sound insulated, the full height of the wall. Provide blocking & strapping @ areas with no gyp. bd. per stud manufacturer's recommendation. Rooms w/ gyp. bd. ceiling: run wall to bottom of ceiling.

DOUBLE PLUMBING WALL:
(Gypsum Association Fire Resistance Design Manual, 15th Edition: WP 1073)
3 5/8" x 18 ga. metal studs @ 16" o.c. Align studs & brace across to each other with 5/8"x 12"x 16" gyp. bd. panels. Install brace panels every 3rd stud (48" o.c.), and @ 48" o.c. vertically; finish each side with 5/8" type "X", tile backer board.
Provide sound insulation same thickness as studs.

6" PLUMBING WALL.
(Gypsum Association Fire Resistance Design Manual, 15th Edition: WP 1073)
Same as Wall Type #4 except with 6"x 18 ga. studs @ 16" o.c. Water-resistant gyp. bd. occurs @ toilet room side. Provide sound insulation same thickness as wall.

**(4)-HOUR RATED AREA SEPARATION WALL
UL DESIGN U490**

3 5/8" x 20 ga. metal studs @ 16" o.c. full height to structure above with base layer of 3/4" type X gyp bd applied parallel to each side of studs with 1 1/4" long Type S drywall screws @ 24" o.c. Joints staggered on opposite sides. Face layer of 3/4" type X gyp bd on each side applied at right angles with vertical joints staggered from base layer joints, or applied parallel with joints centered over studs and staggered on opposite sides. fasten with 2 1/4" type S drywall screws @ 12" along perimeter and at intermediate studs and 1 1/2" type G drywall screws midway between framing along horizontal joints. 2" mineral fiber batts, not less than 2 pcf, in stud cavity.

Figure 11.5h Here is an excerpt from the wall legend describing how the various walls are to be constructed. "IW" stands for interior wall, the details of which are shown on the IW pages in the detail book, a separate document on this project.

NOMENCLATURE:

a. Full-height to underneath side of structure above.

b. Provide 3 1/2" unfaced fiberglass batt sound insulation.

c. Use water resistant gypsum board on interior wet side.

d. Use 20 ga. metal studs @ 16" o.c. & 5/8" Dens-Shield tile backer board by Georgia-Pacific @ tiled wall conditions.

e. Use 8" studs in lieu of indicated stud per wall type.

f. Use 6" studs in lieu of indicated stud per wall type.

g. Provide and install 3/4" fire treated plywood on wall from floor to ceiling/deck above on telephone room side. Fasteners to be countersunk so fastener head does not protrude above plywood surface. Plywood to be painted white.

GENERAL NOTES:

1. Contractor shall provide solid blocking in walls for handrail brackets, fire extinguisher cabinets, equipment, fixtures, hardware, accessories, wall door stops, etc., verify.

2. Dimensions shown on floor plans are to centerline of column grid, face of concrete or face of stud, unless otherwise indicated. Notify the Architect should any discrepancies occur, typical.

3. Where interior 1-hour walls meet interior 2-hour walls in same plane, face layer of gypsum board to be flush on Corridor, Lobby, Stair or Restroom side. Notify Architect of any possible discrepancies.

4. For floor and wall penetrations see details: IW

5. For exterior wall penetrations & surface mounted fixtures, see installation details.

6. Unless otherwise indicated in this wall schedule, all framing and substrate fastening systems shall comply with 1994 UBC Tables 25-G and 25-H, and substrate manufacturer's instructions, verify

7. Rated wall assemblies shall be constructed in strict accordance with the "UL" and "FM" Fire Resistance Directories per the designs specified.

8. Metal stud tracks set on concrete floors/walls shall be anchored with powder-actuated "Hilti DN Pins" .145"Ø x 1 1/2" minimum embedment @ 16" o.c., unless otherwise noted.

9. All metal stud head and sill tracks shall be 16 ga minimum.

10. 16 GA studs to be used at the following:
    a. Door openings
    b. Base cabinets
    c. Upper cabinets
    d. Full height cabinets
    e. Toilet partitions
    f. Restroom grab bars

11. EXTERIOR FINISH SYSTEM:
    E.I.F.S. System , Class PB, consisting of an adhesive, insulation board, base coat with reinforcing mesh(es) and finish. See specifications.
    a. Use impact-resistant reinforcing mesh to +9'-0", typical.
    b. Stud size varies, see details and structural drawings.

12. for fire rated wall penetrations see detail IW6.

Figure 11.5i Of the 17 different wall types on this floor, several are further distinguished by notes correlated to the wall type number. For example, the wall separating the telephone room and the women's restroom (rooms 108 and 109, Figure 11.5g), are type 4 walls, with notes c and g added. Note c calls for water-resistant gyp board, and note d calls for fire-resistive plywood on the telephone room side. Notice under General Notes that the contractor is responsible for installing backing for handrails, fire extinguisher cabinets, grab bars, and the like—none of which is shown on the drawings but is nevertheless required.

NEW FLOOR PLAN

SCALE: 1/4" = 1'-0"

PROJECT
NORTH

Figure 11.5j The imaging room shown in this drawing was renovated after the hospital was constructed. The walls are nonload-bearing. Three details are referenced on the plan (9, 11, and 12 on A5). Note also the interior elevation identifier: the circle inside the square at 45 degrees, denoted as detail 1A4.

NEW PLAN NOTES

1. NEW 2 LB. LEAD LINED GYP. BD. FROM FIN. FLOOR TO +7'-0" MIN. AT WALL OF ROOM 1752. SEE (8/A5)

2. NEW 6 LB. LEAD LINED GYP. BD. FROM FIN. FLOOR TO +7'-0" MIN. SEE (8/A5)

3. NEW SHEET VINYL FLOOR TO MATCH EXISTING.

4. RELOCATED MEDICAL GAS OUTLET. SEE PLUMBING DRAWINGS.

5. EXISTING MEDICAL GAS OUTLET TO REMAIN.

6. COLLIMATORS

7. CONTROL CONSOLE/MONITOR. ANCHOR TO CASEWORK COUNTERTOP. SEE INTERIOR ELEVATIONS.

8. PATCH EXISTING FLOORING.

9. PATCH EXISTING GYP. BD. FINISH.

10. NEW CEILING CURTAIN TRACK. SEE REFLECTED CLG. PLAN.

11. PROVIDE 6" X 6" X 3" DEEP JUNCTION BOX. REMOVABLE COVER SHALL CONTAIN A 1" X 2" GROMMETED NOTCH. BOTTOM OF BOX TO BE 1'-0" A.F.F. SEE ELECTRICAL DRAWINGS.

12. PROVIDE 6" X 6" X 3" DEEP JUNCTION BOX W/ RREMOVABLE COVER. SEE ELECTRICAL DRAWINGS.

13. PICKER MAIN DISCONNECT (C.F.C.I.) LOCATE C.L. OF BOX 5'-0" A.F.F. SEE ELECTRICAL DRAWINGS. ACTUAL LOCATION TO BE DETERMINED BY THE UNIVERSITY'S REPRESENTATIVE.

14. PICKER 120VAC / 20 AMP QUAD RECEPTACLE (C.F.C.I.) W/ CONT. GROUND FROM LOAD CENTER. S-20R (2 EACH) BOTTOM OF BOX 1'-0" A.F.F. SEE ELECTRICAL DRAWINGS.

15. FLOOR WAVINESS OF GANTRY PAD AREA SHALL NOT EXCEED 1/16" WHEN MEASURED WITHIN 1'-0" FROM ANY POINT WITHIN THE GANTRY PAD AREA. FLOOR LEVELNESS OF THE GANTRY PAD AREA SHALL NOT EXCEED 1/8" WHEN MEASURED BETWEEN ANY TWO POINTS WITHIN THE ENTIRE GANTRY PAD AREA.

16. EXISTING DOOR OPENING TO BE INFILLED W/ 1-HR. RATED WALL TO MAKE 1-HR SMOKE PARTITION. WALL TO MATCH ADJACENT CONSTRUCTION.

17. NOT USED

18. NEW FURRED MECH. CHASE. SEE MECH. DWGS FOR DUCT SIZE.

19. INSTALL NEW RUBBER BASE TO MATCH EXISTING ADJACENT BASE.

20. PROVIDE (N) WALL BUMPER. SEE (B/A5)

21. PROVIDE (N) CORNER GUARD. SEE (14/A5)

22. EXISTING HANDWASHING SINK PER U.B.C. SEC. 4204.18.1 (6)

NEW PLAN LEGEND

——————— EXISTING WALL CONSTRUCTION TO REMAIN.

▓▓▓▓▓▓▓ NEW WALL CONSTRUCTION TO MATCH EXISTING ADJACENT WALL T1 RATING AND FINISHES).

—·—1—·— EXISTING 1-HOUR RATED WALL

—··—2—··— EXISTING 2-HR RATED WALL

—·—1S—·— EXISTING 1-HOUR SMOKE PARTITION

—·—2S—·— EXISTING 2-HR SMOKE PARTITION

—·—1O—·— EXISTING 1-HOUR OCCUPANCY SEPARATION

——————— NEW LEAD LINED GYP. BD. ON (E) MTL STUDS

▽ (E)   EXISTING PHONE JACK

▼ (E)   EXISTING DATA JACK

▽   NEW PHONE JACK

▼   NEW DATA JACK

▽ P   NEW PICKER DEDICATED PHONE JACK FOR SERVICE NETWORK SYSTEM.

Figure 11.5k Plan notes and a legend accompanied the drawing in Figure 11.5j. Notice the appearance of detail identifiers within the text of the notes—an unusual location that increases the possibility of the detail being overlooked. The case for consistency in locating subject matter is made here; at least three different details are referenced in this manner.

REFLECTED CEILING PLAN

SCALE: 1/4" = 1'-0"

PROJECT
NORTH

Figure 11.5I The reflected ceiling plan shown here depicts the ceiling as if one were looking down at a floor covered with mirrors. Detail 7 A5 is a section through the curtain rod that is typical of many hospital rooms.

CEILING LEGEND

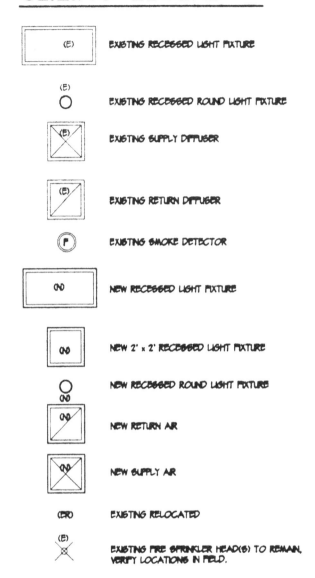

(E)	EXISTING RECESSED LIGHT FIXTURE
(E) ○	EXISTING RECESSED ROUND LIGHT FIXTURE
(E) ⊠	EXISTING SUPPLY DIFFUSER
(E) ◻	EXISTING RETURN DIFFUSER
Ⓕ	EXISTING SMOKE DETECTOR
(N)	NEW RECESSED LIGHT FIXTURE
(N)	NEW 2' x 2' RECESSED LIGHT FIXTURE
○ (N)	NEW RECESSED ROUND LIGHT FIXTURE
(N)	NEW RETURN AIR
(N)	NEW SUPPLY AIR
(ER)	EXISTING RELOCATED
(E) ⊗	EXISTING FIRE SPRINKLER HEAD(S) TO REMAIN. VERIFY LOCATIONS IN FIELD.

Figure 11.5m A ceiling legend and ceiling notes (next figure) accompany the reflected ceiling plan.

CEILING NOTES

1	NEW SUSPENDED CEILING - HEIGHT TO MATCH EXISTING. SEE (1 / A5)
2	PROVIDE NEW CEILING TILE WHERE (E) CLG MOUNTED EXAM LIGHT WAS REMOVED. NEW TILE TO MATCH EXISTING.
3	NEW CURTAIN TRACK
4	EXISTING CURTAIN TRACK TO REMAIN.
5	EXISTING CEILING SYSTEM AND FIXTURES TO REMAIN.
6	NEW MECHANICAL CHASE. SEE MECHANICAL DRAWINGS.
7	EXISTING FIRE SPRINKLER SYSTEM TO REMAIN, IF NEEDED RELOCATE AS REQUIRED TO ALLOW (N) LIGHT FIXTURE INSTALLATION.

Figure 11.5n Notice again the detail reference (1 A5) within the notes adjacent to the drawing.

Figure 11.5o Still more information pertaining to interior construction appears in interior elevations. Here the west elevation is revealed. Notice the detail references in the elevation as well. This interior elevation is 1 of 4 produced by the architect and initially referenced in Figure 11.5j.

(E) STRUCTURE ABOVE

METAL EDGE TRIM, TYPICAL

CONTINUOUS DEEP LEG TRACK WITH .145" DIA. x 1 1/2" POWDER DRIVEN STUDS AT 24" O.C. MAX (1 1/4" MIN. PENETRATION)

6"

FINISH CEILING WHERE OCCURS

3 5/8" x 20 GA. MTL STUDS @ 16" O.C. MAX.

(N) 5/8" TYPE "X" GYP. BOARD, TYP

NOTE:
SHEAR CAPACITY
2000 PSI.
CONCRETE IS 225 LBS.
WHICH IS GREATER THAN
80 LBS REQUIRED FOR
12'-0" HEIGHT.

CONTINUOUS FLOOR TRACK WITH .145" DIA. x 1 1/2" POWER DRIVEN STUDS AT 32" O.C. MAX (1 1/4" MIN. PENETRATION), TYP.

SEALANT, TYP.

FLOOR SLAB

1/2"

O9111f

FURRED WALL
SCALE: 1 1/2" = 1'-0"

9

Figure 11.5p This and the next two details were first introduced on the floor plan shown in Figure 11.5j. Notice how the architect accounts for the movement of the floor above the wall.

(E) STRUCTURE ABOVE

(E) TOP TRACK

(E) DOUBLE METAL STUDS TACK

(E) METAL STUD CRIPPLES

(E) METAL RUNNER CHANNEL

(N) UPPER TRACK, ATTACHED TO (E) RUNNER W/ (1) #10 S.M.S. @ 24" O.C.

(N) MTL. STUDS @ 16" O.C. TO MATCH (E) STUDS IN SIZE AND GAUGE

(N) CONTINUOUS FLOOR TRACK V .145" DIA. x 1 1/2" POWER DRIVEN STUDS AT 32" O.C. MAX (1 1/4" MIN. PENETRATION), TYP.

(E) CONTINUOUS FLOOR TRACK

(E) CONCRETE FLOOR SLAB

(E) DOOR OPENING
SEE FLR. PLAN

DOOR OPENING INFILL FRAMING   11

SCALE: 1/2" = 1' - 0"

Figure 11.5q In this detail the architect is instructing the builder to fill in the space where a door formerly existed; the connections at the head and sill are apparently important enough that yet another detail (12 A5) was created.

(E) STRUCTURE ABOVE

(E) TOP TRACK

(E) METAL STUD CRIPPLES

(E) GYP. BD.

CEILING LINE

(E) METAL RUNNER CHANNEL

(N) UPPER TRACK, ATTACHED TO (E) RUNNER W/ (I) #10 S.M.S. @ 24" O.C.

(N) 5/8" TYPE "X" GYP. BD.

(N) MTL. STUDS @ 16" O.C. TO MATCH (E) STUDS IN SIZE AND GAUGE

(N) CONTINUOUS FLOOR TRACK W .145" DIA. x I I/2" POWER DRIVEN STUDS AT 32" O.C. MAX (I I/4" MIN. PENETRATION), TYP.

SEAL BOTH SIDES

(E) CONCRETE FLOOR SLAB

GYP. BD. SCREWING CONFORM TO U.L. DESIGN # U465

(E) DOOR OPENING

1-HOUR WALL DOOR INFILL

SCALE: I I/2" = I'-0"

12

Figure 11.5r The architect was quite specific about how the space occupied by the door should be filled. Notice how the connection at the top of the wall allows the floor to move without damaging the partition.

(E) STRUCTURE ABOVE

CEILING LINE SEE REFLECTED
CEILING PLANS FOR TYPE
AND ELEVATION

(E) MTL STUDS TO REMAIN

(E) GYP. BD.
TO REMAIN.

(N) 2 LBS. OR 6 LBS. LEAD LINED
GYP. BD., SEE NEW FLOOR PLAN
AND INTERIOR ELEVATIONS.

7'-0" MIN

NOTE:
VERIFY WITH PHYSICISTS
REPORT FOR LEAD LINING
HEIGHT REQUIREMENTS.

O91111

LEAD WALL

SCALE: 1 1/2" = 1'-0"

8

Figure 11.5s Hospitals are complicated projects that make use of a wide variety of products and systems. Here is a simple example of a specialized material (lead-lined wallboard), which is used to prevent radiation from escaping from the room. Wallboard is commonly cut with a razor knife; how do you suppose this material is cut?

1 3/16"

SEE INTERIOR ELEVATIONS

CONTINUOUS TEXTURED HIGH IMPACT SNAP ON COVER. SEE SPECS.

CONTINUOUS ALUMINUM RETAINER

METAL BACKING PLATE SEE

5

CONTINOUS BUMPER CUSHION

(2) 1/4" TEK SCREWS AT 18" O.C. MAX

FINISH WALL

10260b

WALL BUMPER
NO SCALE

13

Figure 11.5t This detail refers the user back to the interior elevations for the vertical and horizontal location of the bumpers, shown in detail here. See note 20 in Figure 11.5k and the left-hand side of Figure 11.5o.

FINISH WALL

5 1/8"

3"

CONTINUOUS CUSHION SHOCK

CORNER GUARD WHERE OCCURS AT 45 DEGREE WALL

3/8"

TEXTURED HIGH IMPACT SNAP-ON COVER

CONT. ALUMINUM RETAINER

CORNER GUARD

NO SCALE

14

Figure 11.5u This and the preceding two details were first referenced in the floor plan notes. (Refer to note 21 in Figure 11.5k.)

FIXED END SEE (6 / -)

#12 GA. VERT. WIRE HANGERS AT 4'-0" O.C. MAX BOTH WAYS. FOR COMPRESSION STRUT DETAIL SEE (4 / -)

CROSS TEES AT 2'-0", TYP. TO MATCH EXISTING

TYP. 12 GA. VERTICAL WIRE HANGERS ON A 4' x 4' GRID AND AT OTHER LOCATIONS AS SHOWN

MAIN RUNNERS AT 4'-0" O.C., TYP. TO MATCH EXISTING

(4) - #12 GA. SPLAY WIRE BRACING MAX 2" FROM CROSS TEES AT 8'-0" O.C.

SPACER BAR TYP.

FREE END SEE (5 / -)

09204a

DIAG. CEILING PLAN
LAY IN TILE   SCALE: 1/4"=1'-0"

1

Figure 11.5v The reader is initially referred to this detail in the reflected ceiling plan notes adjacent to the plan itself. It spawned three additional details (4, 5, and 6), which describe how the suspended ceiling should be attached to the structure.

COMPRESSION STRUTS: STEEL SECTION WITH L/R RATIO OF 200 MAX. ATTACH TO MAIN RUNNERS WITH 1/4" MACHINE BOLT AND TO STRUCTURE WITH 3/16" DIA. POWER-DRIVEN STUD. COMPRESSION STRUT SHALL NOT REPLACE HANGER WIRE

#12 GA. VERTICAL HANGER WIRE AT 4'-0" EACH WAY W/ MIN. 3 TIGHT TURNS IN 1 1/2" BOTH ENDS OF WIRE, TYP.

FOR TYP. CONNECTIONS OF BRACING AND HANGER WIRES TO STRUCT. ABOVE SEE

2

#12 GA. BRACING WIRE W/ MIN. 4 TIGHT TURNS IN 1 1/2" BOTH ENDS OF WIRE, TYP.

45° MAX. TYP.

45° MAX. TYP.

2" (MAX) FROM BRACING WIRES TO CROSS RUNNERS

MAIN RUNNER TO MATCH EXISTING

CROSS RUNNER TO MATCH EXISTING

O9204f

COMPRESSION STRUT

NO SCALE

4

Figure 11.5w The connection of the ceiling to the structure is shown in increasing detail. This is a marvelous example of an isometric drawing.

CROSS TEE 24" O.C.
TO MATCH EXISTING

VERTICAL HANGER, TYP.
FOR DESCRIPTION
SEE

2
—

8" MAX.

2"

FIN WALL

1/2"

1" x 1" x 18 GA. SPACER BAR
W/ ANNULAR RING NAILS

MAIN RUNNER AT 4'-0" O.C.
TO MATCH EXISTING

09510a

FREE END - SUSP. CLG.

SCALE: 3" = 1'-0"

5

Figure 11.5x This detail further explains how the architect wants the free end of the suspended ceiling to be connected to the structure. It is referred to in the compression strut detail shown in Figure 11.5w as detail 2.

(4) 45 DEG. #12 GA. GALV. SPLAY WIRE BRACING AT 8'-0" O.C. WITH MIN. 4 TIGHT TWISTS TYP. MAX 4'-0" FROM SURROUNDING WALL FOR ANCHORAGE ABOVE SEE
③ —

CROSS TEE 24" O.C. TO MATCH EXISTING

VERTICAL HANGER, TYP. FOR DESCRIPTION SEE
② —

LAY-IN ACOUSTICAL CEILING SYSTEM. TO MATCH EXISTING. FOR TYP. INFO. SEE
① —

8" MAX.

2"

FIN. WALL

CONT. I" x I" WALL ANGLE W/ 1/8" POP RIVET TO MAIN RUNNER AND CROSS TEE, TYP. FASTEN WALL ANGLE TO WALL W/ #6 S.M.S. AT 16" O.C. (EA. VERTICAL STUD) TO MATCH EXISTING.

MAIN RUNNER AT 4'-0" O.C. TO MATCH EXISTING

09510b

FIXED END - SUSP. CLG.
SCALE : 3"=1'-0"

6

Figure 11.5y Each detail spawns more details until the architect is satisfied that the component or system is adequately described. This detail leads the reader to three more details.

METAL DECK

TYPICAL ANCHORAGE

90 DEGREE BENT 12 GA. MTL.
ANGLE W/ 3/16" DIA. HILTI
CC27232 EXP. ANCHORS
(MIN. 1 1/8" EMBEDMENT)

4 TIGHT TURNS W/ IN
1 1/2", TYP.

METAL DECK

TYPICAL DIAGONAL
BRACE ANCHORAGE

1 1/2" x 1 1/2" x 1 1/2" 45
DEGREE BENT 12 GA. MTL. ANGLE
IN LINE W/ SPLAY WIRE BRACING
W/ 3/16" DIA. HILTI CC27232 EXP.
ANCHOR W/ MIN 1 1/2" EMBEDMENT
(NO SHOT-IN ANCHORS
ALLOWED)

ALSO SEE $\frac{3}{—}$

4 TIGHT TURNS W/ IN
1 1/2", TYP.

NOTE:
PROVIDE FIELD TEST LOADS
FOR ANCHORS: 200#
FOR VERTICAL AND 440# FOR
DIAGONAL

O9204c

SUSP. CLG. ANCHORAGE

SCALE: 3" = 1'-0"

2

Figure 11.5z Detail 2 A5 describes how the ceiling connects to the floor above.

(E) METAL STUDS

(E) GYP. BOARD

4" x 18 GA. UNPUNCHED "C" STUD MIN. LENGTH TO SPAN 3 STUDS

#10 SELF-TAPPING SHEET METAL SCREWS. (2 TOTAL EA. STUD)

SPLAYED CEILING WIRE THROUGH HOLE DRILLED IN "C" STUD

09204d

SPLAYED WIRE CONN.
SCALE: 3" = 1'-0"

3

Figure 11.5aa Detail 3 A5 describes the connection of the support wires to nearby partitions. It is referenced in the preceding figure (Figure 11.5z).

MTL. TRACK SUPPORT SYSTEM
ARE C.F.C.I, CONTRACTOR MAY
SUBMIT ALTERNATE SYSTEM
FOR UNIVERSITY'S REPRESENTA-
TIVE APPROVAL.

#12 GA. G1 WIRE HANGER AT
4'-0" O.C. FOR TYP. CONNS.
SEE

②
—

CONTINUOUS 4" X 18 GA.
UNPUNCHED STRUCTURAL STUD.

LAY-IN ACOUSTICAL CEILING
SYSTEM. TO MATCH EXISTING.

SURFACE MOUNTED CURTAIN - LV
TRACK BOLTED TO STUD W/
1/4" DIA. MACHINE BOLT AT
16" O.C. THRU PIPE SPACER,
TYPICAL.

LIGHTWEIGHT
FABRIC CURTAIN

10150b

CURTAIN - I.V. TRACK   7

SCALE: 6" = 1'-0"

Figure 11.5bb Elements in suspended ceilings that bear light loads are often reinforced, as shown in this detail.

SEE FLOOR PLAN

CONT. RUNNER TRACK CUT FLANGE AND BEND WEB.

(A)

HEAD-SILL

RECESSED MOUNTED UNIT

(B)

JAMB

.5/8" TYPE "X" GYP. BOARD (2 LAYERS AT 2 HR. WALL)

DOUBLE METAL STUDS, TYP.

INSULATION, TYP.

O9110f

RECESSED PANEL

SCALE: 3" = 1'-0"

10

Figure 11.5cc Detail 10, shown here, is derived from interior elevation D. It is worth noting that neither the elevation nor this detail gives the dimensions of the recessed area.

CONTROL CONSOLE, MOUNT TO COUNTERTOP WITH "Z" CLIPS, SEE 18

3/4" TOP WITH P-LAM FINISH, TYP.

(3) #0 WOOD SCREWS W/ MIN. 1/2" EMBEDMENT.

"Z" CLIPS 2 EA. SIDE W/ #0 WD SCREWS

1 1/2"

1"

DELUXE TUCK-AWAY KEYBOARD SHELF BY "PC ACCESSORIES" STOCK NO. C533TP. PROVIDE 3/4" PLYWOOD PLATE FOR BACKING. INSTALL PER MANUF. REQUIREMENTS.

END PANEL W/ P-LAM FINISH EXPOSED SURFACES, SEE 17

2 1/2" x 2 1/2" x 3/16" CLIP AT 16" O.C. MAX W/ 2 - #0 S.M.S. TO BACKING PLATE, TYP.

2'-1"

4"

2'-6"

BACKING PLATE, SEE 15

FINISH WALL

FINISH FLOOR

2

06222

WRITING DESK (W/ KEYBD. SHELF)

SCALE: 1 1/2" = 1'-0"

16

Figure 11.5dd Casework—counters, cabinets, and the like—may represent a significant amount of work; very specific requirements for supporting a monitor are set forth here. Reference to three other pertinent details is made.

FACE OF STUD. SEE FLOOR
PLAN FOR SIZE

WALL FINISH

CONT. 16 GA. x 6" MTL. PLATE
3 STUD MIN. SPAN

(2) #10 x 5/8" S.M.S. @ EA.
STUD

6"

BACKING PLATE TYPE A

SCALE : 6" = 1'-0"

05500a

15

Figure 11.5ee Casework, chair rail, baseboard, and grab bars must all be supported by backing within the wall. This detail pertains to the casework as well as the bumper described in Figure 11.5t.

3/4" PANEL W/ P-LAM FINISH

1 x 2 WOOD EDGE BAND

3/4" x 1 3/16" x 2 1/2" LONG ANGLE DAP INTO BOTTOM OF END PANEL

1/8" F.H.W.S. COUNTER SUNK, MIN. 2 AT 1 1/2" O.C.

FINISH FLOOR SEE SCHEDULE

(2) - 3/16" DIA F.H.W.S. ONTO EXPANSION ANCHORS

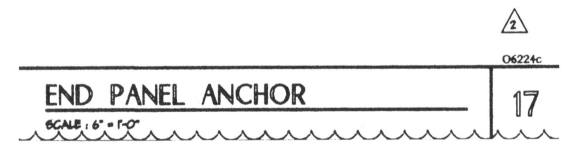

O6224c

END PANEL ANCHOR

SCALE : 6" = 1'-0"

17

Figure 11.5ff This detail shows the connection of the casework to the floor. Consider the work involved just to make this connection.

PLAN
N.T.S.

— MONITOR BASE

(4) 2" x 2" MOUNTING "Z" CLIPS
ATTACHED TO CASEWORK
COUNTERTOP W/ (1) #10 WOOD
SCREWS. SECURE CLIPS TIGHT
TO MONITOR BASE.

— MONITOR BASE

— COUNTERTOP

"A"   "A"

SECT. "A"-"A"
N.T.S.

MONITOR ANCHORAGE DETAIL
N.T.S.

18

Figure 11.5gg The architect is specific about how the monitor connects to the desktop as well.

Subtleties in the Drawings

To fully appreciate and account for all of the tasks required to construct a wall or ceiling or similar assembly, one must integrate what the drawings require with building code requirements and the practices of the indigenous construction community. Building code and similar requirements are often noted on drawings, although they may not appear in the graphic depictions in the drawing set, and practices develop in an area that become the standard of behavior for the area. In a quick review of the portion of the floor plan shown in Figure 11.5g, and the wall type legend and general notes shown in Figures 11.5h and 11.5i, the careful reader will notice a variety of requirements imposed on the builder by the architect. The shopping list shown in Table 11.2 describes the operations and materials associated with all the wall types used in partial floor plan in Figure 11.5g.

TABLE 11.2 Shopping List—Interior Partitions

05400 – cold-formed metal stud framing				
07800 – fire and sound insulation				
09200 - gypsum wallboard				
09820 - acoustical insulation and sealants				

Component or Assembly	Operations	Materials	Comments
05400 Metal stud framing			
Wall types 4, 4cg, 4d, 4dg	Frame walls 3⅝" " × 20-ga. studs @ 24" o.c. to deck or structure (14' AFF)	3⅝" × 20-ga. × 14'; 16 ga. sill track, 16-ga. deep leg head track Hilti DN pins, 16" o.c.	Note 4, wall legend, p. A4.1; Note 9, General Notes, p. A4.1, Detail IW-1 Note 8, General Notes, p. A4.1
	Install blocking, backing, and bracing.	⅝" × 12" × 16" GWB; 16-ga. sheet metal	Note 1, General Notes, p. A4.1
	Install special studs.	16-ga. studs	At doors, base cabinets, upper cabinets, full height cabinets, toilet partitions, grab bars per note 10, General Notes
Wall type 5	Frame double plumbing walls 3⅝" × 18-ga. studs @16" o.c.	3⅝" × 18-ga. × 8'–6" high metal studs; 3⅝" × 16 ga. sill track, 16 ga. deep leg head track Hilti DN pins, 16" o.c.	Note 5, wall legend, p. A4.1; note 4, wall legend for stud height Note 9, General Notes p. A4.1
	Install blocking, backing, and bracing.	⅝" × 12" × 16" GWB 48" o.c. both ways; 16-ga. sheet metal	
	Install special studs.	16-ga. studs	At doors, base cabinets, upper cabinets, full height cabinets, toilet partitions, grab bars, per note 10, General Notes
Wall type 6	Frame plumbing wall, 6" × 18-ga. studs 16" o.c. to deck or structure.	6" × 18-ga. × 14' long studs; 3⅝" × 16-ga. sill track, 16-ga. deep leg head track, Hilti DN pins, 16" o.c	Note 6, wall legend, p. A4.1; see notes for wall type 4 as well as Note 9, General Notes, p. A4.1
	Install special studs.	16-ga. studs	At doors, base cabinets, upper cabinets, full-height cabinets, toilet partitions, grab bars, per note 10, General Notes
Wall type 7	Frame 4-hour wall, 3⅝" × 20-ga. studs 16" o.c. to structure.	3⅝" × 20-ga. × 14' studs;16-ga. sill track, 16-ga. deep leg head track Hilti DN pins, 16" o.c.	Note 7, wall legend, p. A4.1 Note 9, General Notes, p. A4.1
	Install special studs.	16-ga. studs	At doors, base cabinets, upper cabinets, full-height cabinets, toilet partitions, grab bars, per note 10, General Notes

Component or Assembly	Operations	Materials	Comments
Wall type 9, 9d	Frame pump room wall, 3⅝" studs to structure, staggered in 8" channel.	3⅝" × 20-ga. × 14' studs 8" × 16-ga. sill channel; 8" × 16-ga. deep leg head track Hilti DN pins, 16" o.c. Resilient clips on pump room side.	Note 9, wall legend, p. A4.1 Note 9, General Notes, p. A4.1
	Install special studs.	16-ga. studs	At doors, base cabinets, upper cabinets, full height cabinets, toilet partitions, grab bars, per note 10, General Notes
Wall type 11	Frame 2-hour walls, 3⅝" studs 24" o.c. to structure.	3⅝" × 20-ga. studs 3⅝" × 16-ga. sill channel; 8" × 16-ga. deep leg head track Hilti DN pins, 16" o.c.	Note 11, wall legend, p. A4.1 Note 9, General Notes, p. A4.1 Note 8, General Notes, p. A4.1
11d	Frame 2-hour wall, 3⅝" studs 16" o.c. to structure.	3⅝" × 20 ga. studs 3⅝" × 16-ga. sill channel; 3⅝" × 16-ga. deep leg head track Hilti DN pins, 16" o.c.	Note d, "Nomenclature," Note 9, General Notes, p. A4.1, Note 8, General Notes, p. A4.1
11g	Frame 2-hour walls, 3⅝" studs 24" o.c. to structure.	3⅝" × 16-ga. sill channel; 3⅝" × 16-ga. deep leg head track Hilti DN pins, 16" o.c.	Note 9, General Notes, p. A4.1 Note 8, General Notes, p. A4.1
	Install special studs.	16-ga. studs	At doors, base cabinets, upper cabinets, full-height cabinets, toilet partitions, grab bars, per note 10, General Notes
Wall type 12, 12d, 12g	Frame 1-hour wall, 3⅝" studs 24" o.c. to structure (wall 12d studs 16" o.c.).	3⅝" × 20-ga. studs 14' long 3⅝" × 16-ga. sill track, deep leg head track Hilti DN pins, 16" o.c.	Note 12, wall legend, p. A4.1 Note 9, General Notes, p. A4.1
	Install special studs.	16-ga. studs	At doors, base cabinets, upper cabinets, full-height cabinets, toilet partitions, grab bars, per note 10, General Notes
Wall type 13	Frame column enclosures, 3⅝" × 20-ga. studs 24" o.c. to 6" above ceiling.	3⅝" × 20-ga. studs 10' long 3⅝" sill and head track Hilti DN pins, 16" o.c.	Note 13, wall legend, p. A4.1, Note 9, General Notes, p. A4.1
	Install special studs.	16-ga. studs	At doors, base cabinets, upper cabinets, full-height cabinets, toilet partitions, grab bars, per note 10, General Notes
Wall type 14, 14c	Frame chase walls, 3⅝" × 18-ga. studs to 6" above ceiling.	3⅝" × 18-ga. studs 10' long 3⅝" × 16-ga sill and head track Hilti DN pins, 16" o.c.	Note 14, wall legend; Note 9, General Notes, p. A4.1
	Install special studs.	16 ga. studs	At doors, base cabinets, upper cabinets, full-height cabinets, toilet partitions, grab bars, per note 10, General Notes
07800 – Fire and Smoke Protection; 09800 – Acoustical Insulation			
Wall type 4	Insulate bath walls for sound full height.	Check specs, probably batts.	Note 4 not clear; Gyp board runs to 9' only—14' high wall; Securing insulation above the gyp board line not addressed in notes?
Wall type 5	Insulate plumbing walls for sound full height.	3⅝" insulation same thickness as studs to ceiling (8'-6")	Note 5 not specific
Wall type 6	Insulate 6" plumbing walls.	6" sound insulation	Same as wall type 4

(Continued)

TABLE 11.2 (*Continued*)

Component or Assembly	Operations	Materials	Comments
Wall type 7	Insulate 4-hour walls.	2" mineral fiber batts, 2 pcf min.	Height not specified in notes, presume full height
Wall type 9	Insulate pump room walls.	"sound attenuation blanket"	See spec section
Wall type 11	Insulate 2-hour wall.	Check specs	No insulation specified in note
Wall type 12	Install fire stopping 1-hour wall.	Not specified in note, check specs	Wall penetrations and deck flutes
Wall type 13	Insulate column enclosures.	"sound insulation"	See spec section
09200 – Gypsum wallboard			
Wall type 4	Install wallboard to 6" above ceiling.	⅝" type × both sides to 9'; fasteners not identified	From wall legend, p. A4.1; Note 4, wall legend
4cg	Install wallboard to structure, janitor room.	Water-resistant board one side, type × on the other; ¾" fire-treated plywood over ⅝" gyp board telephone room	Note G, "Nomenclature," p. A4.1
4d	Install bathroom wallboard to structure.	⅝" Dens-Shield tile backerboard (Georgia Pacific) bathroom side	Note D, "Nomenclature," p. A4.1
4dg	Install bathroom wallboard to structure.	⅝" Dens-Shield tile backerboard (Georgia Pacific) bathroom side; ⅝" type × gyp board other side; ¾" fire treated plywood, telephone room side over ⅝" gyp board	Note D, "Nomenclature," p. A4.1 Note G, "Nomenclature," p. A4.1
Wall type 5	Install wallboard to 8'–6", double plumbing wall.	⅝" type × tile backerboard	
Wall type 6	Install wallboard (to structure), plumbing wall.	Water-resistant wallboard one side; Type × gyp board on the other	
Wall type 7	Install wallboard full height, 4-hour wall.	Two layers ¾" type × gyp board parallel to studs, staggered joints on second layer 1¼" type S screws 24" o.c. first layer; 2 ¼" type S screws 12" o.c. second layer, with 1½" type G screws midway between studs in board joints	
Wall type 9	Install wallboard, pump room full height, one layer other side.	Two layers ⅝" type × gyp board with #6 × 1" drywall screws 8" o.c. seams, 12" field pump room side; one layer ⅝" type × other side	Fasteners and spacing not specified for second layer; same as type 11 wall?
9d	Install two layers wallboard full height, pump room, one layer on bathroom side.	Two layers ⅝" gyp board pump room side; ⅝" Dens-Shield tile backer board (Georgia Pacific) bathroom side; #6 × 1" drywall screws 8" o.c. seams, 12" field	Fasteners and spacing not specified for second layer; same as type 11 wall?
Wall type 11	Install wallboard to structure, two-hour wall.	Two layers ½" type × gyp board both sides; #6 × 1" drywall screws 8" o.c. seams, 12" field first layer; #6 × 1⅝" drywall screws 9" o.c. on vertical joints; 12" field, 24" o.c. on runners.	Note 11, wall legend, p. A4.1

Component or Assembly	Operations	Materials	Comments
11d	Install wallboard to structure, two-hour wall at bathrooms.	Two layers ½" type X gyp board; ⅝" Dens-Shield (Georgia Pacific) tile backer board bathroom side	Fasten as in wall type 11; apply board vertically, offset joints between layers
11g	Install wallboard to structure, two-hour wall at telephone room.	Two layers ½" type X gyp board each side; One layer ¾" fire treated plywood on telephone room Side	Note g, "Nomenclature"
Wall type 12	Install wallboard to structure, one-hour wall.	One layer each side, ⅝" type X gyp board, #6 × 1" drywall screws 8" o.c. seam, 12" field nailing	
12d	Install wallboard to structure, one-hour wall at bathroom and pump room.	⅝" type X wallboard: ⅝" Dens-Shield tile backerboard (Georgia Pacific) at bathroom	Fasteners and spacing not specified in note
12g	Install wallboard to structure, one-hour wall at telephone room.	One layer ⅝" type X gyp board on each side	As above
Wall type 13	Install wallboard on column enclosures.	One layer ⅝" type X gyp board to 6" above ceiling #6 × 1" drywall screws 8" o.c. seams and 12" o.c., studs 24" o.c.	Note 13, wall legend, p. A4.1
Wall type 14	Install wallboard on chase walls to 6" above ceiling.	⅝" moisture resistant gyp board room side only to 6" above ceiling, or to underside of soffit	Fasteners and spacing not specified in note

What is significant in this list is the variety of requirements imposed on the builder (in the walls alone) in what amounts to a 2,000-square-foot (186 m²) area on one floor of this four-story building. At issue as well is the sequencing of the work. Although some general contractors use a single subcontractor for metal stud framing, wall and ceiling insulation, (fire, acoustic, and thermal), fireproofing (safing and caulking), wallboard, and suspended ceilings—thus streamlining the interior construction process—others do not. In the case of the example project, a contractor who uses separate subs for the aforementioned tasks would have to coordinate the work of some eight subcontractors (when plumbing and electrical are included). There is the quality-control issue as well: Fire caulking and insulation, for example, may affect several subs, and if each is given the contractual responsibility for fire protection (which is frequently the case), the likelihood that the work will be inconsistently performed is greater. It is worth mentioning, too, that deviations from common practice merit a close look and even closer control on the part of the supervisors of the construction. Skilled tradespeople want to do things "the way it's done all the time."

The notion that a builder must evaluate how a project will be constructed while reviewing drawings is well illustrated in Figure 11.6, which is proximate to the partial floor plan shown in 11.5g.

Here the architect has created a chase between two walls, apparently with the idea of concealing the structural columns while separating the building into distinct fire zones, thus the description "(4) hour-rated area separation wall"[1] (see wall type 7 in the legend shown in Figure 11.5h and note 7 under "General Notes" in Figure 11.5i). Wall type 7 is full height, that is, it extends from the slab-on-grade to the underside of the second floor (some 14' [4267 mm] in this project). The frame consists of 3⅝" [92 mm] × 20-gauge[2] metal studs 16" (406 mm) on center in 16-gauge runners, with two layers of ¾" (19 mm) type X gypsum wallboard (a fire-rated wallboard) on each side, applied parallel to the studs (that is, the long edge of the wallboard is parallel to the framing) and fastened with type S drywall screws of two different lengths. The second layer may be applied perpendicular to the studs, with the butt joints staggered (so that they do not line up with the joints in the base layer), or parallel to the studs (like the base layer), with vertical joints staggered. The joints in sheets on one side of the wall must be

---

[1] Building codes set forth minimum standards for safety and health for building projects, among them fire safety standards, which govern a host of building requirements such as building size and construction type. Most large buildings are segregated into distinct fire zones by area separation assemblies (primarily walls and ceilings) to preclude a fire starting in one area from spreading to another.

[2] Gauge refers to the thickness of metals—the smaller the number, the thicker the metal.

Figure 11.6 The chase formed by the walls on both sides of these two structural steel columns creates unusual challenges for the builder.

staggered in relation to the joints on the other side of the wall as well. Where joints occur between framing members, the builder has been instructed to use type G drywall screws. On the left side of the wall (where the lockers and showers are), the builder must install a layer of ⅝" (16 mm)- thick tile backerboard over the second layer of ¾" (19 mm) gypsum wallboard. The wall on the right is similarly constructed, except the stud spacing is 24" (600 mm) on center, and only one layer of ⅝" (16 mm) type × gypsum wallboard is required on each side. The space between the walls is approximately 18" (457 mm).

The normal process for constructing metal stud walls is simple: The walls are laid out (located) on the floor and ceiling (in the case of a full-height wall), the runners are detailed and installed (frequently with powder-actuated pin guns), and the studs are cut to length and installed, along with any blocking or backing within the wall that may be required. Any rough electrical, plumbing, or mechanical requirements within the walls are executed, and the wall is inspected. Following an inspection, the insulation (if required) is installed and inspected. After the insulation has been approved, the wallboard is installed and inspected, it is taped, topped, and textured, and finally it is painted or otherwise finished. For the standard wall, unfettered workspace all around is assumed.

The walls forming the chase mentioned pose an interesting problem for the builder, namely, how to install the layers of wallboard on the inside faces of the walls. The standard installation procedure cannot be followed. Needless to say, an 18" space is inadequate to work in,

particularly when you consider that the weight of one 4' × 8' sheet of ⅝" gypsum wallboard (1219 mm × 2438 mm × 16 mm) is about 72 pounds (33 kg); not to mention its bulk or the limited light that is likely to be available. Even if it were possible to work in the space, at some point the worker would have to exit the chase, and would be forced to install the last piece of wallboard from outside the wall.

It might be possible to frame the wall on the ground, install the wallboard on the inside face, and raise the wall into place, as in platform framing in wood light construction, but a 4' (1219 mm-) wide section of wall type 12 (the lighter of the two walls by a significant amount) will weigh on the order of 250 pounds (113 kg), and at 14' long (4267 mm), would be very awkward to install, if it could be done at all (there are structural steel beams overhead to consider). There is the added difficulty of the requirement in many jurisdictions that a wallboard fastening inspection be done for each layer and taped joints, particularly where firewalls are concerned. It would be impossible to show an inspector the nailing when the wall is completed, except by remote camera (a sort of arthroscopic surgical technique would be required).

A similar condition exists at some elevator shafts in multistory buildings, where the work is accessible from only one side. Known generically as *shaft wall*, the framing system consists of metal studs (called CT studs because of their shape in section) with special channels on one edge (see Figures 11.7a–b) and a runner system that features channel with different-sized legs.

INACCESSIBLE SIDE

Figure 11.7a Shaft wall is a system for framing firewalls that are accessible from only one side.

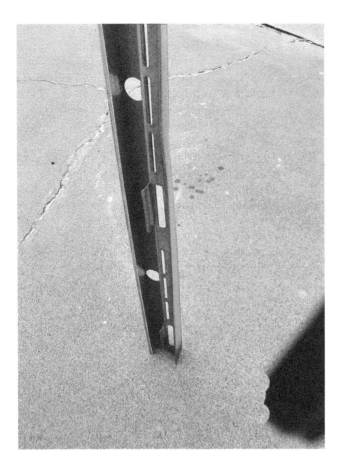

Figure 11.7b This photograph shows a section of CT stud; the slotted section of the channel is the 1" space into which the gypsum wallboard is inserted. (Photo by the author.)

A starter stud is installed between runners (the U- or J-shaped track into which the studs are placed), and the wallboard is stood on its short edge and installed in the vertical channel on the stud. A second stud is then installed in the runners, and its channel is fitted around the remaining uncontained edge of the wallboard; it is then secured to the runners. The installation continues this way until the wall is completed.

The architect has created an interesting situation in this project, however, by specifying that the studs in wall type 7 be installed 16" (406 mm) on center. This means that every sheet of 3/4" (19 mm) wallboard installed in the chase must be cut to 16" (406 mm) in width (wallboard of this thickness comes in 2' × 12' sheets). It is conceivable that the standard installation procedure for shaft wall—sheets 2' wide attached to studs 2' (640 mm × 3658 mm) on center—could be followed; however, it would then be necessary to retrofit studs 16" on center after the wallboard and studs 2' on center were installed, then make the best attempt to attach the wallboard to the retrofitted studs by inserting the screws through the stud into the wallboard—not the best of connections. Gypsum wallboard relies heavily on the head of the screw or nail trapping an adequate amount of paper under the head of

the screw, which is not possible if the screw head does not come into contact with the paper (which would be the case if the screw were installed through the stud first). This is clearly a situation where value engineering is appropriate: The constructor investigates standardized alternatives that do not compromise either the structure or the esthetics of the assembly, then consults with the architect to arrive at the best solution. The point here is that the estimator who evaluates the cost of this particular wall assembly as if it were a standard wall, or even a standard shaft wall, is putting job profitability at risk.

Other conditions that are not revealed by drawings abound. Examples of the typical conditions that framers face in the field are illustrated in the photographs in Figures 11.8a–h.

The Takeoff

With the foregoing in mind, how would one approach the task of developing a material quantity takeoff for interior construction? Focusing just on wall construction for the moment, once the wall types had been distinguished, a list of operations like those listed in Table 11.2 would be developed (it is at this point that the method of construction would be determined in sufficient detail to apply costs to the various tasks). "Hidden" tasks would be revealed; for example, the method of retaining sound insulation in full-height walls that also require sound insulation full height, but on which the wallboard only reaches to a point just above the ceiling (see wall type 4 in Figure 11.5h). Another hidden task, installation of draft stops required by at least one building code on 10' (36 576 mm) centers horizontally and vertically must be accounted for as well. Once the materials have been sorted, they would be quantified and a summary of quantities established. A crew would be selected, its costs determined, and production rates for the various wall types estimated. Unit costs would then be determined for each wall type. Table 11.3 is a matrix of common units used in interior construction.

Summary

Understanding how interior construction is performed—or any other type of construction for that matter—is essential not only to estimating the cost of work, but also to managing it in the field. The sources of the necessary information are often numerous, necessitating an imaginative mind-set on the part of the drawing user.

To as great an extent as possible, good superintendents and project managers review drawings carefully, more

Figure 11.8a This photograph shows a typical condition where partition walls and a large HVAC plenum interface. The top of the walls is free-floating, necessitating the installation of a diagonal brace and ceiling tie. When partition walls are full-height walls, such bracing is unnecessary. If a gypsum wallboard ceiling is called for here, the framer faces another challenge: figuring out how to support the ceiling under the duct. Notice that the underside of the trapeze for the duct is in contact with the top runner. A dropped ceiling is an easy solution, except if the architect calls for a ceiling height that is the same elevation or higher than the underside of the duct. The point here is that the distance that the duct encroaches into the interstitial space is not specified in the design drawings; only the mechanical sub knows with any certainty (duct elevations are recorded on shop drawings, and the work is coordinated with affected trades). (Photo by the author.)

Figure 11.8b It is common for partition walls to be located partway under a structural member; here is one method of securing the top runner to the framing. Bear in mind that this requires cutting and installing additional pieces. Notice how this top runner allows the beam to deflect without compressing the wall; the slots in the runner, through which screws are drilled into the studs, permits vertical but not horizontal movement. This system requires that the gypsum wallboard not be attached to the head track, which it normally is. The gypsum wallboard must be cut short of the beam, and, in a rated assembly, the gap filled with fire sealant. (Photo by the author.)

Figure 11.8c The work reflected in this situation is not described in the drawings, with good reason. The architect may not recognize what is required, and in such conditions, the solution is left up to the specialist, who comes to know what is expected in the construction. It is a fairly clean separation of professional responsibility. The architect has an interest in the wall location and height; it is the framer's responsibility to determine how to accomplish the work. It goes without saying, though, that the framer who overlooks the work inevitably pays for it out of the profit margin. (Photo by the author.)

Figure 11.8d The architect addresses the condition evident here in generic details of predictable conditions, such as in the next figure. (Photo by the author.)

"HILTI" POWDER ACTUATED FASTENER
.145"Ø x 1 1/2" MIN., OR APPROVED
EQUAL @ 16" O.C., MAX.

TYPE "S" DRYWALL SCREWS PER
WALL LEGEND, TYPICAL

CONT. SOUND DEADENING FIRE
CAULKING EACH SIDE OF WALL,
TYPICAL TOP & BOTTOM

CONCRETE OVER METAL DECK
PER STRUCTURAL DRAWINGS

WALL TYPE PER
PLAN & WALL
SCHEDULE

16 GA MIN. CONT
TRACK, TYPICAL
TOP & BOTTOM

SAFING INSULATION
BETWEEN DECK
FLUTES, TYPICAL

1/2" TOP OF
WALL STUDS.

PROVIDE FLEX
TRACK @
TOP OF WALL.

(1) HILTI POWDER
ACTUATED FASTENER
.145"Øx 1 1/2"
EMBED EACH SIDE.

CONTINUOUS 16 GA. METAL PLATE,
WIDTH COMPATIBLE WITH DECKING.

#10 SDSTS @
16" O.C.

TYPICAL SILL & HEAD TRACK
ANCHORAGE DETAIL   3"=1'-0"

Figure 11.8e This detail instructs the builder how to treat wall partitions under a corrugated metal deck.

Figure 11.8f The typical frame often involves duct penetration detailing. Knowing where these penetrations occur will result in a more accurate cost estimate of the work. (Photo by the author.)

Figure 11.8g This curved soffit represents a considerable amount of work. Notice the bracing between the two walls of the soffit, the connection to the corrugated deck, and especially the articulated runners top and bottom. (Photo by the author.)

Figure 11.8h What is interesting about this photograph is the extent to which the outcome described by the architect belies the extent of the work. A rather innocuous detail on a reflected ceiling plan turns out to be an enormous amount of work. In the four curved tracks alone, there are hundreds of screws to loosen and then retighten once the desired radius has been established, after which the tracks must be connected to the ceiling. How would you do it? (Photo by the author, with the permission of the Sacramento County Office of Education.)

TABLE 11.3 Sample Unit Designations for Materials and Work*— Interior Construction

	LS lump sum	EA each	Opng Opening	PR pair	M 1000	LF lineal foot	SF square foot	SY Square yard	FBM foot board measure	CF cubic foot	CY cubic yard	Ga Gallon	PCF Lbs/cf	
Wood light framing	X					X			X					Wall Systems
Metal stud framing						X	X							
Wall backing		X				X								
Gypsum wallboard		(sheet)					X							
Plywood														
Thermal insulation	X													Insulation
Acoustic insulation		X	X				X							
Fire-resistant safing		X				X	X						X	
Fire caulking											X			
Wood trim	X					X								Finishes and Trim
Paint		X					X							
Wallpaper							X	X						
Floor coverings							X	X						

*Metric drawings require different units: liters (liquids), kilograms (weight), lineal meters (length), cubic meters (volume), square meters (area).

or less in the order of construction, with the intention of revealing difficulties and unusual circumstances involving construction and resolving them far in advance of the time they become crises on the project site.

CHAPTER 11 EXERCISES

1. What structural characteristics do most interior walls in a commercial building exhibit?

2. What other important function do interior walls play, besides partitioning space?

3. What is meant by an area separation assembly?

4. List several "hidden tasks" that a builder must consider when installing interior walls.

12 Mechanical Systems

Key Terms

Angle valve
Bends
Butterfly valve
Cleanout
Compressor
Condenser
Couplings
Duct coil
Elbows
Evaporator
Expansion Loop
Flange
Galvanic action
Gate valve
Globe valve
Heat exchanger
Hub
Lift check valve
Plenum
Reducer
Refrigerant
Sleeve
Swing check valve
Tees
Trap
Union
Variable air volume (VAV) box
Water hammer
Wye

Key Concepts

- Exploring mechanical system drawings from two basic viewpoints—source-to-terminus and terminus-to-source—is an effective approach to assimilating the information about the systems.
- Plumbing supply systems are pressurized systems, whereas drainage systems depend largely on gravity to conduct fluids in one direction or another.

Objectives

- List the fundamental components of an HVAC system.
- List the fundamental components of a plumbing system.
- List common graphic symbols in HVAC and plumbing systems.
- Identify the various specialty contractors who become involved in HVAC and plumbing specialty work, and the parts of the system for which they are responsible.
- Identify potential sources of conflicts in mechanical system drawings.

Construction Graphics: A Practical Guide to Interpreting Working Drawings, Third Edition. Keith Bisharat.
© 2025 John Wiley & Sons Inc. Published 2025 by John Wiley & Sons Inc.

Purpose of Mechanical Systems

The term *mechanical systems* refers primarily to HVAC (heating, ventilating, and air conditioning) and plumbing systems in building construction. In industrial construction projects, such as wastewater and water treatment plants, mechanical drawings may constitute the bulk of the design professional's work.

HVAC systems circulate a mixture of fresh and conditioned air throughout the building, and plumbing systems provide hot and cold potable water, sanitary waste disposal, and system vent piping. Mechanical systems are among the more complicated and costly systems in the typical commercial building, representing close to a fifth of the direct (project-specific) construction costs. For hospital projects, the figure approaches a third or more of the costs.

What to Expect in the Drawings

In building construction projects, it is common for mechanical drawings to be segregated into mechanical (M sheets) and plumbing (P sheets), even though the same mechanical engineer may design both systems. The conventional drawing set organization calls for the M sheets to precede the P sheets. Whatever the organization of the plans, it helps to develop an understanding of the drawings by exploring the systems independently.

HVAC Systems

To serve their purpose, HVAC systems must have some means of heating, cooling, conditioning, distributing, filtering, and controlling the speed and temperature of the air in a building. The methods for fulfilling the basic system requirements are numerous and varied, from simple fans to highly controlled environments like clean rooms in silicon chip manufacturing. The temperature of the ambient air, relative humidity, air movement, and the mean radiant temperature of a space conspire to create what people sense as temperature in a building. By various means, HVAC systems seek to affect these four comfort factors by conditioning the air.

Conditioned air refers to the adjustments to the air circulated in a building, which includes temperature adjustments (addressed in the foregoing paragraph), adjustments to the amount of moisture in the air (humidification and desiccation) and air purity (filtration). Humidifiers of varying design may be installed in systems, and when they are, they are most commonly located in the supply air plenums of the system. These systems either add water in the air supply or remove heat at points of distribution. The desiccating process in a residential air conditioning system is a by-product of the cooling cycle. When heat is taken out of the air blowing across the evaporator, the air releases moisture, which collects in the condensate pan under the equipment and is drained off. In industrial applications, other methods such as liquid desiccant dehumidification may be used. Filters capable of extracting dust and allergens take a variety of forms, from simple mechanical systems (air is blown through fine-meshed filters, which trap the dust) to more sophisticated ionizing electronic filters, which charge the particles and trap them electrically. Filters are commonly located in return air grilles, at points where used air is taken in, or very near the air-handling equipment in the return air plenum, in front of the blower.

Elements of HVAC Systems

Heat Generation

Hot water, steam, and warm air systems, fired by natural gas, fuel oil, or various types of coal, and electric resistance strips, or a combination of these methods, commonly heat buildings. The heat is created in one or several locations and is then distributed to terminus points in rooms in the building through conduit such as metal and wire-framed plastic ducts and piping systems made of copper or steel pipe (or a combination of the two). Hydronic systems use boilers of one sort or another to heat water or create steam, and then distribute it throughout the building, where it fills finned coils or convectors of some sort.

Cooling

The refrigerant medium in cooling systems—commonly a gas—is cycled through three components in air conditioning systems, the compressor, the condenser, and the evaporator. The compressor converts the gas into liquid, which results in heat being released from the gas. The condenser, aided by a fan and or other device such as a cooling tower, transfers the heat given off by the compressed gas to a surrounding medium (the atmosphere or a body of water). The converted gas is then sent under pressure through a refrigerant line to the evaporator (a heat exchanger), which allows the gas to expand, thus extracting heat from the surrounding medium (air or water). The cooled medium is then pumped through a distribution system—ducts, when air is the medium, and insulated pipes, when water is the medium. In most systems, the compressor and condenser are close to one another; in fact, in smaller systems they are often components of the same unit, whereas the evaporator is commonly located in the air-handling equipment in or on top of the building.

Controls

The systems that control air temperature, humidity, and air volume and speed are low-voltage systems for which the mechanical subcontractor typically takes contractual responsibility. The electrician commonly furnishes and installs the high-voltage power required to operate the system equipment, the air-handling equipment, pumps, and motors. In the typical residential installation, a mechanical thermostat or a thermostat with an imbedded computer chip that activates the heating or cooling equipment, or that allows the user to operate the fan by itself, is installed at some centralized location. In medium-sized to large buildings, control systems now commonly manage numerous temperature zones and a variety of thermostats, dampers, and other equipment through a centralized computer program and signal wiring. It is predominantly the controls contractor who installs the wiring for this system (frequently under subcontract with the mechanical sub). Cable trays (ladderlike, open metal frames installed horizontally in the interstitial space) or electrical conduit carry the wiring for the control system to the various control panels in the equipment. As with the maintenance of heating and cooling equipment, access to control panels to adjust the system is essential, and it affects the orientation of the equipment being controlled.

Fire Safety

Parts of both HVAC and plumbing systems require alarm circuitry. Fire dampers in ducts, and (depending on the designer of the system) flow switches or other fire response devices connected to the fire system underground, standpipe, sprinkler heads, or elsewhere are required by the fire code.

Common System Types

Most heating and cooling systems in buildings are air or water (hydronic) systems, or some combination of the two—the variety of system combinations is considerable.

Air Systems

Single-zone air systems use a single air handler unit to distribute conditioned air (filtered, heated or cooled, humidified or dehumidified) throughout the building. The volume and temperature of the supply air is constant; to adjust the temperature in the building, the duration of supply air is controlled. Designers of residential systems commonly select the single zone system for both heating and cooling.

Multiple-zone systems produce heated and cooled air simultaneously in the air-handling unit and distribute it to the various rooms through ducts according to the requirements of each. As in the single-duct system, the volume of air is constant; the temperature is regulated by dampers that allow for mixing hot and cold air in the supply ducts of each zone. Since the air is concurrently heated and cooled, multiple-zone systems are not particularly economical.

Reheat systems make the most of the single- and multiple-zone systems by producing conditioned air in the air-handling equipment at the lowest required temperature, then adjusting the temperature by heating the air traveling to zones requiring warmer air. Electric resistance strips or duct coils heated by water or steam raise the temperature of the duct air. Reheat systems are effective systems in buildings where considerable temperature variation within the building is expected.

Variable air volume (VAV) systems are constant-temperature systems that adjust the temperature in a room by controlling the mixture of cool and warm air in the VAV box and discharging it to the specific spaces. They may be used with reheating coils, and are designed for spaces in which humidity is not a critical issue. Office buildings are commonly heated and cooled by this method.

Dual-duct systems have similar disadvantages to the multiple-zone system in that the conditioned air is simultaneously heated and cooled in the air-handling equipment, then distributed in separate duct systems to the spaces, where it is mixed. Reheat coils or VAV boxes may be used in the individual zones. The operating cost of dual-duct systems is higher, as is the initial cost, both from the material and labor standpoints. Figure 12.1a shows a roughed-in VAV box in a dual-duct system.

Figure 12.1a This photograph shows a VAV box installed during the HVAC rough-in stage. The air from the air-handling equipment enters on the side with two duct connections; the discharge to the rooms is on the left. (Photo by the author.)

Figure 12.1b This sheet metal worker is connecting the duct runs to the individual spaces on the ground. Once the duct is connected, it is lifted into place and attached to the hangers that are installed in the metal deck. The marks on the slab to the right of the duct just below the center identify the size (100" or 254 mm) and location of supply diffusers (marked with an X). (Photo by the author.)

Return Air

Returning the air to the air-handling equipment for reconditioning usually requires a separate duct system, referred to as the *return air system* (for certain kinds of equipment—window AC units, for instance—a duct system does not exist; the supply and return air dampers in the machine provide the supply and return air). Return air grilles, where stale air is drawn back into the system, are located in the same rooms as the supply registers or at other collection points. Return air ducts may be fitted with filters and dampers, depending on the system employed.

Ducts may be made of galvanized iron, aluminum, and stainless steel in various geometric configurations (square, rectangular, round), as well as coiled steel wire encased in insulated plastic, as is the case with flexible duct. Figures 12.1b–e show, respectively, flexible duct being connected by an HVAC worker, hanger wires being installed in the deck, the hanger wire assembly, and the tool used to install the hangers.

Hydronic Systems

Hydronic heating and cooling systems depend on hot and chilled water distributed to coils within duct systems, across which air is blown. There are a number of systems in use, but each comprises a supply and return line, or lines, to circulate the water through the boiler and chiller. In systems that both heat and cool, the hot and cold water may travel through the same coil with valves at the duct coils controlling whether hot or cold water is pumped through the coil. Other systems use separate coils for heating and cooling.

Air Purity

Filtering systems cleanse the air of particulate matter, and there are a variety of ways to accomplish that task. Common ones include simple fabric filters or ionizing filters placed in the path of the air returning to the air-handling equipment, which remove dust particles and, depending on the filter type, allergens. The filters are cleaned or replaced as a part of the routine maintenance of the system, which brings to mind issues of regular maintenance.

System Maintenance

A variety of building systems require regular maintenance, and one of the primary responsibilities of the design professional team is to arrange equipment and space in such a way that the equipment can be safely and easily accessed by maintenance personnel. This is certainly true of HVAC equipment. When equipment is mounted on rooftops, a passageway to the roof—a ladder and hatch, or stairway, penthouse, and door must be created. Paths, catwalks, platforms, bridges, and the like—replete with railings—may link the access point to the equipment; and of course there must be enough space around the maintenance panels to temporarily store tools or supplies and provide a worker with adequate workspace. Access lids in suspended ceilings and access into attics (where air-handling and furnace equipment are frequently located, particularly in residential structures) perform the same function. Catwalks and workspace are also required in attic-mounted equipment and for equipment mounted on sloped roofs.

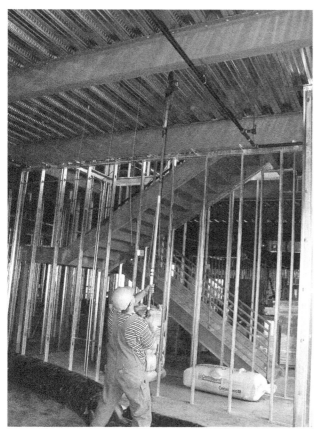

Figure 12.1c The hangers for ducts are installed in a variety of ways. Here, a powder-actuated pin gun on an extension is being used to install hanger wires that are integrated with pins for easy installation. (Photo by the author.)

Figure 12.1d The pins that are shot into the floor deck are integrated with a bracket and hanger wire. (Photo by the author.)

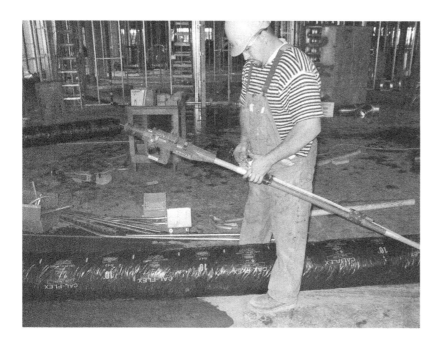

Figure 12.1e Under the hand of the worker in this picture is a section of copper tube (one of two) attached to the pin gun extension by tape, which is used as a holster for the hanger wire/brackets/pin assembly shown in Figure 12.1d. (Photo by the author.)

Controls

The final subsystem to evaluate in HVAC systems is the one that controls air speed, volume, and temperature. Mechanical contractors that furnish and install HVAC equipment also have the responsibility of start-up and testing, to determine whether the system functions as planned. Two building spaces of the same dimension with equally sized windows, one in a west wall and the other in a north wall, will behave differently from one another in the course of a day. Making adjustments in the amount, speed, and temperature of air delivered to a space is part of the system balancing exercise. System controls (thermostats, motor controls, and the like), which are at the center of the balancing effort, are often subcontracted to companies that specialize in this increasingly complicated, technical work. Control systems can be pneumatically, electrically, or electronically operated, or some combination of those methods.

Deciphering the Drawings

Analyzing complicated systems from the builder's standpoint requires some forethought and planning. One useful technique for evaluating an HVAC system is to start with the equipment that heats and cools the space. As noted earlier, heating and cooling is generated in a variety of ways in building projects, from a variety of locations. Some projects are designed with central plants that generate the heating and cooling for a small family of buildings or for very large projects; others use combination systems, where some of the equipment is in a central plant and other pieces of equipment are located on the roof behind mechanical screens. Still others locate the equipment entirely on the roof. One common residential and light commercial application calls for the condenser to be located outside the building, with an air handler, furnace, and evaporator in the attic or in a room in the building.

Once you have grasped the type of system being employed in a project, and the location of the equipment has been identified, it helps to review how the system is powered.

Heating and cooling equipment requires energy to function—commonly natural gas and electricity—so anyone reviewing HVAC drawings should expect to see the piping that transports the energy from where it is supplied—municipal utility departments, quasipublic utilities and power companies, or storage tanks—to where it is used, as well as the wiring for the panels, meters, pumps, switches, and valves used to regulate, transport, control, and account for the energy. Steel pipe is commonly used for fuel, and metal and plastic conduit for the wiring.

When the part of the system responsible for generating heat and cooling has been reviewed, it makes sense to evaluate the distribution system (the duct work or piping)

from the source to the terminus points (the conditioned spaces in the project). The advantage to this approach is that it is logical—after all, you are tracing the path of the heated or cooled air through the system from its point of origin to its end user—and it causes you to focus on high-priority, costly components. In pressurized systems such as HVAC or plumbing systems, the largest, most expensive and difficult-to-install components are closest to the source of the conditioned air or piped fluid. A review of the conditioned air distribution system must also take into account the design zones that have been set aside by the architect specifically for this system. In a multistory project, for instance, horizontal and vertical zones (interstitial space and vertical shafts, respectively) are created to accommodate the ductwork and piping required by the systems. Needless to say, accommodating particularly the ductwork presents significant challenges to architect, design consultant, and builder alike. Ducts as large as 9' (2743 mm) in one dimension may be required even in small midrise structures. The architect has to list the vertical shafts in the summary of design space, and work around them in laying out walls on floors. Ceilings that depart from the ordinary 9' flat ceiling in office buildings (coffered or sloped ceilings, for example), limit the amount of interstitial space that the ceiling can occupy, or the components that otherwise would occupy the space have to be rerouted. Structural engineers must account for the large holes in primarily horizontal members (floors, walls, beams) that are created for the ductwork. Contractor and subcontractor alike must be particularly cognizant of the grave danger to workers that a shaft, open from the ground floor to the roof, creates. The skilled tradespeople working for the mechanical contractor expose themselves to danger, since it is they who suspend very large ducts in the shaft itself. The carpenters who install shaft walls spend their days working adjacent to the shaft, lifting heavy, awkward wallboard into place and fastening metal studs to the underside of floors that are 7' to 10' (2134 mm to 3048 mm) above their heads. Figures 12.2a and b depict this condition.

Components that reside in the interstitial space can be hoisted into their design positions by equipment located directly underneath them on a flat, smooth surface, so safety is less of an issue than in the vertical shafts; however, there are challenges here, too. Numerous other components and assemblies occupy the same space; in fact, this is one zone in which conflicts frequently occur, and construction sequencing—the order in which components are installed in the interstitial space—is an issue. In a very tall building, a difference of 1' (305 mm) in the interstitial space can result in millions of dollars in additional cost in cladding and glazing systems, in the structural frame, in air conditioning equipment, and in long-term operating costs. Consequently, there is considerable incentive to keep the interstitial space to an absolute minimum.

Figure 12.2a This photograph shows a large duct penetrating the floor. Notice the scaffolding that has been erected around the space, to provide workers with a safe platform from which to install the large ducts. (Photo by the author.)

Figure 12.2b This is a photograph looking up through the floor at a duct penetrating the floor. (Photo by the author.)

As noted more than once in this text, conceptualizing in three dimensions from two-dimensional depictions of space is challenging, and when a large number of components share the same zone, there is bound to be some interference. It is worth noting that mechanical contractors seem to be leading the charge to use 3D computer modeling technology to manufacture, install, and account for the mechanical components that are used in a building (a very healthy development in the industry). This technology represents a quantum leap in the depiction of architectural and engineering projects.

Worth mentioning too is the interface between ductwork, design zones, and fire separation assemblies (walls and floors). The shaft walls previously mentioned are fire separation assemblies themselves; they serve to isolate fires that may occur in vertical shafts (and on a smaller scale, vertical and horizontal chases) from the rest of the building. Ducts that penetrate either vertically or horizontally oriented fire-rated assemblies are required to have automatic fire dampers connected to the alarm system; the damper actuator closes the duct when it senses smoke, heat, or both. As with other equipment, dampers must be accessible for maintenance and testing; access doors in the duct adjacent to the damper are common.

A reviewer of mechanical drawings will benefit tremendously by keeping in mind the location of vertical shafts, the structural composition of the floors, the floor-to-floor and finished ceiling heights, and the location of area separation assemblies in the building.

Following close on the heels of supply duct system review is the return air system. Air that has been pumped into a room by the supply system must find its way back to the machinery for reconditioning. Some of the air is exhausted to the exterior of the building (as occurs in bathrooms, janitors' closets, and electrical rooms, for example), and some is returned, mixed with fresh air, and heated or cooled again.

There are normally fewer return air registers than there are supply registers, and they are positioned, as much as possible, at maximum distance from the supply register and from exterior walls. It is at exterior walls where natural convection activity is highest, which explains why supply air registers are commonly located there.

For projects that employ a hydronic system (hot and chilled water for heating and cooling a building), a review of the piping is essential. Some buildings are designed with hot water systems that pump hot water to heat exchangers located in the ductwork. Chilled water is pumped to air-handling equipment on the roof, where it cools air through a heat exchanger. The cooled air is delivered to rooms in the building through ducts. Having done its job of cooling the air, the water is pumped back to the chiller for conditioning. In another iteration, hot and chilled water can be pumped to individually controlled duct units equipped with fans, which heat and cool the space in response to demand.

ENERGY-EFFICIENT DESIGN

Since the early 1970s, public awareness of energy efficiency and conservation has increased dramatically, with the consequence that reasonable alternatives to the standard heating and cooling systems have been developed.

The thoughtful positioning of a building on its site, better quality control in construction, heat sinks (thickened concrete floors, or stone basins under the floor, for example), on-demand water heaters, photovoltaic cells integrated into a building's architecture that produce electricity from the sun's rays, insulated glass, and special glass coatings that selectively block the sun's energy are just a few energy-conserving technologies that have developed and are increasingly being used in buildings. Architects are currently employing passive solar techniques in the cladding systems of skyscrapers, which have historically depended upon costly energy-consuming systems for heating and cooling. Cogeneration facilities, which use the by-products of some process to generate energy, have become more widespread.

Whatever the system, it helps to bear in mind that the piping must form a loop of some kind, such that conditioned water departing the source of heating or cooling can be returned for treatment again. Galvanized iron, stainless steel, and aluminum are the common metals used in ductwork. Flexible ducts are composed of a coiled steel wire frame encased in a plastic sheath, with fiberglass insulation surrounding the duct, which is itself surrounded by a second sheath of plastic.

Plumbing Systems

Elements of Plumbing Systems

Conceptually, plumbing systems are simple. Occupants of a building need potable water for drinking and personal hygiene, and for certain kinds of project—a hospital, for example—many other mechanical services. Water that is used as a conduit for human waste need not be potable, but once used needs to be collected and sent to wastewater treatment plants, where it is treated and released back into the environment. Drinking fountains, restroom facilities, lockers, and kitchens are examples of the terminus points to which potable water in various volumes and at various temperatures must be delivered. The terminus points of the pressurized systems are also the waste generation points of the drainage system, a useful transition point that helps segregate plumbing drawings into supply and drain, waste, and vent components.

Piping

The fluid (or gas) conveyed by a piping system determines the material used and coupling method. Steel pipe, for example, is commonly used for steam, hot water, chilled water, compressed gas or air, and oil transportation. Vitrified clay pipe is used in underground sewer drainage systems, due to its durability and resistance to corrosion. Soldered copper pipe resists mineral build-up and does not leach into the system water and is widely used for domestic water supply. In commercial buildings, cast-iron pipe with no hub connections is very common for drain, waste, and vent piping. Acrylonitrile butadiene styrene pipe (ABS) and polyvinyl chloride (PVC) pipe are the most common materials for drainage in residential buildings.

Pipe thickness correlates to size and pressure—the larger the pipe diameter, the thicker the walls; the higher the pressure, the thicker the walls. Piping comes in standard lengths, depending on size and material, and retains its nominally designated outside diameter regardless of pipe wall thickness (in most cases). Steel pipe lengths, for example, are similar to lumber lengths in that one can order standard lengths, random lengths, or lengths cut to order. The standard steel pipe length is 21' (6400 mm);

random-length pipe varies from 6' to 22' (1800 mm to 6700 mm) in length. The outside diameter of schedule 40 (standard thickness) 6" steel pipe is $6^5/_8$" (168 mm), with a .280" (7 mm) thick wall. A standard section weighs around 398 pounds (181 kilograms)! Installing a fire sprinkler system in the interstitial space of a building takes on a new dimension when pipe weight is taken into consideration.

The end finishes of piping are determined by the connection method: plain ends require a clamp or grip-type connection; beveled ends are required for piping systems that will be welded (and systems that use bell and spigot connections); threaded ends are used in threaded-and-coupled, or flanged connections; and cut or roll-grooved ends are created for bolted grip-type connections.

Fittings

Fittings are used to join sections of pipe, to change direction, reduce or increase pipe size, to convert to another material, or to split or terminate (plug) the system. The fitting material is in most cases matched to the piping material, to simplify matters and avoid the problems associated with dissimilar metals. Cast-iron soil pipe is a common exception to this rule; the sections of pipe are often connected with stainless-steel jacketed rubber connectors known as no-hub connections. Fittings may be:

- *Threaded and coupled.* Smaller lines in fire protection systems, domestic water, low-pressure gas, irrigation, hydronic systems.
- *Flanged.* Steel pipe systems, particularly where the piping connects to pumps or other machinery.
- *Grip-type.* Larger lines in fire protection systems, storm and sanitary drainage, domestic water supply, and hydronic systems.
- *No hub* (jacketed clamp). Soil piping.
- *Fused* (welded, soldered, heat-fused, or glued fittings). Domestic water supply, fire service, gas, and soil pipe systems. Both steel and plastic systems can be welded, the former with arc welding and the latter with special irons.
- *Bell and spigot.* Underground vitrified clay sewer lines, fire system underground, cast-iron soil pipe systems in buildings.
- *Compression.* Refrigeration lines, instrumentation systems and supply lines for oil burners.

Valves[1]

Valves are designed to control the flow, volume, level, and temperature of liquids, to regulate pressure and to facilitate routine maintenance in piping systems. They are available in cast iron, bronze, copper, and plastic,

and may be attached to pipe in most of the ways just listed. The variety of sizes and designs is considerable. A few are described here.

- *Gate valves.* Gate valves are used generally as shut-off valves. A brass gate is lowered by a screw mechanism into the pipe, obstructing the flow of the fluid. They do not impede the flow of the fluid much when fully opened. (See Figure 12.3.)
- *Globe valves.* Globe valves consist of two right-angle turns in body of the valve. When a screw mechanism lowers the seat onto one of the angles, it reduces flow or closes it completely. Even when fully opened, globe valves restrict flow in the valve body, which results in a pressure drop on the downstream side of the valve. (See Figure 12.3.)
- *Angle valve.* Similar to a globe valve, an angle valve controls the flow and direction of fluid in a system by forcing it through one right-angle turn. A screw-operated plunger closes against a seat at the midpoint of the turn. This valve is less restrictive than the globe valve when fully opened, and it offers the advantage of changing the direction of the piping system without an elbow.
- *Ball valves.* In a ball valve, spherical seats contain a drilled globe. When the drilled shaft in the globe is aligned with the pipe using a lever connected to the ball, the fluid is free to move. As the lever is rotated, the ball restricts the flow of the fluid in the pipe.
- *Swing check valve.* Swing check valves are simple plates hinged on one side that allow fluid to travel in one direction. When the force is greater on the downstream side of the valve, the valve swings open. When the pressure equalizes, the valve swings shut against a seat, where it is held closed by the hydrostatic head of fluid on the upstream side of the valve. (See Figure 12.3.)
- *Lift check valves.* Similar in concept to swing check valves, lift check valves—horizontal and vertical—allow fluid (steam, air, gas, water) to travel in one direction only. Rather than being hinged like a swing check valve, lift check valves are axially supported.
- *Butterfly valves.* Butterfly valves make use of a disk attached to shafts installed in the center of the pipe, within the confines of the valve body. When rotated closed, the disk fully obstructs the pipe. As the valve is opened, increasing volumes of fluid are allowed past the disk, to the point it is fully open, when the disk is parallel to the flow of the fluid. These valves control volume well, they are lighter than many other types, and they create little turbulence when fully open.

Figure 12.3 is a comparison of double- and single-line drawings illustrating various fittings and valves. The dual-line drawing is fairly realistic, but it is somewhat less informative than the schematic drawing below it, which makes quite clear what type of valve or fitting is expected.

---

[1]*Means Mechanical Estimating*, 2nd ed., R. S. Means Company, Inc., 1992.

Figure 12.3 This drawing compares the schematic single-line drawing to the more realistic double-line elevation drawing. (From *Technical Drawing*, 12/E, by Giesecke/Mitchell/Spencer/Hill, Pearson Education, Inc., 2003. Reprinted by permission of Pearson Education, Inc., Upper Saddle River, NJ).

Water Sources

City or county municipalities almost always supply building occupants in urban areas with the water they need to function. These producers may get their water supply from rivers, lakes, reservoirs, and wells, or from all four. Untreated water is diverted to treatment plants, where it is mixed with chemicals, filtered as much as possible of particulate matter and pumped through a pressurized regional piping system. In suburban and rural areas, private wells are common. Water extracted from private wells, depending on the quality of the sub-surface soil, will be treated to various extents by the property owner prior to its being used. Water supply lines, being pressurized systems, are commonly steel pipe, or increasingly, thick-walled plastic.

Fire Service Lines

One of the prime responsibilities of a municipality is to provide police and fire protection to its citizens. The input of both police and fire departments is required when a building is being planned, so that each has the opportunity to influence the final design. For both, access to the site is a critical issue. For police departments, exterior lighting layout, illumination, and fixture height and type, for example, are critical issues. Fire departments must have accurate records of the location of fire hydrants, hose cabinets, system connections and general building layout, so that little time is lost to locating occupants and critical life-saving systems. Within buildings themselves, the fire departments frequently play the critical role of final arbiter as to the size and disposition of fire corridors and fire suppression (sprinkler) systems, both of which they evaluate very carefully.

Commercial building projects are normally required to have stand-alone fire service systems, commonly called *standpipe systems*, to which the fire department can connect when fighting a fire. Consisting of steel piping, fire hose valves, hose cabinets, and various fire department connections, standpipe systems are valuable time-saving aids. When combined with automatic fire sprinkler systems, standpipe systems offer maximum fire protection.

Class I standpipe systems are used by fire departments and personnel with specialized training. Class II systems use a small hose and connections, and may be used by building occupants until such time as the fire department arrives. Class III standpipe systems combine the two.

Standpipe systems may also classified by how water is delivered to the system. For example, a Type 1 standpipe system uses an open supply valve, so that all parts of the system are pressurized at all times. Type 2 systems automatically admit water to the system when a hose valve is opened. Type 3 systems use remote-controlled devices at hose stations to manually operate supply valves. Type 4 standpipe systems are dry standpipe systems.

As to whether a fire suppression system will be installed within the building, there are several possibilities. First, an owner may choose to install a sprinkler system. Insurance companies, not surprisingly, view properly selected and designed fire suppression systems favorably, which is manifested in the premiums they charge for fire coverage on a property. For a building owner then, choosing to install a system might be a business decision.

Building departments also view suppression systems favorably. Limits on the floor size and building height set forth in building codes are relaxed for buildings that have a fire sprinkler system. For example, in one code, the floor area in structures within the nonhazardous use group can be tripled. Height limitations can be exceeded as well, and if area separation assemblies are used, an architect can increase the size of a given project substantially by using a fire sprinkler system. So the architect might decide that it is in the building owner's best interests, given the design program, to install a system.

There are several common fire sprinkler system types:[2]

- *Wet pipe systems*, the most common type, are fire sprinkler systems that are connected to a municipal water supply and, like those systems, are under constant pressure. When fire or heat opens one or more sprinkler heads, water is released onto the fire.
- *Dry sprinkler systems* are used in environments that are subject to freezing. The pipes contain pressurized air; a valve maintains the static condition of the water by balancing air pressure on the sprinkler system side against the pressure of the water on the supply side. When a fire starts, the sprinklers relieve the air pressure and water flows into the system. These systems require dry pendant (hanging) heads, which are more expensive than those installed in wet systems, and the actuating valve must be protected from freezing. The dry sprinkler system is slower to respond than the wet pipe system, but is the best system under certain circumstances.
- A *fire cycling system* is used where water damage due to a fire suppression system must be kept to a minimum. These dry systems sense fire, release water to the sprinklers, and then turn the water off. If the fire has not been extinguished, the system will sense the heat and release the sprinklers again.

- *Preaction systems* are dry, unpressurized systems that use conventional sprinkler heads. These systems take advantage of more sensitive heat-detecting devices that will open the water supply valve when they sense heat, fill the sprinkler system with water, and sound an alarm. If even greater heat is generated, the sprinkler heads will be actuated.
- *Deluge systems* are dry systems connected to a water supply that employ heat-sensing valve controllers. Heat causes the valves to admit water into the system, which is released through nozzles rather than sprinkler heads. As their name suggests, deluge systems count on significant volumes of water to control the spread of fire.

Fire suppression systems, when used, are frequently designed by engineers who specialize in the work, using guidelines established by the National Fire Protection Association (NFPA). These systems, unlike many others,

BUILDING CODES AND USE GROUPS

Building codes are widely accepted minimum standards for health and safety in building construction. Municipalities periodically adopt one or another of several codes used in the United States, which raises the adopted code to the level of local law. While codes vary in some details, they begin by defining use groups for buildings. Among the use groups are buildings in which people congregate, conduct business, manufacture chemicals, machines and other commodities, attend school, provide health care, and reside (among other uses). Once the use group has been determined, the building construction type may be selected. Construction type runs the gamut from highly fire-resistive structures built of masonry, reinforced concrete, or fire-protected structural steel to vulnerable buildings such as unprotected wood-framed structures. The height and size limitations in building codes take into consideration building use and construction type, proximity of other structures to the building being proposed, open space surrounding it, and fire-separation assembly use. It should come as no surprise that the more highly resistive construction types are more costly. The architect, therefore, is challenged to arrive at the optimum conglomeration of construction type, floor size, building height, and area separation assembly use, while balancing these design factors against the project budget.

---

[2] *Means Mechanical Estimating*, 2nd ed., R. S. Means Company, Inc., 1992.

are highly sensitive to modification, which makes them particularly challenging systems to design and construct. Fire department personnel (representatives of the state fire marshal) who are familiar with sprinkler system design review and approve the systems as designed, if fire requirements are met. When the inevitable conflict arises, the system designer must demonstrate to the approving authority that the change will not compromise the effectiveness of the system. Fire department personnel inspect and test both underground and above-ground systems, and until their approval has been granted, the building may not be occupied.

Underground fire systems consist of steel pipe or special plastic. The part of the system that is above the ground (pipe, valves, hydrants) is normally steel pipe with cast-iron valve bodies and fittings. Adequate pressure to extinguish fires comes from the municipal water system, reservoirs high in the building or in towers on site, from booster pumps, or through connections to which firefighters connect their pumper trucks. The sprinkler systems in the building are normally steel, fitted with special sprinkler heads as called for on the drawings. While the plumbing contractor may furnish and install the standpipe system, specialty contractors frequently install the different fire sprinkler systems supported by the standpipe.

Control of Water Sources

A user operating a valve of some kind produces potable water at some terminus point. Since the more or less constant pressure in the water supply system is greater than the medium resisting it (ambient air in the case of a faucet valve), water flows out of the pipe, into a vessel or container of some sort (a lavatory, for example). Once the water has been used, or as it is being used, it enters the drainage system, where it is transported through the building's soil piping to the site underground, where it connects to the municipal sewer system. Eventually, it arrives at a wastewater treatment plant or other destination.

Controls for fire systems operate in similar fashion, that is, when a valve is opened, water flows to that point. In fire systems, of course, this may indicate that a sprinkler head has ruptured in a nascent fire, or perhaps that the system is being tampered with. In either case, when water starts flowing in a fire system, an alarm is usually activated, and the fire department (and any people occupying the space) are informed of the impending danger.

Drain, Waste, and Vent Piping

Drain lines transport liquid other than sewage to the municipal sewer system or, in some residences, a septic tank. The pipes that drain lavatories, showers, bathtubs, laundry rooms, and rooftops are examples of drain lines. Waste piping refers to the pipes that drain sewage-producing fixtures such as urinals, water closets, and bidets,

as well as those fixtures that produce significant amounts of organic matter, such as commercial kitchen sinks. *Soil piping* is a generic term for drain and waste piping.

Vent piping is the part of the plumbing system that allows the drain and waste lines to function by providing a source of air to replace the liquid escaping from the system. Without a source of air, turbulence in the pipes occurs, and draining water will siphon water out of traps. Traps are U-shaped bends in drain and waste pipes, situated close to the fixture, that trap water in the drain pipe, thus sealing the pipe against the intrusion of sewer gas into the building. Stacks are vertical pipes, usually contained in walls or plumbing chases, that rise from various plumbing fixtures up through the structure to 18" (457 mm) or so above the roof of the building. Condensate drains—a common interface between the HVAC and plumbing subs—transport condensation generated by air conditioning equipment from collection basins under the equipment to the storm or sanitary sewer system, or in some cases, from rooftop equipment into gutters and rainwater leaders.

Drainage systems depend on gravity to transport waste out of the system to a distant municipal wastewater treatment facility or to collection points, where it is pumped to a higher elevation and allowed to flow to the treatment facility or, in certain cases—as noted in the section on maintenance of plumbing systems—where the waste is destined to attract attention. Consequently, all of the horizontal pipes in the system are sloped. Depending on the application and the pipe size, 1/8" to 1/4" per foot (10 to 23 mm/meter) is the range of slope in drain lines in buildings. Slope and pipe size are correlated; the larger the pipe, the flatter the slope can be.

Fixtures

The term *fixture* refers to the devices that ultimately are used by the occupants of the building—lavatories, water closets, and drinking fountains are examples. Fixtures are typically installed late in the construction process, in the finish stages of a project. Fixtures are common interfaces between the supply and drainage systems of buildings.

Though they are not considered fixtures, drains and interceptors (used to trap grease, sand and oil) can nevertheless be costly to install, from both the material and labor standpoints. Drains are used in numerous locations in a building; in the roof, in restrooms, in lockers, in kitchens, in floors, and of course in the numerous fixtures that are part of the system. The typical roof drain, for example, is a large cast-iron assembly that must be installed in the rough-in stage, prior to the installation of the roof covering. Although the plumber sets it, parts of the roof drain are left to the roofer to install, so that the roof covering and drain are integrated. Figures 12.4a–d shows the architectural detail and photographs of a common roof drain.

Figure 12.4b The part of the drain under the roof structure can be awkward to install. It is heavy cast iron, and is installed overhead in the 12' to 15' range (4 to 5 meters). Any work performed overhead is more costly. (Photo by the author.)

Figure 12.4a Roof drains are heavy and require installation above and below the roofline. (Drawing courtesy of Comstock Johnson Architects.)

Figure 12.4c This photograph shows the portion of the drain that is installed above the roofline. (Photo by the author.)

Figure 12.4d The roof drain (below) and overflow (above) are shown in this photograph of the installed roof drains. (Photo by the author, with the permission of the Sacramento County Office of Education.)

Pipe Hangers, Supports, Carriers, Seismic Restraint, Thermal Expansion, Sound Attenuation, Water Hammer, and Galvanic Action

Piping systems undergo changes as they are being used—they are hardly static systems—and architects and engineers must provide support for piping systems and address several common phenomena involving piping systems.

Hangers support pipes while simultaneously allowing them to move (depending on the hanger type) in seismic events and when thermal expansion takes place, which is a regular occurrence in hot water lines. The thrusting that occurs when water lines change direction or branch must be contained, and in seismic events, large, rigid supply lines that penetrate vertical or horizontal structural elements must be segregated from each other to avoid damage to the pipe. This is frequently accomplished by creating a hole in the member several inches larger in diameter than the pipe penetrating it, which creates anchoring challenges. A variety of hooks clamps, brackets guides, "rolls," sleeves, and trapezes, either surface-mounted or embedded in concrete members, exists to support piping of all sizes and weights. Several types of hanger are illustrated in Figures 12.5a–d.

The term *carrier* refers to the components that provide structural support for cantilevered fixtures such as water closets, lavatories, and drinking fountains. Fixtures mounted on carriers are sanitary and easy to maintain—the fixture is suspended above the floor, supported by the floor or wall. To save time and effort in the rough-in process, drainage components may be integrated into the carrier; in other words, in addition to providing support for the fixture, carriers may also function as the drain and vent bodies. Depending on the carrier type, support comes from floor or wall-mounted pipe or brackets. When the wall is used to support the fixture, wall backing, stiffer studs, closer stud spacing, or all three may be required. Such requirements do not often show up in the drawings themselves, except perhaps in details; however, they are noted on the drawings or in specifications. Figures 12.6a and b are photographs of one type of floor-mounted carrier system.

Expansion joints (typical of drainage piping) or expansion loops in pipe runs (typical of supply piping) are designed to accommodate thermal expansion, which is significant in copper and plastic piping, both of which have high coefficients of expansion. Expansion loops are U-shaped pipe diversions perpendicular to straight pipe runs; as the pipe expands, the ends of the straight runs—which are clamped but not so tightly that they cannot

Figure 12.5a Hangers, clamps, and other hardware of remarkable variety are used to support installed pipes. Here is a sampling. (*After Means Mechanical estimating*, 2nd ed., R. S. Means Company, Inc., 1992.)

Figure 12.5b This clamp supports the fire sprinkler pipe from one side of the bottom flange of the wide flange beam. (Photo by the author.)

Figure 12.5d Domestic water is connected to the frame in much the same way as fire sprinkler systems, except here plastic material separates the copper pipe from the iron hanger, to avoid the problems caused by electrolysis. (Photo by the author.)

Figure 12.5c This anchor attaches to the side of a glued-laminated beam using a lag screw. This pipe is connected by a grip-type fitting. (Photo by the author.)

Figure 12.6a This and the following photograph show a floor-mounted water closet carrier in a plumbing wall. This assembly relies heavily on the orientation of back-to-back plumbing fixtures for structural support. (Photo by the author.)

Figure 12.6b Carriers, drain, waste and vent piping, and hot and cold water supply are all shown in this photograph of restroom walls in an office building. (Photo by the author.)

O-RINGS

SLIP JOINT

FIRM CONNECTION

Figure 12.7 An expansion joint allows pipes to expand and contract due to thermal expansion without destroying the supply line. An expansion loop builds flexibility into a supply line by giving the system room to grow. (After National Standard Plumbing Code, 1993.)

move—are allowed to extend into the space created by the loop. Figure 12.7 is a sketch of an expansion joint and an expansion loop.

Fundamental to expansion relief methods in piping, as is the case with thermal expansion in other systems, is restraint-avoidance solutions.

Fluid moving through pipes—pressurized and non-pressurized systems—generates sound, which, if the pipe is in direct contact with framing members, may be transmitted from space to space in a building (particularly true of wood-framed construction). The common method of mitigating the transmission of sound is to use various isolation devices, including plastic "doughnuts" and offset pipe clamps, whose purpose is to absorb the vibration created in the pipe. Doughnuts surround the pipe and fit within an oversized hole in the framing member, thus isolating the pipe from the framing. Offset pipe clamps are clamps set on legs that attach to the framing or substrate.

A phenomenon known as "water hammer" occurs when an open valve closes quickly, which is common with many sink faucets. When swiftly moving water is stopped, a shock wave in the system occurs, which can cause a momentary rise in pressure many times that which the system was designed for. The shock itself can cause damage, as well as an unpleasant clunking sound.

Air chambers (vertical sections of pipe mounted above the fixture that trap air) and water hammer arrestors (devices containing a bellows and gas-filled chamber) are used to counteract water hammer.

Corrosion in metals[3] (which are generally negatively charged) is a naturally occurring phenomenon involving the loss of an electron in a metal atom to an available ground, thus positively charging the atom momentarily. Water molecules in contact with the metal attract the charged metal atom, wrenching it from its crystalline matrix into an electrolyte solution, where it combines with a dissolved negative ion, forming an ionizable salt that precipitates as a metallic salt on the surface of the metal. If the concentration of metallic salts is high enough to precipitate, crystals will grow on the base metal. Those crystals are the visible evidence of corrosion.

Localized corrosion can occur in the absence of a ground when water droplets form small electric cells on the surface of the metal. The flow of electrons in metals, and a similar counterflow in the electrolyte solution, results in the occasional "lifting" of metal atoms within the small electric cell, resulting in localized corrosion, or pitting.

Combining water and two metals that are far apart on the electromotive chart of metals results in galvanic action, which is an accelerated form of corrosion. A copper pipe in contact with an iron hanger, for example, will result in the iron corroding more rapidly than the copper, due to iron's propensity to give off electrons at a greater rate than copper. A copper hanger or clamp, or an iron hanger with a device on it that isolates the two metals (plastic or rubber coating for instance), is called for. This phenomenon exists for valves, couplers, and fittings as well; dissimilar metals are isolated from one another, or simply not used together.

Insulation

Mechanical equipment and piping is insulated either after it is installed, often by subcontractors who specialize in the work, or is manufactured as an integrated system, that is, with the insulation encasing the pipe: just the ends, where connections will be made, are exposed pipe. The manufacturer frequently insulates mechanical equipment, thus relieving the builder of insulating it in the field. Fiberglass, cellular glass, rock wool, polyurethane foam, closed-cell polyethylene, flexible elastomeric, rigid calcium silicate, phenolic foam, and rigid urethane are all used. Insulation comes in blankets of different thicknesses and lengths and as rigid board. Pipe insulation comes in lengths varying from 3' to 6' (1000 mm to 2000 mm), depending on

[3]*Building Pathology,* Samuel Harris, New York: John Wiley & Sons, Inc., 2001.

Figure 12.8 This photograph shows a common method of insulating pipes. (Photo by the author.)

the material. Insulation may be integrated with the pipe itself; refrigeration lines for smaller air conditioning systems, for instance, commonly come encased in foam. After their ends are gas-welded, the exposed sections are separately wrapped. Pipes intended for use with chilled water are wrapped with insulation and encased in a hard plastic or sheet metal sheath, which acts as a shield or as a vapor retarder. Figure 12.8 is a photograph of mechanical piping that has been insulated and wrapped.

Maintenance, Service, and Preservation of Plumbing Systems

Part and parcel of plumbing drainage systems are the components that allow for maintenance and service; for example, clean-outs (which allow maintenance personnel to access the drainage system with machinery that unclogs soil pipe) are installed periodically at convenient points of access (through access doors or plates in walls below fixtures, immediately outside the building, over hallways in interstitial space). Shut-off valves are used to shut down parts of the plumbing system for repair or fixture replacement. Unions allow threaded and coupled lines, as well as soldered or welded lines, to be separated temporarily to replace plumbing equipment such as water heaters, without having to disassemble and reassemble the pressure lines. Although they are not thought of as maintenance fixtures in the normal sense, trap primers maintain water seals in infrequently used drains such as floor drains and sinks. They periodically deposit water into floor drains and sinks to maintain the trap seal, which is what blocks methane gas from emanating into the building from waste lines. Were the traps not primed, the trap seal would evaporate and no longer provide the protection they are designed for.

Similarly, water hammer arrestors—devices that absorb the concussion occurring when rapidly flowing water is halted by a rapidly closed valve—are installed above fixture valves in the wall. Interestingly, the overflow condensate line— a separate line from the primary condensate drain—terminates in a place where it is bound to attract attention, for example, at a point in the ceiling directly above the lavatories in a restroom. Roof overflow drains are similarly situated, although because of the quantity of water that might be deposited if the roof floods, they terminate outside the building, for example, on the underside of an entrance canopy. When those lines produce water, people take notice and alert facility personnel of the abnormal condition, which is exactly the point.

Common Graphic Conventions in Mechanical Systems

Order of Information

Mechanical drawings, like many other drawings, begin with generalized information in small scales and proceed to specialized information in larger scales. The first few pages of mechanical drawings—both HVAC and plumbing—frequently consist of equipment schedules and the legend of graphic symbols and abbreviations. After the introductory pages, the two disciplines part company, insofar as addressing the drawing user is concerned.

HVAC Drawings

Following the introductory sheets, mechanical engineers superimpose the supply and return air duct system, registers, exhaust vents, and duct coils (if used) over the architectural floor plan, floor by floor, all the way to the roof. If a hydronic system is used, the plan view of the plant that generates the hot and cold water may be shown next. What often follows are drawings of the hot and chilled water distribution system, floor by floor, from the ground to subsequent floors. Plans of rooftop equipment conclude the hydronic drawing set. Schematic piping diagrams may then follow, which in turn are followed by myriad details of the mechanical equipment, including mounting instructions, piping schematics of duct coil piping (hot and cold), duct/register connections, and rooftop curb design, among others. Mechanical control drawings—mostly schematic depictions of equipment, mechanical system, and fire and alarm controls, complete the HVAC drawings. Figures 12.9a–u provide a tour of the HVAC drawings for a portion of a research office building.

VAV BOX SCHEDULE – 2ND FLOOR

SYMBOL	MANUFACTURER	MODEL	SERVED BY	INLET SIZE	MAX. CFM	LOW MIN. CFM	HIGH MIN. CFM	COIL ROWS	GPM	MAX. WPD	MBH	VALVE (TYPE)	MAX. APD (IN. WG.)	REMARKS
VAV 2-1	TITUS	DESV	AHU-1	10	840	380	610	2	2	.65	30.1	2-WAY	.42	ROUND OUTLET WHENEVER FEEDING SINGLE OUTLET, OTHERWISE A PLENUM BY CONTRACTOR, SEE DRAWINGS.
VAV 2-2	TITUS	DESV	AHU-1	9	630	285	460	2	2	.33	22.7	2-WAY	.35	
VAV 2-3	TITUS	DESV	AHU-1	6	240	110	175	2	1	.24	8.6	3-WAY	.31	
VAV 2-4	TITUS	DESV	AHU-1	9	540	245	395	2	2	.84	19.5	3-WAY	.26	
VAV 2-5	TITUS	DESV	AHU-1	10	880	400	640	1	2	.69	24.6	2-WAY	.32	
VAV 2-6	TITUS	DESV	AHU-1	7	420	190	305	1	1	.50	11.7	2-WAY	.29	
VAV 2-7	TITUS	DESV	AHU-1	5	250	115	185	1	1	.50	7.1	2-WAY	.34	
VAV 2-8	TITUS	DESV	AHU-1	9	600	270	438	1	2	.69	16.8	2-WAY	.25	
VAV 2-9	TITUS	DESV	AHU-1	8	540	245	400	1	1	.69	15.4	2-WAY	.32	
VAV 2-10	TITUS	DESV	AHU-1	8	500	175	300	2	1	.33	14.8	2-WAY	.38	

PROVIDE:
A) 120 VOLT/40 VA TRANSFORMER EACH VAV BOX.
B) PRESSURE INDEPENDENT.
C) TCC, DDC CONTROLS FACTORY MOUNTED.
D) CONTROLLER ENCLOSURE

Figure 12.9a Mechanical drawings often start with equipment schedules, graphic symbols, and abbreviations. This schedule–describing the VAV (variable air volume) equipment requirements–may be one of 15 or more types of mechanical equipment listed in the HVAC drawings (M sheets).

HVAC LEGEND

ABC		ABOVE CEILING
ACD	ACD	AUTOMATIC CONTROL DAMPER
AFF		ABOVE FINISHED FLOOR
BDD	BDD	BACK DRAFT DAMPER
BF		BOTTOM FLAT
BOD		BOTTOM OF DUCT
BOP		BOTTOM OF PIPE
CFM		CUBIC FEET PER MINUTE AIR
CLG		CEILING
(D)		DROP, DOWN
DSD	DSD	DUCT SMOKE DETECTOR
(E)		EXISTING
EA		EXHAUST AIR
EC	E	ELECTRICAL CONTRACTOR
FD	FD	FIRE DAMPER
FSD	FSD	FIRE/SMOKE DAMPER
GC		GENERAL CONTRACTOR
LVR		LOUVER
MCC		MOTOR CONTROL CENTER
MCD	MCD	MOTORIZED CONTROL DAMPER
MVD	MVD	MANUAL VOLUME DAMPER
(N)		NEW
NIMC		NOT IN MECHANICAL CONTRACT
OBD		OPPOSED BLADE DAMPER
OPNG		OPENING
OSA		OUTSIDE SUPPLY AIR
PC		PLUMBING CONTRACTOR
POC		POINT OF CONNECTION
(R)		RISE, RISER
RA		RETURN AIR
(RL)		RELOCATE
SA		SUPPLY AIR
SW	$	SWITCH
TF		TOP FLAT
TOD		TOP OF DUCT
S		SUPPLY AIR GRILL
R		RETURN AIR GRILL
E		EXHAUST AIR DIFFUSER
TAG		TRANSFER AIR GRILL
L		GALVANIZED STEEL RECTANGULAR DUCT LINED
	10"x10"	DUCT SIZE (INCHES)
		ROUND WIRE-FLEX DUCT W/ VOLUME DAMPER
	8"ø	DUCT SIZE (INCHES DIAMETER)
	øS	ROUND SPIRAL DUCT
	T	THERMOSTAT
	S	SENSOR
		DUCT TRANSITION
		VAV BOX
		3' LINED PLENUM

Figure 12.9b Abbreviation and symbol legends differ from drawing set to drawing set. Here is a brief one.

MECHANICAL LEGEND

Figure 12.9c This drawing, a partial legend from another drawing set, contains more information.

Figure 12.9d It takes a moment to get oriented in this drawing but you will soon be able to identify how the conditioned air is delivered to the individual spaces in the building (adding color with highlighter or colored pencils is very helpful in deciphering the layout). Two 68" × 20" (1727 mm × 508 mm) supply air ducts coming from the air handler on the roof are shown in the upper quarter of the drawing, directly over the center. Supply air ducts are designated in this drawing by a rectangle with diagonals from each corner (one shaded side means that the duct penetrates the horizontal plane of the drawing—note how supply air diffusers at the ends of duct runs are clear). Return air ducts have a single diagonal line from one corner to the opposite corner. A 64" × 40" (1625 mm × 1016 mm) return air duct is visible close to the center of the drawing. It is possible to track one duct run in this drawing from the source to the terminus. The source is the 68" supply duct from the roof, which ties into 30" × 16" (762 mm × 406 mm), 24" × 16" (610 mm × 406 mm), 12" (305 mm), and 10" round ducts (look in the center of the drawing). Refer to the symbols in Figure 12.9c for more information on the symbols themselves.

Figure 12.9e After the supply and return air systems have been described for each floor, mechanical drawings may then display the hydronic system, starting with the source of the heated and chilled water. This drawing shows most of the mechanical plant for a four-story office building.

Figure 12.9f On the left edge of this drawing is the hot and chilled water supply from the mechanical plant. Once in the building, the hot water lines supply VAV boxes, located in duct runs in office spaces. The chilled water is pumped to air handlers on the roof. Subsequent plans show similar information for each of the floors of the structure.

Figure 12.9g Air-handling equipment commonly inhabits the roof structure of building projects. Note the catwalks that are required for equipment service and maintenance on the two air-handler units.

Figure 12.9h The hot and cold water and condenser lines schematics often follow the mechanical equipment plan. This drawing, of a portion of the chilled water piping, is typical of schematics. Note that the drawing is not to scale; the purpose of these drawings is primarily to identify the relationship of components.

Figure 12.9i Mechanical drawing details are a mixture of schematic drawings, sections, and elevations. This single-line drawing shows how the piping and the valves to the reheating coils in the duct runs are supposed to be connected. Unions allow parts of the system to be worked on without shutting off the whole system.

Figure 12.9j The details in mechanical drawings depict critical system components and how they are to be secured to the structure, among other things.

1¼"X#10 TEK SCREWS
EVERY 4' O.C.

1½x¼" WOOD SCREW
W/ WASHER 4' O.C.

NOTE: CAT WALK BY STRUCTURAL
AND SEPERATE FROM CURB.
SEE ROOF PLAN

18 GA SHEET
METAL FLASHING

SINGLE PLY
ROOFING

RIGID INSULATION

REFER TO ARCH
DRAWINGS

AC
1&2

AHU C-CHANNEL

2"X¼" GALV PLATE
STEEL 4' O.C. (TIE DOWN)

CUSTOM FABRICATED EXTENDED
CURB W/ FRAMING/INSULATION
(CUSTOM CURB CRC-1)

2X8 HEADER

2X4 STUDS FRAMING/SUPPORT
PER CUSTOM CURB. MANUFACTURER
DESIGN

INSULATION

2X8

RIGID ROOF
INSULATION

METAL CURB

METAL DECK

3/8"ØX4" BOLT, NUT
& WASHER EVERY 4' O.C.

2½X2½X¼"X2' LONG ANGLE

42"

5"

AHU−1&2 SPECIAL HIGH CURB, BY EQUIPMENT MANUFACTURER

(A) NO SCALE

NOTE:
CUSTOM EQUIPMENT CURB BY MANUFACTURER.
MANUFACTURER SHALL PROVIDE CONSTRUCTION
DETAIL AND STRUCTURAL CALCULATION FOR
EQUIPMENT CURB SUPPORT TO STRUCTURE.

Figure 12.9k Sections are used in mechanical details also. Here, the manner of connecting the air handler to the curb on the roof, as well as how it should be protected against the weather, is made clear.

5/16"Ø HKB II
21/2" MIN. EMBED
(TYP 4)

2"X20 GA. STRAP (TYP 4)

#10 SELF TAPPING SM
SCREWS, 2 EACH STRAP.
(1) ON BOTTOM (TYP.)

CONTROL BOX
OPPOSITE SIDE

MOUNT VAV BOX TO ALLOW FOR ADEQUATE
ACCESS TO CONTROLS AND ON BOTH SIDES
SEE F/M4.2 FOR INSTALLATION WITH METAL DECK NO CONCRETE.

VAV BOX MOUNTING − CONCRETE DECK

(G) NO SCALE

Figure 12.9l Details as to the connection between VAV boxes and the structure are set forth in this drawing.

VAV BOX
INLET DUCT CONNECTIONS
H
NO SCALE

Figure 12.9m This detail describes the requirements for connecting the ducts to one another, as well as the location of critical components within the ducts.

Figure 12.9n This detail shows the connection between a typical duct run and the diffuser in the ceiling of the room.

DIFFUSER @ T-BAR
I
NO SCALE

NOTES:

1.
PIPE SIZE	NOMINAL JACKET THICKNESS
UP TO 6"	0.080"
8" TO 12"	0.100"
14" TO 22"	0.120"
24" TO 36"	0.150"

2.
SERVICE PIPE SIZE	L
2-1/2" & BELOW	20' MAXIMUM
3" & ABOVE	40' MAXIMUM

3. FOR 2" AND SMALLER PIPE: SQUARE ENDS
   FOR 2-1/2" AND LARGER PIPE: BEVELED ENDS

Figure 12.9o As noted in the text, some piping comes insulated by the manufacturer. The instructions in this detail pertain to hot and chilled water pipes running underground from the mechanical plant to the building. The proper thickness pipe jacket for various sizes of pipe, the maximum allowable pipe lengths, and the end treatment for pipes of various diameters are listed. The jacket referred to in Note 1 is a seamless high-density urethane polymer. On the same sheet in this project, details of butt joints, elbows, and the manner of insulating the pipe when a round steel anchor supports it are drawn. Accompanying the details is a long list of written specifications; among them the admonition to backfill the jacketed pipe by hand!

(A) **STEEL PIPE STRAIGHT LENGTH**
NO SCALE

(D) **BOILER FLUE THRU ROOF DETAIL**
NO SCALE

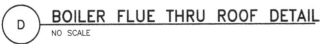

Figure 12.9p Furnace flues require clearance to other elements in the supporting structure and good flashing.

BALANCE VALVE
"B & G" CIRCUIT SETTER

BUTTERFLY VALVE
ON SUCTION AND DISCHARGE

LINE SIZE CHECK
VALVE

F.C.

REDUCER

PUMP

F.C.

SUCTION DIFFUSER

PUMP RAIL

4" HIGH LEVEL
CONCRETE
HOUSEKEEPING PAD

GATE VALVE

(6) 3/8"∅x4" LONG
WEDGE ANCHORS
W/2-1/4" MIN. EMBED.
EQUALLY SPACED.
MAINTAIN MIN. 6"
FROM EDGE OF SLAB

6" MIN. DISTANCE
FROM ANCHOR TO
EDGE OF SLAB
(TYP. ALL ANCHORS)

FLOOR
SLAB

SECURE MOUNTING CHANNEL TO PAD
W/(3) 1/2"∅ "HILTI" KWIK BOLT II (EACH SIDE)
2-1/2" MIN. EMBEDMENT. MOUNTING SHALL BE
LEVEL PER MFR'S RECOMMENDATIONS.
SHIM AS NEEDED.

B ## PUMP MOUNTING DETAIL
NO SCALE

Figure 12.9q Pump mounting details are the subject of mechanical details, too. Here, the engineer is calling for a "housekeeping" pad under the pump. The pad is connected to the floor slab by expansion bolts. The drawing suggests, intentionally or not, that the anchors be installed in the floor slab, followed by the casting of the housekeeping pad. The pump skids are connected to the housekeeping pad with expansion bolt anchors. Note how the pump is isolated from the piping by the flexible coupling (denoted by "FC").

9"∅
OUTLET

10' ABOVE
ROOF

H
M5.4

(3)PVC COATED
12 MIN SPLAY WIRES
PAINTED ORANGE.
CONNECT TO SUPPORTS.

12"∅ 316 STAINLESS STEEL
SPIRAL DUCT, PRIME AND
PAINT WHITE

FUME HOOD
EXHAUST FAN

FC

MASON MODEL
SSLFH MIN 1"
DELECTION. 1
PER MOUNTING
POINT. ISOLATOR
SHALL BE OSHPD
PRE-APPROVED.
(R-0198)

4 # 10 SMS
DUCT TO
BASE

(2)3/8"∅ DRAIN HOLES @
BOTTOM OF DUCT.

1 1/4" PW FIRE TREATED

20 GA. GALV. S.M.
PLATFORM CAP

ROOFING UP
& OVER RAIL

CANT STRIP

SINGLE PLY
ROOF

FIN. ROOF
ROOF INSUL.

SEE ARCH.
DWG'S.

(E) 3 1/2" LIGHTWEIGHT CONC.
OVER 1 1/2"x20ga METAL DECK

3/8"∅ "HILTI" HKBII 2 1/2" EMBED.
(MIN. 4 EACH)

G ## EF-3 MOUNTING DETAIL
NO SCALE

Figure 12.9r The exhaust fan shown here is designed to extract air from certain spaces in the building—janitor's closets, bathrooms, and the like. Note the guy wires used to stabilize the 10'- high (3050 mm) flue. It is worth noting that the exhaust flue from within the building connects to the fume hood on the side away from the viewer, which explains why the end of the flue attached to the mounting pad is sealed (other than two 3/8" diameter [10 mm] drain holes). Based on what is shown, what would the likely sequence of construction for this assembly be?

1/8" PLATE STEEL
OVER TOP OF
PIPE

SPLAY WIRE

1/4"ø BENT STEEL ROD

3/16

2"ø SCHED 40 PIPE
PAINT 2 COATS, PRIMER

CLAMP

1/8" PLATE

SINGLE PLY
ROOF

12"

3/16

6"

8"

PIPE JACK

(E) 31/2" LIGHTWEIGHT CONC.
OVER 11/2"x20ga METAL DECK

3/8"øx2" BOLTS, WASHERS & NUTS
(MIN. OF 4)

)

T INTO STRUCTURAL STEEL

IIGH

(H) SPLAY WIRE SUPPORT
NO SCALE

Figure 12.9s This is the guy wire anchor for the flue described in Figure 12.9r. Unlike similar roof penetrations, this one does not define how the flashing is sealed around the anchor.

PUMPS   BOILER
APLC   APLC

AHU-1   AHU-2
APLC   APLC

PUMPS   CHILLER
APLC   APLC

VAV   VAV
TUX   TUX

SEE SCHEDULES
AND PLANS FOR
QUANTITIES

ALERTON COMPUTER

NEW ON-SITE COMPUTER WILL
COMMUNICATE WITH EXISTING
ALERTON COMPUTER IN RCE
BUILDING.

GLOBAL CONTROLLER PANEL
INSTALL IN ROOM ???

RS-485 LAN

ALERTON SYSTEM ARCHITECTURE

Figure 12.9t This drawing shows the layout for the local area network that is the heart of the mechanical control system.

120V TO 24V AREA
TX TRANSFORMER(S)
BY EMS CONTACTOR

120V POWER
AREA J-BOX
BY DIV. 16
SEE PLANS

TX

120V

VAV BOX WITH ENCLOSURE

CF

VAV

VAV

DTS

A

RTS

73

MS/TP —— LAN

VAV BOX CONTROL DIAGRAM

Figure 12.9u The controls for all the components of the heating and cooling system need to be detailed as well. Here is the drawing identifying the circuitry for the VAV boxes.

Plumbing Drawings

Drawings describing the drain waste and vent piping routinely follow the introductory sheets in plumbing. It is not unusual for a mechanical engineer to show the architect's floor plan, highlight the restroom, food preparation, and other areas in which plumbing is concentrated, then show enlarged-scale plans of the areas adjacent to the floor plan, which makes it very easy to put the plumbing into context in the building. Subsequent floors are shown the same way. The roof plan, or plan views of ancillary structures—the mechanical plant, for example—are shown next, often at a somewhat enlarged scale (if the normal floor plan is 1/8" = 1'-0" [1:100], this drawing might be at 1/4" = 1'-0" [1:50]). Following these plan views is the equivalent of plumbing elevations. Called *riser diagrams*, these drawings depict the domestic water supply and controls in schematic form (single-line drawings). Following the riser diagrams are the waste and vent schematics (also single-line drawings), which depict the drain waste and vent (DWV) system in an isometric drawings. Following the DWV isometrics are the fire system standpipe schematics, in which the main supply system

is located, followed by detail sheets showing a variety of things, including the manner of connecting pipes to the structure, seismic connections, and pump mounting instructions, to name a few. Figures 12.10a–i comprise a collection of drawings that illustrate the kind of information to be found on plumbing drawings.

Fire System Drawings

Fire system drawings are often listed as "deferred approval" items, meaning that someone other than the architect and design consultants will furnish fire system drawings, submit them, and have them approved by the permitting authorities after the project drawings have been put out to bid. Deferred approval items exist in structural drawings as well; for example, truss systems are often designed by truss manufacturers according to design parameters established by the structural engineer. The truss manufacturer submits shop drawings for review after the material contract for the trusses has been awarded. The mechanical engineer may simply locate the standpipes and main drains, leaving the sizing of the standpipe design and details of the sprinkler system to design-build fire safety specialists.

Figure 12.10a The typical plumbing plan shows drain waste and vent piping, as well as hot and cold water supply and return lines. This plan describes the second-floor DWV and supply lines (the drain lines are actually in the interstitial space of the first floor, and the vent and supply lines are above the second floor).

Figure 12.10b Roof plumbing consists largely of vents rising through the roof from the floors below, water heater flues, roof drains, and condensate drain lines for roof-mounted HVAC equipment.

HOT AND COLD
WATER RISER DIAGRAMS

A SCALE: NO SCALE

Figure 12.10c Riser diagrams are schematic drawings of the hot and cold water supply and return lines, valves, and gas lines. This diagram shows part of the first and second floors of a building.

NOTE:
PROVIDE WALL CLEANOUTS ON ALL
SINKS AND URINALS

B 2nd FLOOR
WASTE & VENT ISOMETRIC
SCALE: NO SCALE

Figure 12.10d Isometric single-line drawings of the drain waste and vent system make visualizing the system easier.
They are not drawn to scale.

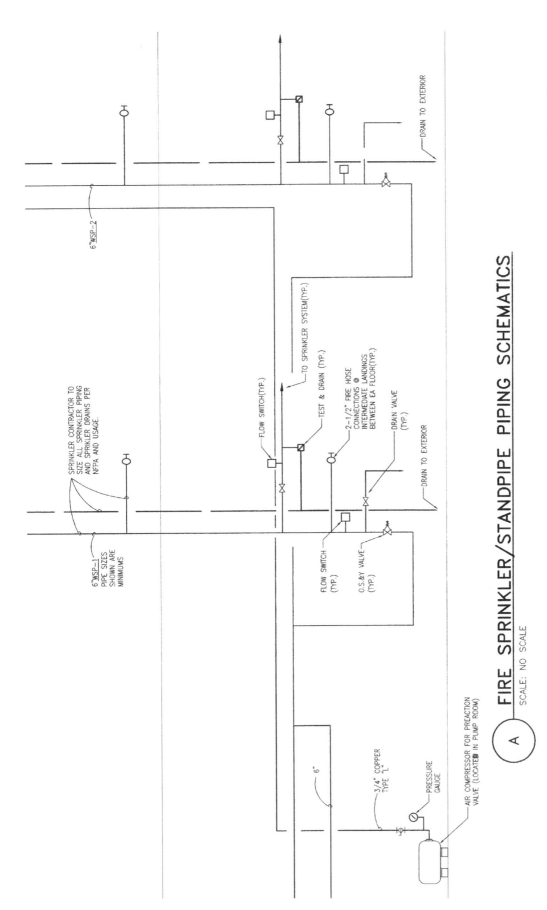

Figure 12.10e Schematic drawings are produced for fire sprinkler systems. This system combines a wet standpipe and preaction system. Note the design professional's comment regarding the sizing for the fire system piping (top, center of drawing).

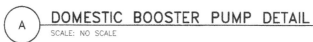

PRESSURE REDUCING/
CHECK VALVE

ISOLATION VALVE
TYP. @ SUCTION
& DISCHARGE

DUPLEX PUMP
SYSTEM

THERMAL RELIEF VLV.
1/2" FPT PIPED
TO FLOOR SINK

MASON SUPER W
PAD ISOLATOR
EA. CORNER

MIN. 4" THICK
CONCRETE PAD
PAINT SIDE OF
PAD GREEN

6" MIN. TO EDGE
OF SLAB (TYP.)

5/8"ØX3" EMBED
HKBII (1)EA, CORNER

BUILDING SLAB

MAINTAIN REBAR MIN. 8"
FROM EDGE OF PAD.

#3 REBAR @12" O.C.
MID SPAN OF PAD

A — DOMESTIC BOOSTER PUMP DETAIL

SCALE: NO SCALE

Figure 12.10f Pump assemblies are frequently shown in elevation, and can be quite realistic.

12" MIN. TO
WALLS

WALL (INTERIOR OR EXTERIOR)

FILL VOID W/ FLEXIBLE
MATERIAL APPROVED
BY THE STATE
FIRE MARSHALL.

6"

CONCRETE SLAB

PIPE PIPE SLEEVE

RISER

RESTRAINT JOINT EBAA IRON 2000PV
(ASTM F 1674-96). RATED FOR MIN.
200 PSI (TYP.)

DUCTILE IRON PIPE TO 5'
OUTSIDE OF BUILDING
MIN. 36" COVER OVER PIPE.

C — FIRE SPRINKLER PIPE RISE THRU SLAB

NO SCALE

Figure 12.10g In certain seismic regions, fire service supply lines must be separated from the structure, so that the two can move independently without damage.

1. CONCRETE SLAB OR CONCRETE OVER STEEL DECK.

2. STEEL OR COPPER PIPE.

3. SPECSEAL SERIES 100 SEALANT INSTALLED TO
   1" DEPTH W/ ANNULAR SPACE 3/8" MIN. TO 2" MAX.

NOTE:
FIELD VERIFY LOCATIONS WHERE PENETRATION APPLIES.
ALTERNATE UL SEALING IS ACCEPTABLE WHERE UL
SEAL MEETS THE APPLICATION.

NOTE: THE PRODUCT USED IN THIS DESIGN SYSTEM HAVE
BEEN TESTED AS FOLLOWS.

ASTM E814 (UL 1479). REFER TO SYS. C-AJ-5021 (479)

ASTM E119 (TIME/TEMPERATURE EXPOSURE)
(COTTON WASTE IGNITION)
ANNULAR SPACE REQUIREMENTS

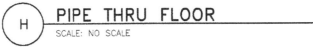

H   PIPE THRU FLOOR
     SCALE: NO SCALE

Figure 12.10h Isolation of the fire system supply occurs in horizontal and vertical planes.

MOUNT GAS PIPING
TO CHANNEL WITH
"SUPERSTRUT" PIPE
CLAMP

GAS PIPING, SEE
PLUMBING PLANS
FOR SIZE AND LOCATION
PAINT 1 COAT PRIMER 2 COATS
GREY. PREPAINT AS MUCH AS
POSSIBLE PRIOR TO INSTALLATION.
NO OVER SPRAY @ JOB SITE.

MINIMUM 8"
ABOVE SLAB

"SUPERSTRUT"
J-844 CHANNEL
ANGLE CONNECTION

"SUPERSTRUT" A-200
METAL FRAMING CHANNEL
(ZINC PLATED)

MOUNT CHANNEL TO
CONCRETE BASE WITH
3/8"ø HKB II & EACH END,
2-1/2" EMBEDMENT.

GAS/LPG
PIPING SUPPORT @ YARD

I   SCALE: NO SCALE

Figure 12.10i Other useful details, such as this gas pipe mounting
instruction, are shown in plumbing detail sheets.

Applicable Line Types

There is a limit to the number of ways a line can be drawn and still be meaningful, and there is considerable variety in mechanical system elements, so annotated lines of a single thickness and density are used extensively. Figure 12.11 offers a sampling of line types.

Symbols

Symbols are extensively used in mechanical drawings. What distinguishes mechanical drawings from architectural, structural, and interior construction drawings particularly is the extensive use of object symbols and the incidental use of material symbols (mostly in details).

Figure 12.11 Common line types in mechanical drawings.

Figure 12.12a Symbols for HVAC drawings vary from office to office. Here is one legend.

Figure 12.12b This legend shows one firm's plumbing symbols.

Object symbols, which, except for scale, approximate the appearance of the objects they are describing, are commonly used. Fortunately for drawing users, design professionals identify the symbols and abbreviations they use in legends that accompany the equipment schedules in the early sheets of the mechanical drawing set. Figures 12.12a–b show a collection of common mechanical symbols taken from a variety of drawings.

Projections

Mechanical drawings consist largely of plan views and schematic drawings, although in the details sheets, sections and elevations are common. Elevations in the detail sheets often take the form of double-line drawings, which are reasonably realistic (see Figure 12.10f). Isometric line drawings are used in drain waste and vent drawings, which are schematic in nature. These drawings are not done to scale. The more pictorial isometric drawings,

of flashing and anchors for example, are to be found in details.

■ The Takeoff

Effective mechanical system estimating requires specialized, intimate knowledge of the work, and when a detailed estimate is called for, every nut, bolt, hanger, fixture, and assembly must be accounted for. There are, however, reasonable approaches to understanding the drawings and identifying work for people who do not have experience doing the work. It was suggested earlier in this chapter that one reasonable approach to breaking the work into manageable subcategories is to first segregate the HVAC and plumbing systems. The HVAC system further divides into heating and cooling equipment and associated hardware and mounting elements, fuel supply piping; supply and return air systems, duct-mounted equipment, anchors, mounting brackets; trim packages (grilles and diffusers, for example), fire safing, caulking, and finally, controls.

A plumbing takeoff is easily segregated by subsystem. Separating the water supply from the drain waste and vent system is logical and, since the two systems are fundamentally different (one is pressurized, the other is not), the materials, connections, and quality-control requirements are different. The character of the fittings and valves echoes the piping system, so they should be taken off as part of the system, as should the pipe hangers, sound attenuation devices, and insulation. It is reasonable to take off the fixtures separately; in fact, since the supply and DWV system frequently interface at the fixture, it is a logical demarcation for the two takeoffs. Included in a fixture takeoff would be the ancillary hardware and equipment—mounting brackets, bolts and nuts, seals, escutcheons, and the like.

In general, a prudent approach to any estimate is to begin with the most important or expensive items and areas of greatest risk or uncertainty first, and proceed to the least costly and risky. Although it takes experience to do that, common sense goes a long way. Within a given system—say the water supply system—bigger items are usually more costly since they are heavier and more difficult to transport and install without special equipment or support, and they cost more to buy. Higher or deeper is more costly than lower or shallower (for instance, piping that is installed over 10' (3048 mm) high increases costs by about 1 percent per foot starting at 10 percent, so a ceiling over 50' (1500 mm) would result in an increase in cost of around 50 percent.[4] The degree of assembly of components and the manner in which they are connected to each other affects costs, too. An integrated carrier package, although more costly than comparable plumbing assembly without a carrier, saves rough-in costs. It is worth noting that the decisions to use one or another installation method should be evaluated on a systems basis—what makes things easier and less expensive for one trade may well translate into higher costs for another trade. Quality requirements affect work as well, since what is concealed by finishes of one sort or another is treated differently from that which is on display when the project is complete.

"Costing" a project—a term used in the construction industry that describes the process of determining project costs—takes a variety of forms, but in mechanical contracting, the worker-hour factor is a popular method. This method is effective with labor-paced operations; it involves identifying the cost of work by correlating the weighted cost of composite crews (for instance, a journey-level sheet metal worker, an apprentice, and a laborer who are all involved in a certain operation) to the amount of time

the crew takes to do one unit (the productivity factor), multiplied by the number of units. Accurately identifying numbers of units of various parts of mechanical systems in varying locations in a building is, therefore, critical. It goes to the heart of the takeoff process, and it underscores the relevance of the construction drawings to determining construction costs. The units that are commonly used in mechanical work are listed in Table 12.1.

Summary

The graphic portrayal of mechanical systems—conceptually simple but complicated systems in reality—can be challenging. Using a limited graphic "vocabulary" (primarily single-line drawings in plan view, and elevation and section views in details), mechanical engineers must conceive of and describe:

- The general requirements of the mechanical systems, within the parameters established by the architectural program, plumbing, and mechanical code requirements, and an established budget.
- The proper mix of basic HVAC and plumbing systems and equipment, and the manner in which they interrelate and are supported.
- How the fuel for HVAC systems is to be connected between the source and the equipment that requires it.
- How potable water or other required fluids and gases are produced at the project and how they connect to the supply side of the system.
- How the water and other fluids and gases are disposed of once they have been used.
- The distribution system for conditioned air (heated, cooled, humidified, dehumidified, filtered, and freshened) and the manner in which it is supported.
- Which fixtures are required for an effective plumbing system and how they are connected to each other and supporting structures.
- The nature of the control systems (operational and alarm) for both HVAC and plumbing.
- How all these components correlate to other systems; for example, the structural system and any interior architecture used in the project, and electrical work and mechanical systems.

Using the drawings and text developed and produced by the mechanical engineer, the builder must conceive of the plan that will result in all the work being done. What are the categories of required work? What operations must be performed? Who will execute them? What is the nature of the labor and type of equipment involved? What is the prudent sequence of work? How long will it take? How can we improve on what is suggested in the drawings, or make it more economical?

---

[4] *Means Mechanical Estimating*, 2nd ed., R. S. Means Company, Inc., 1992.

TABLE 12.1 Sample Unit Designations for Materials and Work*— Mechanical Systems

	LS lump sum	EA each	PR pair	M 1000	In. inch	LF lineal foot	SF square foot	SY Square yard	CF cubic foot	Ga Gallon	Lbs. pounds	Other	
Pipe						X							Plumbing
Fittings		X			X (welded fittings)								
Cutting oil										X			
Welding materials	X										X		
Pipe dope										X			
Valves		X											
Supports, anchors, guides		X											
Insulation						X							
Equipment		X					X					Ton, BTU- (equipment cooling, heating capacity);	HVAC
Access doors		X											
Ducts						X							
Fittings		X											
Registers		X											
Grilles		X											
Diffusers		X											
Insulation							X						
Controls		X				X (wiring and conduit)							

*Metric drawing require different units: liters (liquids), kilograms (weight), lineal meters (length), cubic meters (volume), square meters (area).

CHAPTER 12 EXERCISES

1. What is the principal difference between plumbing supply and drainage systems?
2. List several ways that heat may be created in HVAC systems, and describe the different media used to distribute it.
3. What are the principal components of a cooling system?
4. What is the principal drawback to drawings that describe HVAC systems?
5. List a couple of similarities between HVAC and plumbing systems.

13 Electrical Systems

Key Terms

Branch circuitry
Bus bar
Cable tray
Conduit
Devices/loads
Disconnect
Feeder
Load centers
Motor control centers
Panel boards
Raceway
Raised floors
Surface metal raceways
Trench duct
Underfloor duct
Wire duct
Wireways

Key Concepts

- Electrical drawings leave more to the imagination than other kinds of drawings, since they frequently identify a source and a destination but not necessarily a specific path.

Objectives

- Identify the elements of an electrical system.
- Identify commonly used graphic conventions in electrical drawings.
- Describe the graphic shortcomings of electrical drawings.

Construction Graphics: A Practical Guide to Interpreting Working Drawings, Third Edition. Keith Bisharat.
© 2025 John Wiley & Sons Inc. Published 2025 by John Wiley & Sons Inc.

Purpose

Electrical systems play a critical supporting role in the daily lives of people in thousands of cities and rural areas alike. The most fundamental needs of society—warmth, water, fuel, and light—are met and controlled by electrical systems. One need endure only one power failure to have contemporary society's dependence on electrical power emphasized. The entire power system from electrical generation to distribution to points of consumption is designed by electrical engineers and put into place by industrial, underground utility, and electrical specialty constructors, whose work represents a significant portion of the overall costs of construction projects. In building projects, the costs for electrical work ranges from 8 to 12 percent of the hard construction costs, a percentage that is likely to increase in the future as systems, particularly control systems, become more sophisticated.

Elements of Electrical Systems

There are similarities between mechanical and electrical systems in that both consist of loops of sorts. In HVAC systems, air is conditioned, pumped to remote spaces, partly exhausted, and returned to the air handler for reconditioning. Plumbing systems deliver conditioned water (heated and perhaps purified water) under pressure through conduit (pipes) to remote spaces where it is used; and although it is not returned for reconditioning, the drainage system completes the conceptual loop. Electrical systems move electrons through conductors housed in conduit to remote spaces under a sort of pressure (voltage being the corollary to pressure in plumbing systems, and amperage being the corollary to flow) caused by the imbalance of electrons generated at the utility company. As with mechanical systems, there is the source, a supply and return system, and controls.

Power Sources, Distribution, and Controls

The electrical power source for most projects is the service provided by the utility company operating in a given region. Overhead or underground lines in easements on private and public property form the power grid, into which an owner taps when construction of a project commences. The power is extended to an owner's parcel of land, and may be further distributed to various locations on the site itself, according to a site plan. The utility company may perform the underground work on a new site, or subcontract the work to a responsible subcontractor. The interface between electrical specialty contractor and the utility company frequently occurs at the transformers, where the electrical power is stepped down from its transportation mode (very high voltage,

low amperage) to its consumption mode (lower voltage, high amperage). The utility company provides the power to the transformer through aluminum or copper conductors housed in conduit, and the electrical specialty contractor connects the transformer to the main switchboard in the building, where it is distributed to smaller panelboards and transformers situated at various locations in the building. Panelboards, also called *load centers*, house the bus bars and circuit-protective devices (fuses or breakers). Branch circuitry, the electrical wiring connecting the circuit breakers to devices such as switches, light fixtures and receptacles, is the final leg in the distribution system. All circuits have at least two wires, one to carry current to the devices from the power source and the other to return it; most have three.

Controls occur wherever they are needed in a circuit. They take a variety of forms, from rheostats that control the flow of electricity to a fixture to switches that turn lights and equipment on and off to motor control centers—large metal cabinets, common in commercial buildings, where motor starters and disconnects and associated electrical circuitry are centralized.

Auxiliary Power

Without reasonable safeguards, the occasional failure of a system upon which society is so dependent can be catastrophic. Although residents in houses are left to their own devices when it comes to power failures, building codes, which establish minimum standards for health and safety, require public buildings to have back-up electrical systems or emergency devices. These may take the form of battery-operated lights strategically located at exit points or complete auxiliary electrical systems. Hospitals, for example, because of the nearly continuous life-threatening drama occurring in them, must have extensive auxiliary systems that immediately provide critical power during a failure.

Maintenance and Modifications to the System

Electrical systems must be inspected and maintained, and for building projects, be flexible enough to be adjusted. Businesses that occupy commercial space expand, contract, and disappear altogether, so a variety of measures are undertaken in the design phase to provide access to equipment and wiring. Cable trays, trench duct, raised floors, underfloor duct, wireways, surface metal raceways, and regular conduit all belong to the component family known as *raceways*, channels created to house and protect electric cabling.

Building codes require that certain kinds of electrical equipment be wired through devices such as disconnects and safety switches, so that while maintenance or repair work is being performed, the worker is protected from being inadvertently shocked. Disconnects for HVAC equipment and fans, for example, must be close to the equipment itself.

What to Expect in the Drawings

Electrical drawings reveal myriad combinations of the various elements just mentioned. In building construction projects, the information pertinent to electrical systems is found in the electrical drawings, designated by the letter E. Electrical drawings occur most commonly toward the rear of the construction drawing set, just ahead of landscape drawings. The electrical work is the last of the 16 specification divisions of MasterFormat, at least until the Construction Specifications Institute, the developer of MasterFormat, expands the number of divisions.[1] In drawing sets for civil work such as highway and bridge construction, electrical drawings may play an incidental role in the construction. In a wastewater treatment plant, electrical drawings constitute a significant number of the drawing sheets and project costs. Whatever the project, it is helpful to segregate the electrical work into subcategories.

In building construction, service refers to the utility-provided feeders (conduit and electrical wiring) from the transformer to the building (the source of the electricity). Distribution pertains to the switchboards, panelboards, switchgear, feeders, and raceways that emanate from the panels. Branch circuitry consists of the conductors and raceways between the fuse or breaker and the power outlet or light fixtures the circuit supports. Devices/loads describe the receptacles and switches that provide or control the power and the fixtures and equipment that consume it. Segregating projects into these subcategories is certainly not the only way to make understanding electrical drawings and quantifying electrical components and labor easier, but it does possess a certain logic that is appealing. Tracing the origin of the electrical power for a project to where it is distributed and finally to where and how it is used (or backwards from where it is used to where it is distributed to where it originates) is logical and makes assimilating the information easier.

Correlation to Other Drawings

Electrical drawings are largely schematic in nature, no doubt due in part to the linear character of the system, the small size of the components, the vast number of them required in the typical system, and the limited benefit to describing them realistically. With the exception of the detail sheets, in which elevations, sections, and the occasional pictorial drawing are used, most electrical drawings (like plumbing floor plans) consist of single-line drawings, accompanied by a wide variety of object symbols, superimposed over the architectural drawings. Electrical drawings of buildings reveal virtually nothing about the system in its critical third dimension (between floors). Understanding the vertical component of the system and how it relates to other systems in the interstitial space is derived from architectural or structural drawings (building sections, interior, and exterior elevations). Electrical engineers may stipulate the horsepower of a motor in a motor control center, but not the size of the conduit and conductor. That information may have to be determined by the reviewer of the drawings.

Other architectural drawings help tell the story of electrical systems; the type and location of light fixture is often shown in reflected ceiling plans—views of the ceiling as if one were looking down on a mirrored floor.

Landscape drawings and electrical utility (site) drawings should be consulted for potential interference between electrical equipment, such as transformers, and site improvements that may be described in the landscape drawings, as well as for the location of power distribution and branch circuitry.

The mechanical drawings are also critical to understanding the electrical work, since air conditioning equipment and fire system alarms are electrically powered. Figure 13.1 is a matrix of work responsibility on one building project that shows how extensive the interfaces between electrical and other specialties are.

Common Graphic Conventions in Electrical Drawings

The design-build project delivery system, a form of project delivery in which a single entity takes responsibility for both design and construction (either for the project overall or for specialties in it), is increasing in popularity, especially in the private sector, but in public works construction as well. The explanation for this phenomenon is a matter of spirited discussion in the design and construction communities, and is a complex issue more appropriate to a treatise on construction contracting; but for reviewers of construction drawings, the rise in popularity of design-build portends that variety in construction graphic work is likely to persist or even increase. Getting the "pure" design community to shift to a single graphic standard is one thing; getting designers and increasing numbers of contractors, whose graphic work under the design-build delivery system tends to focus on complying with local building requirements and internal communication, is another. (The design-build delivery system has the effect of narrowing the audience for the drawings. The drawings are produced for the contractor by the contractor, or by a design professional over

---

[1]CSI has developed another iteration of MasterFormat, consisting of 49 divisions (see Appendix F). Electrical work is now listed in Division 26.

SUBCONTRACTOR COORDINATION ⚠

NOTE 1: IF DISCONNECT IS BUILT-IN, NO ADDITIONAL PROVIDED.
NOTE 2: "MECH" "PLUG" "MECH/PLUG", OR "GENERAL" INDICATES ITEM NOT PROVIVED BY ELECTRICAL. REFER TO CONTRACT FOR PROVIDER.
NOTE 3: "FIRE" PROVIDED BY ELECTRICAL WHERE INCLUDED IN ELECTRICAL CONTRACT.
NOTE 4: N/A = NOT APPLICABLE.

CEILING DESCRIPTION	FURNISHED BY (NOTE 2, 3)	INSTALLED BY (NOTE 2, 3)	POWERED BY (NOTE 2, 3)	DISCONNECT BY (NOTE 1, 2)	STARTER BY (NOTE 1, 2)	CONTROL WIRING (ALL VOLTAGES) (NOTE 2, 3)
ITEMS						
CEILING OR FRACTIONAL HP EXHAUST FAN	MECH	MECH	ELECT	ELECT	N/A	N/A
DUCT SMOKE DETECTORS	MECH	MECH	ELECT	N/A	N/A	FIRE
FANS ("AHU", EXHAUST, RETURN, SUPPLY)	MECH	MECH	ELECT	ELECT	MECH	MECH
FIRE ALARM SYSTEM	FIRE	FIRE	ELECT	N/A	N/A	FIRE
FIRE / SMOKE DAMPER	MECH	MECH	ELECT	N/A	N/A	FIRE
FIRE SPRINKLER FLOW AND TAMPER SWITCH	FIRE	FIRE	N/A	N/A	N/A	N/A
HOT WATER HEATER / BOILER	MECH/PLUMB	MECH/PLUMB	ELECT	ELECT	MECH/PLUMB	MECH/PLUMB
HOT WATER CIRCULATING PUMP	MECH/PLUMB	MECH/PLUMB	ELECT	ELECT	MECH/PLUMB	MECH/PLUMB
HVAC UNITS, HEAT PUMPS, FAN COILS, SPLIT UNITS	MECH	MECH	ELECT	ELECT	N/A	MECH
LIGHTING CONTROLS	ELECT	ELECT	ELECT	N/A	N/A	ELECT
MECH OR PLUMBING CONTROL DEVICES	MECH	MECH	MECH	N/A	N/A	MECH
MOTORIZED CONTROL DAMPER	MECH	MECH	ELECT	N/A	N/A	MECH
THERMOSTATS	MECH	MECH	N/A	N/A	N/A	MECH
VARIABLE FREQUENCY DRIVES	MECH	MECH	ELECT	N/A	N/A	MECH
SITE WORK						
CONCRETE PADS (FOR ELECTRICAL EQUIPMENT ONLY)	ELECT	ELECT	N/A	N/A	N/A	N/A
TRENCHING (SOFT DIRT ONLY)	ELECT	N/A	N/A	N/A	N/A	N/A
ELEVATOR						
ELEVATOR PIT SUMP PUMP	MECH/PLUMB	MECH/PLUMB	ELECT	ELECT	MECH/PLUMB	MECH/PLUMB
ELEVATOR POWER SHUNT TRIP	ELECT	ELECT	ELECT	N/A	N/A	FIRE
HEAT AND SMOKE DETECTOR AT ELEV MACHINE RM	FIRE	FIRE	FIRE	N/A	N/A	FIRE
HEAT AND SMOKE DETECTOR AT HOISTWAY	FIRE	FIRE	FIRE	N/A	N/A	FIRE
MBC						
DUCT HEATER HEATING COILS (GAG)	MECH	MECH	N/A	N/A	N/A	MECH
ELECTRICAL VAULTS	ELECT	ELECT	N/A	N/A	N/A	N/A
ELEVATOR HOISTWAY DAMPERS	MECH	MECH	N/A	N/A	N/A	N/A
ELEVATOR RECALL	ELEV	ELEV	ELEV	N/A	N/A	ELEV
ELEVATOR WATER LEVEL SWITCHES	ELECT	ELECT	ELECT	N/A	N/A	N/A
END-SWITCHES (POSITION INDICATORS) (NC)	MECH	MECH	ELECT	N/A	N/A	ELECT
OUTSIDE AIR FAN	MECH	MECH	ELECT	MECH	MECH	MECH
PRESSURE BOOSTER SYSTEM	PLUMB	PLUMB	ELECT	ELECT	PLUMB	MECH
PUMP CONTROL DEVICES	MECH	MECH	MECH	N/A	N/A	N/A
SMOKE CONTROL DAMPERS AIR HANDLERS	MECH	MECH	ELECT	N/A	N/A	MECH
TEMPERATURE CONTROL VALVES	MECH	MECH	MECH	N/A	N/A	MECH

Figure 13.1 This matrix shows the division of work between the electrical subcontractor and several other specialties on one building project. It points out how complex the coordination of subcontracts can be. The project manager would prepare contract language reflecting this division of work. Such a matrix would not exist on drawings created by an electrical engineer; the electrical contractor under a design-build contracting format created this matrix.

whom the contractor, being a teammate, has some influence. In this respect, design-build drawings are like shop drawings.)

Order of Information

The electrical work is one specialty area that lends itself well to the design-build contract form, particularly for smaller and medium-sized commercial building projects in the private sector. As with other disciplines, there is variety in how drawings are packaged (some electrical drawings for building projects follow a similar format to mechanical drawings, in which the preliminary sheets are devoted to a sheet index, energy regulation compliance data, symbols, and fixture schedules). Figures 13.2a–e comprise a sampling of data included on the first few sheets of electrical drawings, including a sheet index.

When a commercial remodeling project is the subject of the drawings, the introductory sheets may be preceded by demolition drawings. Site utilities (which include power and communications conduits, cabling, and related equipment such as transformers and pull boxes), if they are not part of a separate set of drawings produced by the municipal power generator, follow. Site lighting drawings describe the number type and location of light standards in parking lots, driveways, and open space around the building. Figures 13.3a and 13.3b are examples of, respectively, site utilities and site lighting drawings.

The mechanical plant provides heating and cooling in medium to large commercial building projects through a variety of systems. Electrical power plays a critical role in the mechanical plant: Electric motors run the hot and chilled water pumps and myriad other devices, and the auxiliary power generator is typically housed in the plant. Some commercial building campuses, hospital complexes, and industrial projects use cogeneration machinery to optimize electrical efficiency, making the drawings for those projects extensive. Plant drawings are logically placed after site lighting and before building power plans, which follow them. Figures 13.4a–c are excerpted from the mechanical plant drawings for an 85,000-square-foot office building.

On residences and small commercial projects, lighting, power, switches, communication, and signal circuitry are superimposed on a copy of the architectural floor plan. In commercial building projects, power plans are often segregated from power, lighting, and signal circuitry (telephone, computer networks, cable television, clock systems). In a multistory project, these systems may be shown separately for each floor, starting with power plans by floor, followed by signal plans by floor. Figures 13.5 and 13.6 show the same area of the first floor of the office building mentioned above.

Telephone and electrical equipment rooms are generally shown in regular floor plans, which orient the reader as to general locale, as well as in larger-scale plans, where deciphering the specific layout of equipment, cabinets, and cable trays is made easier (see Figure 13.7).

At some point in a drawing set, a schematic drawing of the electrical distribution system (panelboards, feeders,

(text continues on page 321)

DRAWING INDEX

SHEET NO.	DESCRIPTION
E0.1	LEGEND, SCHEDULES, NOTES, AND TITLE 24
E1.0	SITE UTILITY PLAN
E1.1	SITE LIGHTING PLAN
E1.2	LIGHTING AND POWER PLAN — MECHANICAL PLANT
E2.1	POWER PLAN — FIRST FLOOR
E2.2	POWER PLAN — SECOND FLOOR
E2.3	POWER PLAN — THIRD FLOOR
E2.4	POWER PLAN — FOURTH FLOOR
E2.5	POWER PLAN — ROOF
E3.1	SIGNAL PLAN — FIRST FLOOR
E3.2	SIGNAL PLAN — SECOND FLOOR
E3.3	SIGNAL PLAN — THIRD FLOOR
E3.4	SIGNAL PLAN —FOURTH FLOOR
E4.0	ELECTRICAL, TELEPHONE, AND ELEVATOR MACHINE ROOM
E5.1	SINGLE LINE DIAGRAM
E5.2	PANEL SCHEDULES
E5.3	PANEL SCHEDULES
E6.1	LIGHTING PLAN — FIRST FLOOR
E6.2	LIGHTING PLAN — SECOND FLOOR
E6.3	LIGHTING PLAN — THIRD FLOOR
E6.4	LIGHTING PLAN — FOURTH FLOOR
E8.1	DETAILS
E8.2	DETAILS
E8.3	TELECOM DETAILS
E8.4	TELECOM DETAILS
E8.5	TELECOM DETAILS

Figure 13.2a A sheet index (this figure), general notes (b), fixture schedules (c), various compliance requirements (d), and a graphic legend (e) are data commonly shown on the first sheet of electrical drawings.

GENERAL NOTES

I   GENERAL

1.   THESE SPECIFICATIONS APPLY TO ALL ELECTRICAL SHEETS.

2.   ALL WORK SHALL BE IN ACCORDANCE WITH APPLICABLE STATE
     AND FEDERAL CODES, LAWS AND REGULATIONS AND UL LISTINGS.

3.   CONTRACTOR SHALL VISIT SITE AND BE FULLY COGNIZANT OF ALL
     CONDITIONS PRIOR TO SUBMITTING PROPOSAL. VERIFY ALL
     CONNECTIONS TO EXISTING WORK.

4.   CONTRACTOR SHALL OBTAIN AND PAY FOR ALL REQUIRED FEES, PERMITS
     AND INSPECTIONS.

5.   DURING ENTIRE CONSTRUCTION PERIOD, THE CONTRACTOR SHALL
     MAINTAIN ADEQUATE CO_2 FIRE EXTINGUISHERS READY FOR USE IN CASE
     OF FIRE.

6.   PROTECTION OF PUBLIC: THE CONTRACTOR SHALL PROTECT THE PUBLIC
     FROM INJURY DURING PROGRESS OF THE WORK BY POSTING WARNING
     SIGNS, GUARD LIGHTS AND BARRICADES.

7.   COORDINATE ALL CUTTING AND PATCHING WITH GENERAL CONTRACTOR.
     SUBCONTRACTOR SHALL BE RESPONSIBLE FOR ALL CUTTING AND
     PATCHING RELATED TO HIS WORK.

8.   RESTORE ALL DAMAGE RESULTING FROM YOUR WORK AND LEAVE PREMISES
     IN CLEAN CONDITION WHEN FINISHED WITH WORK.

9.   GUARANTEE ALL WORK AND MATERIALS FOR ONE YEAR MINIMUM FROM
     DATE OF FILING NOTICE OF COMPLETION.

10. FURNISH AND INSTALL ALL MATERIALS, EQUIPMENT AND LABOR AS
     SHOWN AND AS NECESSARY FOR COMPLETE WORKABLE SYSTEMS.

11. CONNECT ALL EQUIPMENT FURNISHED BY OTHERS AS REQUIRED.

12. UPON COMPLETION OF THE WORK FOR THIS SECTION AND AT VARIOUS
     TIMES DURING THE PROGRESS OF THE WORK WHEN REQUESTED BY THE
     OWNER, THE ELECTRICAL CONTRACTOR SHALL REMOVE FROM THE
     BUILDING ALL SURPLUS MATERIALS, RUBBISH AND DEBRIS RESULTING
     FROM THE WORK OF THIS DIVISION AND THE INVOLVED PORTIONS OF
     THE SITE SHALL BE LEFT IN A NEAT, CLEAN AND ACCEPTABLE
     CONDITION AS APPROVED BY THE OWNER.

II SCOPE OF WORK

1.   CONTRACTOR SHALL SUPPLY NECESSARY LABOR AND MATERIALS TO
     PROVIDE FOR A WORKING SYSTEM INCLUDING BUT NOT LIMITED TO:

     A.   PROVIDE AND INSTALL NEW EQUIPMENT AS INDICATED.
     B.   PROVIDE ALL CONDUIT, WIRE AND DISCONNECT AS REQUIRED TO
          CONNECT NEW EQUIPMENT.
     C.   MAKE ALL CONNECTIONS TO EQUIPMENT SUPPLIED BY OTHER AS
          REQUIRED, COORDINATE WITH THE EQUIPMENT SUPPLIER.
     D.   COORDINATE ALL WORK WITH OWNER'S REPRESENTATIVE.

III DRAWINGS

1.   KEEP ONE SET OF PLANS AT THE JOBSITE TO RECORD ANY DEVIATIONS
     FROM THE DESIGN.

2.   PROVIDE FOUR SETS OF RECORD DRAWINGS AND FOUR BOUND SETS OF
     ALL OPERATIONS MANUALS, DIAGRAMS, SERVICE CONTRACTS,
     GUARANTEES, ETC.

IV METHODS AND MATERIALS

1.   SEE SPECIFICATIONS FOR SPECIFIC REQUIREMENTS FOR EQUIPMENT
     AND INSTALLATION.

2.   ALL ELECTRICAL EQUIPMENT SHALL BE LISTED, LABELED, OR CERTIFIED
     BY A NATIONALLY RECOGNIZED TESTING LABORATORY.

Figure 13.2b General notes consist of comments and admonitions to the builder on a variety of issues.

LIGHTING FIXTURE SCHEDULE

FIXTURE TYPE	MANUFACTURER	MODEL NUMBER	DESCRIPTION	WATTS	VOLTS	LAMP	MOUNTING	REMARKS
A	LITHONIA	2PM3NFB2U316-16LD 277 GEB 10	2X2 PARABOLIC TROFFER	62	277	(2)FB 31T8	RECESSED	ELECTRONIC BALLAST
A1	LITHONIA	2PM3NFB2U316-16LD 277 GEB 10 EL	2X2 PARABOLIC TROFFER	62	277	(2)FB 31T8	RECESSED	BATTERY BACKUP ELECTRONIC BALLAST
A2	LITHONIA	2PM3NFB32 277 GEB 10	2X4 PARABOLIC TROFFER	89	277	(3)F32 T8 35K	RECESSED	ELECTRONIC BALLAST
B	LITHONIA	2LB-232-277	FOUR FOOT LOW-PROFILE WRAPAROUND	62	277	(2)F32 T-8 35K	SURFACE	ELECTRONIC BALLAST
B1	LITHONIA	2LB-232-277-EL	FOUR FOOT LOW-PROFILE WRAPAROUND	62	277	(2)F32 T-8 35K	SURFACE	BATTERY BACKUP ELECTRONIC BALLAST
B2	LITHONIA	WP232BF277GEB	FOUR FOOT WALL BRACKET	62	277	(2)F32 T-8 35K	WALL	BATTERY BACKUP ELECTRONIC BALLAST
D	LITHONIA	UN 232 277 EB WGCUN	4' OPEN STRIP	63	277	(2)F32 T-8	SURFACE	ELECTRONIC BALLAST WIRE GUARD STEM HANGER +10 FEET
D1	LITHONIA	DMW 2 32 120	4' ENCLOSED STRIP	63	120	(2)F32 T-8	WALL	ELECTRONIC BALLAST WET LOCATION
D2	LITHONIA	TDM 2 32 120	8' ENCLOSED STRIP	126	120	(4)F32 T-8	SURFACE	ELECTRONIC BALLAST DAMP LOCATION
G	LITHONIA	AFV 32TRT 6P 277 EB PLUS 6LD1 FOR SHOWERS	6" FLUORSC. OPEN DOWNLIGHT	35	277	32W TRT	RECESSED	ELECTRONIC BALLAST PROVIDE OPAL DROP LENS FOR SHOWER LOCATIONS
G1	LITHONIA	AFV 32TRT 6P 277 EB -EL	6" FLUORSC. OPEN DOWNLIGHT	35	277	32W TRT	RECESSED	BATTERY BACKUP ELECTRONIC BALLAST
K	LIGHTOLIER	F3454	SCONCE	26	120	(1) F25 T8	WALL	SEE ARCHITECTURAL PLANS FOR LOCATION
L	HALO	ETCHTINGS SERIES H2529	17 3/4" SCONCE	38	277	(2)18W PL-C	WALL	
L1	HALO	ETCHTINGS SERIES H2526	11 1/2" SCONCE	38	120	(1)100W T4	WALL	
M1	FINELITE	SERIES 8	4' FOOT INDIRECT FLUORESCENT	62	277	(2)F32 T8 35K	PENDENT	MOUNT AT 14 INCHES BELOW CEILING
M2	FINELITE	SERIES 8	8' FOOT INDIRECT FLUORESCENT	126	277	(4)F32 T8 35K	PENDENT	MOUNT AT 14 INCHES BELOW CEILING
M3	FINELITE	SERIES 8	12' FOOT INDIRECT FLUORESCENT	189	277	(6)F32 T8 35K	PENDENT	MOUNT AT 14 INCHES BELOW CEILING
M4	FINELITE	SERIES 8	4' FOOT INDIRECT FLUORESCENT	62	277	(2)F32 T8 35K	PENDENT	MOUNT AT 14 INCHES BELOW CEILING EMERGENCY BATTERY
N2	DAY-BRITE	2TG-232-01-277	2X4 PRISMATIC TROFFER	62	277	(2)F32 T8 35K	RECESSED	ELECTRONIC BALLAST
N3	DAY-BRITE	2TG-332-01-277	2X4 PRISMATIC TROFFER	89	277	(3)F32 T8 35K	RECESSED	ELECTRONIC BALLAST
N4	DAY-BRITE	2TG-232-01-277	2X4 PRISMATIC TROFFER	62	277	(2)F32 T8 35K	RECESSED	BATTERY BACKUP ELECTRONIC BALLAST
P1	HOLOPHANE	MA250MH27FFVW BR-1091-WH	AREA LIGHTING	295	277	250W MH	POLE	20FT POLE
P2	HOLOPHANE	MA250MH2779FHW BR-1064W	AREA LIGHTING	590	277	(2)250W MH	POLE	20FT POLE
P3	GARDCO	111 FT 70MH 277	MINI SCONCE	75	277	70 MH	WALL	FINISH TO MATCH BUILDING
P4	SISTEMALUX	S-4158-14	BOLLARD	87	277	70 MH	BOLLARD	
Q	DEVINE	BCW50 SERIES BCW55-100HPS	WALLPACK	100	277	100HPS	WALL	WET LOCATION VERIFY FINISH W/ ARCH.
X	LITHONIA	LQM-S-W-G-277-EL	LED EXIT SIGN	4	277	LED	UNIVERSAL	

Figure 13.2c Fixture schedules are widely used to organize a great deal of information in a simplified format.

LIGHTING MANDATORY MEASURES

BUILDING LIGHTING SHUT—OFF

☒ THE BUILDING LIGHTING SHUT—OFF SYSTEM CONSISTS OF AN AUTOMATIC TIME SWITCH, WITH A ZONE FOR EACH FLOOR.

☒ OVERRIDE FOR BUILDING LIGHTING SHUT—OFF

THE AUTOMATIC BUILDING SHUT—OFF SYSTEM IS PROVIDED WITH A MANUALLY ACCESSIBLE OVERRIDE SWITCH IN SIGHT OF THE LIGHTS. THE AREA OF OVERRIDE IS NOT TO EXCEED 5,000 SF.

☒ AUTOMATIC CONTROL DEVICES CERTIFIED

ALL AUTOMATIC CONTROL DEVICES SPECIFIED ARE CERTIFIED. ALL ALTERNATE EQUIPMENT SHALL BE CERTIFIED AND INSTALLED AS DIRECTED BY THE MANUFACTURER.

☒ FLUORESCENT BALLAST AND LUMINARIES CERTIFIED

ALL FLUORESCENT FIXTURES SPECIFIED FOR THE PROJECT ARE CERTIFIED AND LISTED IN THE "DIRECTORY". ALL INSTALLED FIXTURES SHALL BE CERTIFIED.

☒ TANDEM WIRING FOR TWO—LAMP BALLAST

ALL ONE AND THREE LAMP FLUORESCENT FIXTURES ARE TANDEM WIRED WITH TWO (2) LAMP BALLASTS WHERE REQUIRED BY STANDARDS 132; OR

ALL THREE LAMP FLUORESCENT FIXTURES ARE SPECIFIED WITH ELECTRONIC HIGH—FREQUENCY BALLAST AND ARE EXEMPT FROM TWO—LAMP TANDEM WIRING REQUIREMENTS.

☒ INDIVIDUAL ROOM/AREA CONTROLS

EACH ROOM AND AREA IN THIS BUILDING IS EQUIPPED WITH A SEPARATE SWITCH OR OCCUPANCY SENSOR DEVICE FOR EACH AREA ENCLOSED WITH FLOOR—TO—CEILING WALLS.

☒ UNIFORM REDUCTION FOR INDIVIDUAL ROOMS

ALL ROOMS AND AREAS GREATER THAN 100 SQUARE FEET AND MORE THAN 1.2 WATTS PER SQUARE FOOT OF LIGHTING LOAD ARE CONTROLLED WITH BI—LEVEL SWITCHING FOR UNIFORM REDUCTION OF LIGHTING WITHIN THE ROOM.

☒ DAYLIT AREA CONTROL

ALL ROOMS WITH WINDOWS AND SKYLIGHTS GREATER THAN 250 SQUARE FEET, THAT ALLOW FOR THE EFFECTIVE USE OF DAYLIGHT IN THE AREA, HAVE 50% OF THE LAMPS IN EACH DAYLIT AREA CONTROLLED BY A SEPARATE SWITCH.

☒ CONTROL EXTERIOR LIGHTS

EXTERIOR LIGHTING IS CONTROLLED VIA A DIRECTIONAL PHOTOCELL AND/OR ASTRONOMICAL TIME SWITCH THAT WILL TURN OFF EXTERIOR LIGHTING WHEN DAYLIGHT IS AVAILABLE.

Figure 13.2d Mandatory criteria for energy savings and other considerations may be shown in tabular format.

POWER DEVICES

Symbol	Description
⊖	SINGLE RECEPTACLE AS SPECIFIED ON THE PLANS.
⊕	DUPLEX RECEPTACLE. MOUNT CENTER +18" ABOVE FINISHED FLOOR U.O.N. 15 AMPERE.
⊕	ISOLATED GROUND DUPLEX RECEPTACLE. MOUNT CENTER +18" ABOVE FINISHED FLOOR U.O.N. 15 AMPERE.
⊕	DOUBLE DUPLEX RECEPTACLE. MOUNT CENTER +18" ABOVE FINISHED FLOOR U.O.N. 15 AMPERE.
⊕	ISOLATED GROUND DOUBLE DUPLEX RECEPTACLE. MOUNT CENTER +18" ABOVE FINISHED FLOOR U.O.N. 15 AMPERE.
⊕	ABOVE COUNTER DUPLEX RECEPTACLE. 15 AMPERE.
▤	FLOOR MOUNTED DUPLEX RECEPTACLE
⊢◁	SPECIAL RECEPTACLE, SEE PLANS
⊢◁	SPECIAL RECEPTACLE, SEE PLANS
⊢⊝	CLOCK OUTLET
Ⓙ	JUNCTION BOX
⊢Ⓙ	WALL MOUNTED SURFACE JUNCTION BOX
▣	FLOOR MOUNTED RECESS JUNCTION BOX
⬡ EF/2	MECHANICAL EQUIPMENT. VERIFY CONNECTION REQUIREMENTS WITH DIVISION 15.
⬡	DETAIL REFERENCE. # = DETAIL NUMBER, E-# = SHEET
▬	SURFACE MOUNTED PANELBOARD. MOUNT TRIM TOP 6'6" AFF UON.
▬	RECESSED MOUNTED PANELBOARD. MOUNT TRIM TOP 6'6" AFF UON.
▨	SWITCH BOARD
▭	SURFACE MOUNTED AUXILIARY SYSTEM CABINET. MOUNT TRIM TOP 6'6" AFF UON.
▱	RECESSED MOUNTED AUXILIARY SYSTEM CABINET. MOUNT TRIM TOP 6'6" AFF UON.
⊏⊦	FUSED DISCONNECT. FUSE PER EQUIPMENT MANUFACTURERS NAMEPLATE
$ᴛ	MOTOR RATED THERMAL SWITCH
⊘	MOTOR, # = HORSE POWER
⟍⟋ᴍ	HOMERUN TO PANELBOARD HASHMARKS = NO. OF CONDUCTORS NO HASHMARKS = 2 #12 IN 3/4" CONDUIT, \ = ISOLATED GROUND
— — — —	CONDUIT BELOW FLOOR OR GRADE NO HASHMARKS = 2 #12 IN 3/4" CONDUIT
————	CONDUIT RUN CONCEALED IN WALL OR CEILING U.O.N. NO HASHMARKS = 2 #12 IN 3/4" CONDUIT
— · —	EXISTING CONDUIT OR EQUIPMENT
✕—✕·✕—✕	EXISTING CONDUIT OR EQUIPMENT TO BE REMOVED
•,○	CONDUIT/CIRCUIT DOWN,UP
ℙ	POWER POLE
Ⓜ	UTILITY METER

Figure 13.2e Legends such as this one explain the symbols used by a particular design professional's office.

Figure 13.3a This partial site utility plan shows the source of power that the new structure is tapping into. Part of the auxiliary power line coming from the mechanical plant is shown in the lower part of this drawing.

Figure 13.3b Site lighting includes parking lot standards and walkway illumination.

MECHANICAL PLANT — POWER PLAN
SCALE: 1/4"=1'-0"

2 / E1.2

Figure 13.4a This floor plan shows the location of various pumps with their associated circuits and several panelboards. "MCC" in room M101 stands for motor control center. Directly above this panel bank is an unidentified panel.

MECHANICAL PLANT — LIGHTING PLAN
SCALE: 1/8"=1'-0"

$$\begin{array}{c}1\\\hline E1.2\end{array}$$

N

Figure 13.4b The lighting plan of the mechanical plant was shown in smaller scale than the power plan, on the same sheet.

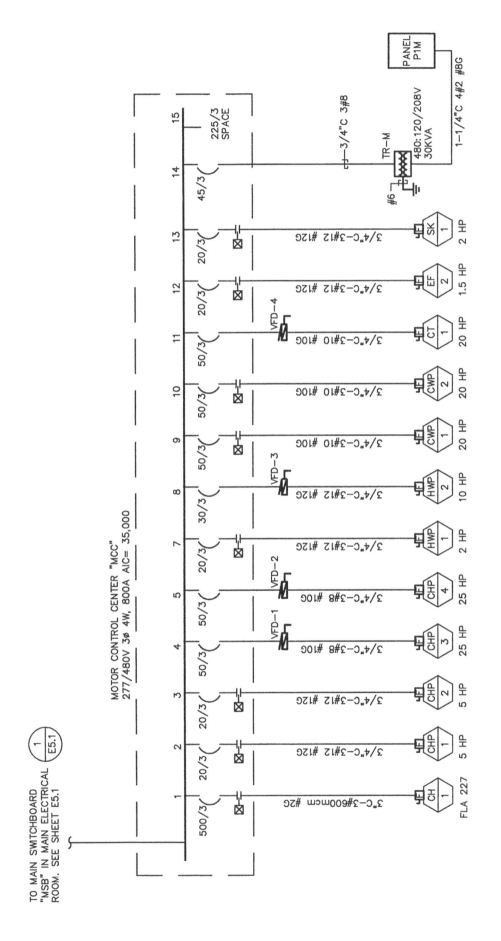

MOTOR CONTROL CENTER "MCC" SCHEDULE
SCALE: NONE

$\begin{array}{c}\boxed{3}\\[-2pt] \hline E1.2 \end{array}$

Figure 13.4c The diagram for the motor control center, which is located in the boiler room in the plant, is shown here.

and conduit) from the source to the panelboards is inserted into the drawing set. These drawings, variously dubbed single-line, one-line diagrams, or power riser diagrams, correlate the main switchboard to numerous panelboards used in the structure. Like most schematic drawings, they are not drawn to scale. The key to getting along with these drawings is to avoid trying to associate any realistic spatial relationship with the many parts described. That the auxiliary power circuit is shown to the right of the transformer in the diagram is no guarantee that it actually is to the right of the transformer, or to the left for that matter. About the only discernible characteristic is that the auxiliary power occupies the same horizontal plane as the transformer (the ground floor). They nevertheless are helpful, since the entire system, at least on a medium-sized project, can be organized and described on a single sheet. Figure 13.8 shows part of a single-line diagram.

Figure 13.5 Power plans show the location of receptacles and junction boxes for the variable air volume boxes, pumps, and various HVAC equipment.

Figure 13.6 The same floor area as that shown in Figure 13.3 is shown here, but the focus is the signal plan. Cable trays (designated by two parallel medium-width lines) are shown reaching over room 106 into room 108 (near the elevators in the lower part of the drawing), and from the hallway into room 108 (adjacent to the door of the room). Notice how the grid symbols have changed from the preceding drawing.

FIRST FLOOR ELECTRICAL/TELEPHONE ROOM
SCALE: 1/4"=1'-0"

Figure 13.7 The reader is first introduced to Detail 2 on E4.0, shown here, in the floor plan in the previous figure. The cable tray system is elaborate in this room.

Figure 13.8 Single-line or riser diagrams are schematic drawings that describe the relationship of panelboards and other components to the main switchboard. The transition from the ground floor to the second floor (shown by a dashed line) is visible left of center, toward the upper edge of the drawing.

Following the single-line diagram is some drawing sets are panelboard schedules, in which specific circuits are identified and organized. Figure 13.9 illustrates a typical schedule.

Lighting for commercial buildings, considering its overall contribution to the architectural statement in some projects, often deserves its own sheets. That said, when lighting drawings are shown by themselves, the issue of conflicts with HVAC supply and return air fixtures arises. For this reason, light fixtures, diffusers, and return air grilles are shown together in some drawing sets. Figures 13.10a–b show one project in which the lighting plan and mechanical drawings were apparently not coordinated.

Projection Types and Applicable Lines

As noted previously, electrical plans are, for the most part, line diagrams of parts of the electrical circuitry, in plan view (looking down from above). The more easily assimilated architectural depictions of electrical objects do exist, however, in the detail sheets, where pictorial drawings may also be used. Several examples are shown in Figures 13.11a–c.

Symbols

When the graphic alphabet is limited to a couple of line types, communicating the variety of elements of the electrical system becomes more difficult. Consequently, symbols combined with ordinary lines and abbreviations are used extensively in drawings of electrical systems. Figures 13.12a–d illustrate the range of lines and symbols used in the electrical drawings of one construction project. (See also Figure 13.2e.)

Subtleties in the Drawings

The abstract character of electrical drawings and the tradition of showing only portions of the electrical system

PANEL P1C

120/208 VOLT 3 PHASE 4 WIRE

TYPE: NQOD
AIC: 10,000

BUS: 125 AMPERE
MAIN: 125 LO

MOUNTING: FLUSH
LOCATION: 1ST FLOOR

NO	DESCRIPTION	KVA	BREAKER A	P	TOTAL LOAD	BREAKER P	A	KVA	DESCRIPTION	NO
1	RECEPTS	0.9	20	1	1.8	1	20	0.9	RECEPTS	2
3	RECEPTS	0.9	20	1	1.8	1	20	0.9	RECEPTS	4
5	RECEPTS	0.9	20	1	1.8	1	20	0.9	RECEPTS	6
7	RECEPTS	0.9	20	1	1.8	1	20	0.9	RECEPTS	8
9	WORK STATION OUTLETS	1.0	20	1	1.9	1	20	0.9	RECEPTS	10
11	WORK STATION OUTLETS	1.0	20	1	1.9	1	20	0.9	RECEPTS	12
13	RECEPTS	0.9	20	1	1.8	1	20	0.9	RECEPTS	14
15	RECEPTS	0.9	20	1	1.8	1	20	0.9	RECEPTS	16
17	RECEPTS	0.9	20	1	0.9	1	20	0.0		18
19	FPB	0.8	20	1	1.0	1	20	0.2	REEL RECEPT	20
21	VAV	0.8	20	1	1.7	1	20	0.9	RECEPTS	22
23	RECEPTS	0.9	20	1	1.8	1	20	0.9	RECEPTS	24
25	RECEPTS	0.4	20	1	1.3	1	20	0.9	RECEPTS	26
27	WORK STATION OUTLETS	1.0	20	1	1.9	1	20	0.9	RECEPTS	28
29	WORK STATION OUTLETS	1.0	20	1	1.8	1	20	0.8	RECEPTS	30
31	ELEV. #1 PIT	0.3	20	1	1.0	1	20	0.7	RECEPTS	32
33	ELEV. #2 PIT	0.3	20	1	1.9	1	20	1.6	VP-1	34
35		0.0	20	1	0.0	1	20	0.0		36
37		0.0	20	1	0.0	1	20	0.0		38
39		0.0	20	1	0.0	1	20	0.0		40
41		0.0	20	1	0.0	1	20	0.0		42
	PHASE A	72	AMPS		9	KVA			NOTES:	
	PHASE B	92	AMPS		11	KVA				
	PHASE C	68	AMPS		8	KVA				
	TOTAL				28	KVA				

Figure 13.9 Panel P1c is shown in the preceding figure, in the lower right corner of the drawing. P1c is one of three first-floor power panels.

Figure 13.10a The reflected ceiling plan for a portion of the first floor is shown here. In the next figure, the same ceiling area is shown occupied by supply air diffusers and ductwork.

Figure 13.10b Conflicts between the lighting and ductwork in a portion of the building are not uncommon. Using a simple measuring device, one can see the conflict in the diffusers on the right side of rooms 136 and 137 and the light fixtures shown in Figure 13.10a.

(the location of elements of the system, for example, and not the path) conspire to present drawing reviewers and estimators alike with a significant conceptualization challenge. The people who have substantial trade experience and excellent visualization skills, not surprisingly, are the most effective at determining the cost of electrical work. The advantage to building from inexplicit drawings is that the specialty contractor is free to determine the most efficient manner of doing the work. That said, whenever drawings are not explicit and the authority to make decisions is not clear, there is greater potential for disagreement.

Translating the Drawings into Work

While it is tempting and often appropriate to approach a quantity takeoff in the same sequence that a project is constructed, identifying the most costly and time-consuming activities, regardless of when they occur in construction, is more prudent. As noted earlier in this chapter, tracing the power from its source to the point of consumption, or vice versa, is an effective method of determining the cost of electrical work.

Starting the takeoff with the lighting makes sense from several viewpoints. It is one of the most costly

WHERE SHOWN ON PLAN MOUNT
WP DUPLEX RECEPTACLE ON UNIT

WP (FUSED)
DISCONNECT

ROOF TOP
MECHANICAL
UNIT

ROOF CURB

FLEXIBLE WIRING
UP THROUGH
CURB OPENING

CONTROL
FBO UON

ROOF

ACCESSIBLE JUNCTION
BOX(ES) AS REQUIRED

NOTE: RECEPTACLE, DISCONNECT & WIRING TO BE
INSTALLED CLEAR OF ALL MAINTENANCE
ACCESS OPENINGS

HVAC UNIT MOUNTING DETAIL
SCALE: NONE

14
E8.1

Figure 13.11a Perspective projections are very rare in construction drawings; here is one describing the mounting detail of an HVAC unit. Can you see where the projection departs from being a true perspective?

Figure 13.11b The typical mounting detail for fluorescent fixtures is shown in this detail.

Figure 13.11c Recessed fixtures require a different connection from the suspended fluorescent fixture.

ABBREVIATIONS

SYMBOL	DESCRIPTION
A	AMPERES
AFF	ABOVE FINISHED FLOOR
AFG	ABOVE FINISHED GRADE
AIC	AMPERE INTERRUPTING CAPACITY
AL	ALUMINUM
BC	BARE CONDUCTOR
BLDG	BUILDING
C	CONDUIT
CB	CIRCUIT BREAKER
CLG	CEILING
CT	CURRENT TRANSFORMER
CU	COPPER
DN	DOWN
(E)	EXISTING
ET	ELAPSE TIMER
FA	FIRE ALARM
FACP	FIRE ALARM CONTROL PANEL
F.B.O.	FURNISHED BY OTHERS
GFI	GROUND FAULT INTERRUPTER
GND	GROUND
HP	HORSEPOWER
IG	ISOLATED GROUND
I_{SCA}	SHORT CIRCUIT AMPERES
KW	KILOWATTS
MAX	MAXIMUM
MT	METALLIC TUBING
KVA	KILO−VOLT−AMPERES
MIN	MINIMUM
(N)	NEW
NIC	NOT IN CONTRACT
NL	NIGHT LIGHT
OC	ON CENTER
PNL	PANEL BOARD
ø	PHASE
(R)	EXISTING TO BE REMOVED
(RL)	EXISTING TO BE RELOCATED
RMS	ROOT MEAN SQUARE
SQFT	SQUARE FEET
TYP	TYPICAL
U.O.N.	UNLESS OTHERWISE NOTED
V	VOLTAGE
WP	WEATHER PROOF

Figure 13.12a An abbreviation legend is essential in electrical drawings, and when it is alphabetized, is much easier to use. In the same drawing set, the abbreviations for mechanical work were not alphabetized.

LIGHTING

SYMBOL	DESCRIPTION
▭◦	2X4 RECESSED FLUORESCENT TROFFER 3 LAMP
▦	2X4 RECESSED PARABOLIC FLUORESCENT TROFFER, 2 LAMP
▦	2X4 RECESSED PARABOLIC FLUORESCENT TROFFER, 2 LAMP EMERGENCY BATTERY BACK-UP.
▣	2X2 RECESSED PARABOLIC FLUORESCENT TROFFER, DOUBLE SWITCH
▣	2X2 RECESSED PARABOLIC FLUORESCENT TROFFER, 2 LAMP EMERGENCY BATTERY BACK-UP.
▭◦	SURFACE MOUNT FLUORESCENT WRAPAROUND
⊢◦⊣	PENDENT OR SURFACE MOUNT FLUORESCENT STRIP
▭	WALL MOUNTED FLUORESCENT FIXTURE
▭	INDIRECT PENDANT FLUORESCENT FIXTURE
▢	RECESSED DOWN LIGHT FIXTURE
▣	RECESSED DOWN LIGHT FIXTURE, WITH EMEGENCY BATTERY
▣	RECESSED WALL WASH DOWNLIGHT FIXTURE
⊗	LED EXIT SIGN WITH DIRECTIONAL ARROW
Ⓢ	LOW VOLTAGE PUSHBUTTON SWITCH STATION TWO CIRCUIT - PUSH ON PUSH OFF SEE DETAIL
\$₃ \$ₐ \$_{D,3}	SWITCH, SPST +42" HIGH U.O.N. SUBSCRIPT: 3=THREE WAY, 4=4WAY, D=DIMMER, K=KEYED, P=PILOT, a,b,c,d=AREA SWITCHED
(#)	LIGHT FIXTURE TYPE. REFER TO FIXTURE SCHEDULE
W₁	OCCUPANCY SENSOR MOUNTED AT SWITCH BOX. AUTO-OFF-ON SWITCH WATT STOPPER WI-300 OR EQUAL.
◄◇2◇►	CEILING MOUNTED OCCUPANCY SENSOR 360° TWO-SIDED PASSIVE INFRARED WATT STOPPER WT-2200 OR EQUAL.
◇6►	CEILING MOUNTED ULTRASONICOCCUPANCY SENSOR 360° ONE-SIDED WATT STOPPER WT-600 OR EQUAL
◁	WALL MOUNTED PASSIVE INFRARED SENSOR WATT STOPPER CX-100 OR EQUAL
◦▭	POLE MOUNTED AREA LIGHTING
⊙	BOLLARD AREA LIGHTING

Figure 13.12b Lighting symbols vary considerably between drawing sets. Here is one example.

SIGNAL DEVICES

SYMBOL	DESCRIPTION
▼	VOICE/DATA OUTLET, 4 11/16" SQUARE BOX AND DUAL DEVICE RING W 1 1/4" CONDUIT TO THE CABLE TRAY SHOWN ON THE PLANS, +18" AFF UNLESS OTHERWISE NOTED. PROVIDE TWO AVAYA M106FR4-246 MODULAR FRAME W/3 AVAYA GIGASPEED MODULAR INFORMATION OUTLETS AND FIVE M20AB-246 IVORY BLANKS. PROVIDE STAINLESS STEEL TWO GANG DUPLEX OUTLET PLATE. INSTALL ONE 2071 CABLE FROM EACH JACK TO THE MTTB OR TTB.
▼	WALL TELEPHONE OUTLET 4 11/16" SQUARE BOX W/SINGLE DEVICE RING +48" AFF UNLESS OTHERWISE NOTED. PROVIDE A 1 1/4" CONDUIT TO THE CABLE TRAY SHOWN ON THE PLANS. PROVIDE ALLEN TEL PRODUCTS AT630B-8 WALL TELEPHONE JACK W (1) 2071 CABLE FROM THE JACK TO THE MTTB OR TTB.
⍖	VIDEO/TV OUTLET 4 11/16" SQUARE BOX W/SINGLE DEVICE RING +78" AFF UNLESS OTHERWISE NOTED. PROVIDE A 1 1/4" CONDUIT TO THE CABLE TRAY SHOWN ON THE PLANS. PROVIDE STAINLESS STEEL PLATE PUNCHED FOR F-TYPE CABLE TV BULKHEAD JACK. PROVIDE TERMINATION ON CHAIN SECURED TO COVER PLATE.
▼+48"	VOICE/DATA OUTLET, 4 11/16" SQUARE BOX AND DUAL DEVICE RING W 1 1/4" CONDUIT TO THE CABLE TRAY SHOWN ON THE PLANS, +48" AFF UNLESS OTHERWISE NOTED. PROVIDE TWO AVAYA M106FR4-246 MODULAR FRAME W/8 AVAYA GIGASPEED MODULAR INFORMATION OUTLETS, IVORY COLOR. PROVIDE STAINLESS STEEL TWO GANG DUPLEX OUTLET PLATE. INSTALL ONE 2071 CABLE FROM EACH JACK TO THE MTTB OR TTB.
ᴾ○	VOICE/DATA OUTLET, POWER POLE PUNCHED FOR TWO DUPLEX OUTLETS. PROVIDE 1 1/4" CONDUIT FROM THE JUNCTION BOX ABOVE THE POLE TO THE CABLE TRAY SHOWN ON THE PLANS. PROVIDE TWO AVAYA M106FR4-246 MODULAR FRAME W/8 AVAYA GIGASPEED MODULAR INFORMATION OUTLETS, IVORY COLOR. INSTALL ONE 2071 CABLE FROM EACH JACK TO THE MTTB OR TTB.
▼ +40"	COIN TELEPHONE OUTLET 4 11/16" SQUARE BOX W/SINGLE DEVICE RING +40" AFF UNLESS OTHERWISE NOTED. PROVIDE A 1 1/4" CONDUIT TO THE CABLE TRAY SHOWN ON THE PLANS. PROVIDE ALLEN TEL PRODUCTS AT630B-8 WALL TELEPHONE JACK W (1) 2071 CABLE FROM THE JACK TO THE MTTB OR TTB.

Figure 13.12c Telephone, computer data, television, and the like fall under the category "signal devices." One legend is shown here.

SECURITY SYSTEM

⊏◁	SECURITY CAMERA
₍ₒₛ◣	OCCUPANCY SENSOR, WATTSTOPPER DT200
Ⓜ	MOTION DETECTOR SECURITY SYSTEM
Ⓓ	DOOR CONTACT, SECURITY SYSTEM-ROUGH IN ONLY
Ⓚ	KEY PAD
⊡	PUSH BUTTON
CR	CARD READER

FIRE ALARM

SYMBOL	DESCRIPTION
▣	FIRE ALARM MANUAL PULL STATION
SD	SMOKE DETECTOR (LETTER 'R' = RECALL)
DSD	DUCT SMOKE DETECTOR
DH	MAGNETIC DOOR HOLD OPEN
H◁	HORN/STROBE
S◁	STROBE
HD	HEAT DETECTOR
FSD	FIRE SMOKE DAMPER

Figure 13.12d Some electrical contractors specialize in security and fire systems; legends such as this provide them with some of the information they need.

items of work in most building construction projects, and performing a quantity takeoff of the lighting helps familiarize the reviewer of the drawings with the overall design of the building. At best, quantifying light fixtures is tedious—it amounts to methodically counting the number of the types of fixture called for, space by space. Within the same takeoff, the branch circuitry for the lighting—the raceways, wiring, fixture whips, switches junction boxes, and required hangers and connectors for each fixture—should be included. Whether branch circuitry for light fixtures is taken off separately or included in a takeoff of all the branch circuitry is influenced by the design. Every project has different

requirements; engineers may devote separate panels to lighting, power, emergency lighting, and other fixtures, which simplifies the takeoff process—the natural demarcation points for work are derived from the drawings. Site lighting is costly as well, but is taken off separately.

Service equipment and panelboard feeders are also costly items, and the information related to them is found on riser diagrams, which include all of the panels that the service equipment feeds, as well as the main switchboard and auxiliary power. Electrical wiring requires termination hardware of some sort: both ends of a wire require them. Termination hardware consists of clamps, clips, lugs, and similar connectors. Panelboards, which

TABLE 13.1 Sample Unit Designations for Materials and Work*—Electrical Systems

	LS lump sum	EA each	LF lineal foot	CLF Hundred lineal foot	SF square foot	SY Square yard	CF cubic foot	CY Cubic yard	Ga Gallon	Other	
Pipe			X								Underground
Trenching and backfill			X					X			
Fittings		X									
Conductors				X							
Access holes, boxes		X									
Duct banks			X								
Supports, anchors, guides		X									
Transformer pad		X									
Cutting, drilling, and patching	X	X	X		X						
Main switchboard		X									Building Systems
Motor Control Center		X									
Panelboards		X									
Feeders				X							
Conduit			X								
Circuit-protective devices		X									
Termination hardware		X									
Lighting		X									
Hangers, clips, supports		X									
Switches, receptacles		X									

*Metric drawings require different units: (liquids), kilograms (weight), lineal meters (length), cubic meters (volume), square meters (area).

contain the circuit-protective devices and bus bars, are costly, as are the fuses and circuit breakers that control each circuit.

Motor control centers, when they are included in projects, represent a significant cost, too. This equipment is used to start electric motors and control their operation, and they can include numerous modular sections that must be installed, connected to one another, and secured to their foundations.

Branch circuitry—the conduit and wiring used to power receptacles and lighting, is one of the least costly items of work from the material standpoint, although there is considerable labor involved in pulling all the wiring and making up the necessary connections. Likewise, special alarm, signal circuits, motorized dampers, and emergency equipment such as battery-powered lighting and signage and the associated conduit, wiring, connectors, hangers, and clamps are important parts to quantify.

A variety of factors affect the cost of electrical work: the size, weight, and degree of assembly of the components; location of and space around the work; and the technical complexity of the work are some of the important considerations. "Costing" electrical work is similar to the technique employed for mechanical work. It involves identifying the cost of work by correlating the weighted cost of composite crews (for instance, a journey-level electrician and two apprentices involved in a certain operation, say installing and making up light fixtures) to the amount of time the crew takes to do one unit (the productivity factor), multiplied by the number of units. Accurately identifying numbers of various parts of electrical systems and recognizing the factors that affect cost is therefore critical. It goes to the heart of the take-off process—accounting for details—and it emphasizes the importance of correlating what is shown on construction drawings to the pictures in the mind's eye of people

accomplishing work. The units that are commonly used in electrical work are listed in Table 13.1.

Summary

Considering the extent to which society depends on electricity, it should come as no surprise that electrical work in construction projects affects the work of numerous other specialty contractors. Electrical work appears in site plans—in the form of source and signal wiring (electrical power and fire system tamper switching) and branch wiring for site lighting—in numerous architectural drawings, including floor and roof plans, interior elevations and reflected ceiling plans, extensively in the mechanical sheets, in building fire system drawings, in the landscape plans, and of course in the electrical sheets. Not only does electrical work affect many other trades, the drawings, being largely schematic, leave more to the contractor's imagination than do other drawings. Hundreds of small components, from termination lugs to circuit breakers to installation clips to hangers to switches and receptacles—many of which are not shown at all in the drawings—are required in the work. For the architect coordinating design work, and the contractor coordinating subcontractors' work, this discipline demands a good imagination and attention to detail.

CHAPTER 13 EXERCISES

1. How are electrical systems similar to HVAC systems?
2. What are the key elements of an electrical system?
3. What is meant by the term *conduit*?
4. How are electrical drawings different from drawings produced for other trades?

14 Fabrication Drawings

Key Terms

Closeout submittals
Construction submittals
Preconstruction submittals
Shop drawings
Submittal process

Key Concepts

- Contracts for design and construction are essentially promises to the owner made by the design professional (for project design) and the contractor (for project construction) that the effort will conform to the buyer's expectations.
- The contractor promises to produce a building or other project according to the graphic and written descriptions provided by the design professional, to the specified quality, within the allotted time and budget.
- The protracted duration of many projects, the significant costs associated with their construction, and the difficulty of foreseeing the project's physical characteristics put the project buyer's intentions to a severe test. The construction process is therefore fraught with potential troubles.
- Solutions to the potential difficulties exist, and they take into consideration the expectations of the involved parties, and seek to educate and communicate effectively.

Objectives

- Identify the purposes of the submittal process.
- Explain the role that fabrication drawings play in the construction process.

Construction Graphics: A Practical Guide to Interpreting Working Drawings,
Third Edition. Keith Bisharat.
© 2025 John Wiley & Sons Inc. Published 2025 by John Wiley & Sons Inc.

Problems Inherent in the Design and Construction Processes

Recognizing the role that the expectations of participants in the construction process play is central to understanding potential areas of dispute. When an owner is asked to sign a construction contract, s/he is being asked to take a leap of faith, that after months of effort by hundreds of people and the expenditure of frequently millions of dollars (with what often seems like modest progress), the project for which the owner has contracted will become a reality, and that it will be what the owner had hoped. After all, an owner cannot drive to a nearby constructor's sales lot, examine a few buildings, and select a model to have made. The exception to this rule is of course production housing projects, where sales models are constructed specifically for that purpose.

For virtually every other kind of project, the buyer must resort to the information developed by the design professional (the written program, renderings or digitally enhanced photographs, computer-generated or scale models of the project, and the drawings and specifications) to help an owner decide what to build. Under the best of circumstances, this information has its limitations, as noted repeatedly in this text, and it can be costly to generate. Consequently, owners do not always avail themselves of all the information that may help them understand what they are purchasing, with the occasional result that the project does not equate to the image that for months or years may have resided in the owner's mind (despite the often exhaustive efforts on the part of the design team to identify the owner's needs and wants). As noted in early chapters, accurately imagining a huge object based mostly on a desk-sized two-dimensional depiction of it or even a scale model is expecting a great deal, particularly from people who do not have design and construction experience. Owners understandably become nervous under these circumstances. The submittal process is designed to inform and properly educate the owner (and the design professional as the owner's representative) as to how the parts of the project for which submittals are required will be integrated into the overall construction, and to document certain decisions.

Submittals Defined

Submittals are data collected by the contractor according to a list created on behalf of the owner by the design professional. They are a contractual requirement on virtually every project of any consequence. In the typical construction contract, the required submittals are listed in the general requirements section (Division 1) of the

IDENTIFYING EXPECTATIONS

The true value of the model home is that nothing is left to the imagination of the buyer—the intangible is made tangible. This approach has measurable benefits for buyer and builder alike. The U.S. legal system generally views home buyers as it does regular consumers—as unknowing, unsophisticated, and vulnerable—so various devices may be required to educate them. Disclosure statements pertaining to the real cost of financing and "hidden" home-ownership costs, such as improvement bonds, are examples. None of these devices is as effective as the model home, however—at least from the builder's standpoint. Envisioning an object of the size, complexity, and importance of a new home as it will be, based on a rendering in a colored brochure, is at best a leap of faith. Model homes clearly explain the environment in which a family will spend much of its time. In this sense, model homes serve as much to protect the builder as they do to promote sales.

But builders must be careful to avoid creating false hopes. Model homes often display the efforts of interior designers, who can actually make a silk purse out of a sow's ear. Consequently, caveats of all sorts abound in the model homes themselves: signs on furniture, countertops, walls with special textures, plumbing fixtures, floor coverings, and elsewhere. These signs are common, all admonishing the buyer that these items are upgrades from the standard model or are auxiliary items.

By contrast, commercial building owners, particularly developers and real estate investors, are considered to be sophisticated and knowledgeable. That is not to say, though, that transactions with these buyers are free from some of the same mysteries that befuddle homeowners, that is, that a project has somehow failed to meet their expectations, even though they may not be able to identify why. The equivalent to the model home is, of course, out of the question with commercial building projects. Efforts to understand whether a client is fully "on board," by using computer-generated models, scale models, mock-ups for critical aspects of the project, and similar educational devices, as well as communicating regularly during the course of the construction work, are well worthwhile for the design professional and builder alike.

specifications.[1] Administrative procedures involving submittals are set forth in the general and supplementary conditions of the contract (see Chapter 2 for an explanation of general and supplementary conditions).

The list of submittals varies for every project, as a consequence of owner and design professional preferences, legal requirements, the product requirements of the project, and the customs of the design profession. Most submittals can be listed under three categories:[2]

- Preconstruction submittals
- Construction submittals
- Project closeout submittals

Preconstruction submittals consist of data required prior to commencement of construction, namely contractor insurance certificates, payment and performance bonds, the contractor's product list (which the design professional uses to determine whether the contractor is complying with specifications requirements, and to compare proposed substitutions), and the progress schedule for the project.

Construction submittals include product data, product samples, mock-ups and fabrication drawings (which require the design professional's approval), and submittals that are filed for informational purposes only (which do not require the design professional's approval).

Product data includes manufacturers' illustrations, instructions for proper installation, finish selection, color charts, test reports, and any applicable limitations on product use. One of the key roles of product data is to provide a standard to which proposed substitutions can be compared.

Product samples are physical samples of products that are intended for use in the project. Tile, carpet, and brick samples are examples of products that are commonly required in the submittal process.

Mock-ups are usually full-scale portions of the project— a section of cladding, for example—constructed specifically for approval by the architect and for use in performance tests of building systems. For a skyscraper, for example, the architect may require the contractor to construct a portion of the cladding with windows, flashing, and pressure equalization system completely installed. The mock-up is then subjected to demanding air and water infiltration tests to determine its efficacy. Additionally, the esthetic characteristics of the cladding are more apparent when a very large sample is constructed. Product samples and mock-ups help make the intangible tangible, and they establish standards against which the final product can be judged.

Among the best known of the commonly required submittals are fabrication (shop) drawings, which derive their name from their origin as instructions given by fabricators to their production shops. The purpose of shop drawings is to inform the design professional as to how the contractor intends to fabricate or assemble certain portions of a project and how that particular system correlates to other systems. Among the common specialty areas for which shop drawings are produced are structural steel, reinforcing steel, cladding (architectural precast concrete and metal and glass systems), vertical transportation (elevators and escalators), and fire sprinkler systems, to name a few.

Informational submittals document the performance and quality of project components and serve to verify that the installed work meets the specified requirements. Design data, test reports, certificates of a product's quality, manufacturer's instructions, manufacturer's field reports, qualifications of certain material suppliers and subcontractors (used by the design professional to determine whether a fabricator, erector, or manufacturer is qualified to perform the work), and progress photographs are examples of informational submittals.

Project closeout submittals consist mostly of record set drawings, operation and maintenance (O & M) manuals, warranties, and documents related to occupancy and final payment. Record set drawings are commonly referred to as *as-builts*, which is a more descriptive title—they are a record of the project as it was actually built. Changes to the original drawings are inevitable in construction. A change of 1' or 2' in the elevation of a sewer system connection, for instance, may result in the contractor having to relocate the sewer line several feet in one direction or other. Such changes are recorded in the as-built drawings which, not surprisingly, are critical documents to facility maintenance personnel; in fact, owners are loath to release final payment without a complete collection of submittals.

In addition to establishing the standards by which the construction work will be evaluated, submittals manifest the contractor's interpretation of the project requirements, particularly the critical tasks such as structural steel detailing and cladding systems. Needless to say, discovering that the contractor had a different interpretation of project requirements after the work at issue has been performed puts both owner and contractor in untenable positions.

Submittal Process

As noted earlier, the submittal procedure is set forth in the contract documents, and it commonly requires the product supplier or manufacturer to review the submittal requirements, prepare the submittal accordingly, and submit it to

---

[1]MasterFormat, a widely used construction indexing system developed by the Construction Specifications Institute devotes Division 1 to general project requirements, in which (among other things) submittal procedures are described (Section 1330). Divisions 2-16 are devoted to the products and materials of the project. Each section is segregated into three parts: Part 1 is devoted to general information, which is where submittal requirements for that section are listed; Part 2 addresses product information; and Part 3 addresses execution (how the work is to be done).

[2]From the Construction Specifications Institute's *Manual of Practice.* 1996.

the buyer (usually a subcontractor or prime contractor). The buyer/contractor then determines whether the submittal complies with the contract requirements, that the dimensions and quantities are correct, and that all affected parties have the information they need. The submittal is then stamped by the contractor, who notes the action taken on the document and sends the submittal to the design professional. Under most contracts, the design professional is given a reasonable period of time to review the submittal based on its conformance with the design intent and the contract documents. (Defining a reasonable period of time is often an issue for both the architect and contractor). Submittals are either approved, approved as annotated by the design professional, returned to the contractor for revision and resubmission, or rejected (which means the submittal is incomplete or does not conform to the contract documents). In few cases may work proceed without an approved submittal. The project owner can take some comfort in the submittal process itself, since several sets of eyes evaluate a given product or system and jointly produce the necessary information, which should reduce errors and oversights, resulting in a more desirable outcome.

Shop Drawing Production

The contractual responsibility for producing shop drawings resides with the contractor (they are not part of the "instruments of service"—design documents—provided by the A/E under the design contract). The contractor frequently delegates the responsibility for producing shop drawings to specialty contractors, who have the expertise to produce the drawings to the exacting detail required.

Shop drawings differ from construction drawings in that they are usually produced by subcontractors for a more limited audience than standard construction drawings. The people who produce the shop drawings must have intimate knowledge of engineering requirements and manufacturing and construction field practices in their particular specialty. Although the contractor and design professional use shop drawings, the principal user group is the workers who are fabricating and assembling parts of the system in the shop and those who install the components in the field. Although all shop drawings are detailed, some are so detailed that they leave nothing to the imagination of the workers assembling the system. The form that shop drawings take on a given project varies according to the submittal requirements, the type of work being described, and the preferences of the subcontractor or material supplier responsible for the work. It should come as no surprise that there is considerable variety in shop drawings even in the same trade.

It goes without saying that shop drawings should exhibit all of the characteristics of good drawings, as described in Chapter 4, but there are several additional caveats:

Figure 14.1 Shop drawings for some specialties depart from convention in some respects. This drawing demonstrates how a beam might be described in structural steel drawings. In design drawings, all critical structural elements are drawn to scale. Take a column, for example: The critical connections are described in detail (base plate, column/beam connection, column/column connection), and the column itself, other than a short portion of it adjacent to each connection, is eliminated. Dual break lines signify to the reader that there is a section of column missing and that the written dimension is to be used. This saves considerable space on the drawings themselves, and nothing of value is lost in the technique. In some shop drawings, the element is drawn to scale, except for its length. Drawing by Tom Krejci after Keith Bisharat, produced from *Technical Drawing*, 12th ed., by Giesecke, Mitchell, Spenser & Hill, Pearson Education, Inc., 2003.)

- Shop drawings should follow the same general layout as the project drawings.
- Single-line schematic drawings may be effective in plan views or orientation drawings for certain systems—structural steel, for example, since they simplify the reader's understanding of the project.
- Every detail in a shop drawing set must correlate and be properly cross-referenced to the applicable source drawing (architectural or engineering) when it exists, to facilitate comparing the two.
- All dimensions should be listed—although scaled drawings are produced, shop drawings depend entirely on written dimensions. In structural drawings, for instance, beams are not commonly drawn to their actual lengths (see Figure 14.1). Other shop drawings use double break lines in the fashion common to construction drawings.
- Every separate element in a system should be clearly identified for use in shipping and erection or assembly.

What to Expect in the Drawings

Shop drawings can be extremely complex, particularly for reinforced concrete and structural steel, and by contrast with architectural drawings, for example, leave no doubt as to what is required. Every element of the system—nuts, bolt washers, screw spacers, struts, ties, bolt holes, the list goes on—must be accounted for and described. The American Institute of Steel Construction (AISC) and the American Concrete Institute produce guidelines to preparing shop drawings for, respectively, structural steel and reinforced concrete ("Detailing for Steel Construction," and the "Manual of Standard Practice for Detailing Reinforced Concrete Structures"). In other specialties, there is less information on drawing protocol, so drafters of shop drawing often develop their own methods of presenting information to the builder. The shop drawings in this chapter pertain to the glazing and curtain wall system for the project described in Chapter 8—a 70,000-square-foot two-story office building consisting of both "storefront" and curtain wall systems.

As to projections, line types, and symbols, shop drawings follow familiar graphic conventions, except perhaps in some specialty areas, where abbreviations and symbols are more esoteric. Fortunately for the user of the drawings, legends typically accompany shop drawings. Shop drawings are drawn to scale, as noted earlier; however, the drawings are not necessarily proportional, particularly where beam and column length are concerned. The double break line used in architectural and engineering design drawings, which alerts the reviewer that a significant portion of the element is not shown, is a convention used less frequently in shop drawings of structural elements (refer to Figure 14.1), although it is used in other shop drawings (glazing systems, for example).

Order of Information

This shop drawing set is formatted similarly to mechanical drawings, in that the first pages are devoted to project information (glass, finish, and hardware schedules, design criteria, graphic legends, sheet index, door handing, and general notes). Figures 14.2a–e comprise a collection of samples from the introductory pages of the drawings.

Subsequent pages show the various glazing systems in plan view, in elevation, and, finally, in details, which take advantage of plan, section, and elevation views. As with other types of drawings, the information goes from general to specific. Figures 14.3a–m illustrate the correlation between architectural drawings (the source documents for cladding and glazing) and shop drawings.

ABBREVIATIONS & SYMBOLS	
D.O.	= DOOR OPENING
F.O.W.	= FACE OF WINDOW (STOREFRONT / CURTAIN WALL)
F.O.S.	= FACE OF SLAB
F.S.	= FRAME SIZE
G.D.	= GLASS DIMENSION
L.O.	= DAYLITE OPENING
M.L.	= MULLION LENGTH
M.D.	= MODULE DIMENSION
M.O.	= MASONRY OPENING
P.D.	= PANEL DIMENSION
R.D.	= REFERENCE DIMENSION
R.O.	= ROUGH OPENING
S.P.	= SPANDREL PANEL
T	= TEMPERED GLASS
W.P.	= WORK POINT TO WORK POINT
(XXX)	= GLASS TYPE MARK
(XXXX)	= DOOR NUMBER MARK
◀	= DEADLOAD ANCHOR
⊕	= DEADLOAD ANCHOR
◁	= WINDLOAD ANCHOR
⊕	= WINDLOAD ANCHOR
◆	= LUG ANCHOR
⊕	= WORKPOINT REFERENCE
⫰	= PERIMETER JAMB REFERENCE
⌐	= CORNER REFERENCE

Figure 14.2a The abbreviations used in these drawings are largely esoteric, which is why a legend is so useful. (Drawing courtesy of Entelechy.)

GENERAL NOTES

1. ALL ROUGH OPENING DIMENSIONS REQUIRED FOR FABRICATION OF WINDOW MATERIALS SHALL BE GUARANTEED AND RELEASED BY THE GENERAL CONTRACTOR PRIOR TO FABRICATION.

2. ALL ELEVATIONS ARE VEIWED FROM THE OUTSIDE UNLESS OTHERWISE INDICATED

3. ALUMINUM, STEEL, AND CONCRETE SURFACES TO BE SEPARATED FROM CONTACT WITH VINYL TAPE OR PLASTIC SHIMS.

4. FABRICATION, ASSEMBLY AND INSTALLATION OF FRAMES AS PER PUBLISHED MANUFACTURER'S RECOMMENDATIONS.

5. COORDINATE LOCATIONS REQUIRED BY CODE TO HAVE SAFETY GLAZING

SHEET INDEX

SHEET	DISCRIPTION
101	COVER
102 – 103	DOOR SCHEDULE
201 – 202	FLOOR PLANS
301 – 310	BUILDING ELEVATIONS
401 – 405	451 CENTER GLAZE DETAILS
501 – 502	DG & FF FRONT LOAD DETAILS
601 – 613	3250 1" C.W. DETAILS
701 – 703	3250 1/4" C.W. DETAILS
38	TOTAL

Figure 14.2b General notes are common on a multitude of drawings. Here the notes consist of admonitions to the builder. (Drawing courtesy of Entelechy.)

GLASS SCHEDULE

(G1) 1" VISION GLASS OF:
1/4" TINTED ANNEALED
1/2" AIR
1/4" CLEAR ANNEALED

(G1T) 1" VISION GLASS OF:
1/4" TINTED TEMPERED
1/2" AIR
1/4" CLEAR TEMPERED

(G1T1) 1/4" VISION GLASS OF:
1/4" TINTED TEMPERED

(G2S) 1/4" SPANDREL GLASS OF:
1/4" TINTED (BLACK-OUT)

(G2S1) 1/4" SPANDREL GLASS OF:
1/4" TINTED TEMPERED (BLACK-OUT)

(G3) 1" VISION GLASS OF:
1/4" CLEAR LOW 'E' ANNEALED
1/2" AIR
1/4" CLEAR ANNEALED

(G3T) 1" VISION GLASS OF:
1/4" CLEAR LOW 'E' TEMPERED
1/2" AIR
1/4" CLEAR TEMPERED

(G3T1) 1/4" VISION GLASS OF:
1/4" CLEAR LOW 'E' TEMPERED

(G4S) 1/4" SPANDREL GLASS OF:
1/4" TINTED (BLACK-OUT)

(G5P) 1/4" SPANDREL PANEL OF:
1/4" ALUM. PANEL TO MATCH FINISH

Figure 14.2c Look at the variety in glass on this very simple project. (Drawing courtesy of Entelechy.)

Figure 14.2d Door handing–labeling the swing of interior and exterior doors–is inconsistent in the industry, and it demands that the builder understand the shop drawing producer's definitions of "swing." In the example shop drawings, this legend exists on both pages that display door information. (Drawing courtesy of Entelechy.)

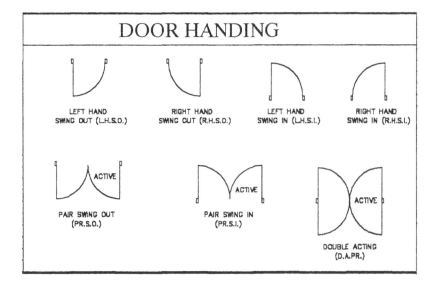

DOOR HANDING

LEFT HAND SWING OUT (L.H.S.O.)   RIGHT HAND SWING OUT (R.H.S.O.)   LEFT HAND SWING IN (L.H.S.I.)   RIGHT HAND SWING IN (R.H.S.I.)

PAIR SWING OUT (PR.S.O.)   PAIR SWING IN (PR.S.I.)   DOUBLE ACTING (D.A.PR.)

DOOR HARDWARE SCHEDULE

DOOR#	ELEV.	SWING	SIZE	HARDWARE GROUP #
105	S2/304	R.H.S.O.	3'–0" X 8'–0"	N/A
106	S4/305	L.H.S.O.	3'–0" X 8'–0"	N/A

MEDIUM DOOR w/ 2" X 4 1/2" FRAMING

QTY.	DISCRIPTION	MANUFACTURER / PART NUMBER		FINISH
1 EA	TOP PIVOT	DORMA	75120	***
1 EA	INTERM. PIVOT	DORMA	75220	***
1 EA	BOTTOM PIVOT	DORMA	75232	***
1 EA	FLOOR CLOSER	DORMA	BTS–80	***
1 EA	PANIC DEVICE	VON DUPRIN	33 X 334	***
1 EA	CYL. DOGGING	VON DUPRIN	***	***
1 EA	CYLINDER	SCHLAGE	***	***
1 EA	THRESHOLD	PEMKO	176A	***
1 EA	DOOR SWEEP	PEMKO	315CN	***
1 EA	RAIN DRIP	PEMKO	346C	***
1 EA	9" OFFSET WIRE PULLS	US ALUMINUM	***	***

Figure 14.2e The parts list for one door is listed in this schedule—one of several in the shop drawing set. Adjacent to the schedules is written, in very large letters; "Verify Door Hardware Schedule." As noted in the text, verifying dimensions, quantities of parts, and compliance with contract submittal requirements is the responsibility of the contractor. (Drawing courtesy of Entelechy.)

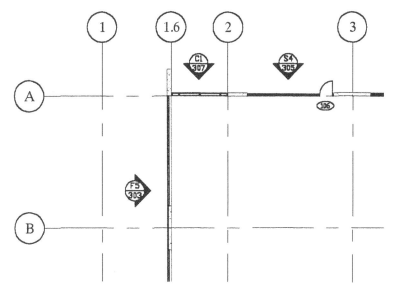

Figure 14.3a This portion of floor plan shows a curtain wall system (C1-307) and two storefronts (F5-303 and S4-305) in the northwest corner of the building. The shop drawing page number, 201, is cross-referenced to the source drawing, page A2.1 of the architectural drawings (see Figure 14.3b). (Drawing courtesy of Entelechy.)

Figure 14.3b Compare this architectural floor plan to the shop drawing plan in Figure 14.3a: You will notice immediately that the shop drawing is used for orientation purposes only. (Drawing courtesy of Comstock Johnson Architects.)

NORTH

Figure 14.3c This is a partial elevation, taken from page A3.1 of the architectural set. The curtain wall is located between gridlines 1.6 and 2, and the storefront between gridlines 2 and 3. (Drawing courtesy of Comstock Johnson Architects.)

Figure 14.3d The framework for the curtain wall is clearly shown in this photograph of the northwest corner of the building. Notice how the aluminum frame sits outside the second-floor beam (center of the photograph). Visible in the background to the right are the structural frame for the curtain wall at the building entrance and a portion of the exterior wall of the building behind that. (Photo by the author.)

Figure 14.3e The outcome first described in Figure 14.3c is visible in this photograph of the same portion of the building. (Photo by the author, with the permission of the Sacramento County Office of Education.)

Figure 14.3f In somewhat greater detail, the curtain wall and storefront for that portion of the building shown in Figure 14.3e is shown here. Notice the seam in the break metal in this picture, referred to in Figure 14.3f as 9-603. (Photo by the author, with the permission of the Sacramento County Office of Education.)

ELEVATION 'C1'

SERIES 3250 (2 1/2' X 7') PRESSURE PLATE SYSTEM
(1) REQUIRED THUS

REF. A3.1

Figure 14.3g Some 19 different details apply to this portion of the curtain wall system alone; 17 are visible here. The other two are shown in Figure 14.3h, which is a section through this wall, viewed from the right side. The different details are required because the specific conditions differ. For example, starting from the bottom of the wall, the glazing changes from a single ¼" pane of spandrel glass to 1" vision glass (two-¼" panes of clear glass with a half-inch spacer between them), then (at 119") back to a single spandrel glass panel, back to vision glass (at 211 ½"), then back to a single pane spandrel panel. The horizontal aluminum elements between these points of transition must accommodate the changes in glass thickness. Notice how this elevation is cross-referenced to the source drawing, in this case, page A3.1 (part of which is shown in Figure 14.3c). (Drawing courtesy of Entelechy.)

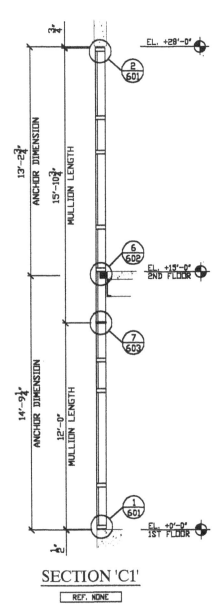

SECTION 'C1'

REF. NONE

Figure 14.3h The section through the curtain wall shows two conditions that the shop drawing producer could not effectively show in the elevation: one was a splice between mullions (7-603), and the other was the windload anchor (6-602), which is designated by a triangle symbol at the approximate midpoint of the curtain wall in Figure 14.3f. The windload anchor is the connection between the curtain wall and the second floor. (Drawing courtesy of Entelechy.)

Figure 14.3i Details 1 and 2 are, respectively, the sill and head details of the curtain wall system. The glazing conditions are identical; in fact, the parts that are called for are the same. Can you identify the one additional part required in the sill assembly? (Drawing courtesy of Entelechy.)

Figure 14.3j Details 4 and 5 show the required details at transitions from spandrel to vision glass. Details 3 and 5 are similar; however, in Detail 3, the panes of glass are reversed. Other details in the curtain wall show still different conditions. (Drawing courtesy of Entelechy.)

Figure 14.3k This drawing describes the connection between the mullions in the curtain wall and the second-floor deck. Notice how this connection accounts for thermal expansion. (Drawing courtesy of Entelechy.)

Figure 14.31 Detail 7 is an elevation of the mullion splice at 12'. One of the critical decisions the detailer must make is where to make joints in a system. Knowing the lengths in which the material is manufactured is critical. (Drawing courtesy of Entelechy.)

Figure 14.3m This drawing is a section through the double mullion within the spandrel panel (refer to the elevation in 14.3f, center of the drawing just under the windload anchors). A different condition in the same mullion is called for where vision glass is used on both sides. Detail 13-604 addresses this condition. (Drawing courtesy of Entelechy.)

Subtleties in the Drawings

As noted a number of times in this text, there are subtleties in drawings that can affect the builder in a variety of significant ways. This applies to virtually every trade, so it behooves the designer particularly to thoroughly investigate the more critical building systems, such as structural systems and cladding and glazing, perhaps in consultation with specialty contractors. There are many different curtain wall and storefront systems, for example, each with advantages and disadvantages, and consulting with specialists (when possible) is a prudent use of the owner's fee. Of particular interest to glazing contractors is the procedure required to install the glazing in the window frame. Some systems are installed from the front (outside the building) and others may be installed from within the building, depending on the design and the manufacturer. The number of steps required to install the glazing is an issue as well—a "three-way tuck" for example, where glazing is installed in the deep jamb, slid into the shallow jamb, and raised up on setting blocks before the sill section of frame can be installed in the sill—takes more time than a two-way tuck—wherein glazing is set in a deep jamb, rotated into a deep sill (first tuck), slid into the shallow jamb (second tuck), and the snap bead installed in the head. The definitions for various frame and glazing systems vary; for some, *front loaded glazing* means a system that locates the glass in the frame close to the outside of the building; *center-loaded glazing* is located in the center of the frame; and *back-loaded glazing* sits close to the inside face of the frame, all of which can affect how the windows are installed. In the project that is the subject of these shop drawings, the architect's decision to use the exterior wall panels to hide rooftop mechanical equipment, and the desire to have the spandrel and vision glass flush with the face of the panel in certain locations, affected the construction of the exterior wall panels and the glazing system (see discussion in Chapter 8 and Figures 8.10a–c). Notice in Figures 14.4a–e how the two front-loaded glazing conditions differ as to installation methods, although they are side by side in the example building.

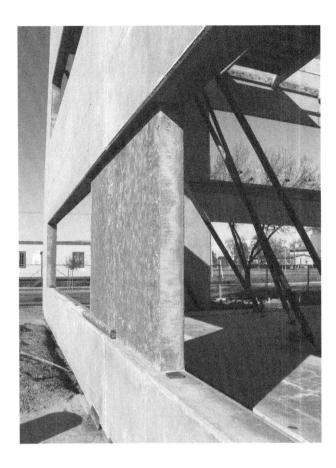

Figure 14.4a This photograph shows the condition that the details in Figures 14.4b-d address. (Photo by the author.)

Figure 14.4b This is a plan view of the glazing system where the spandrel panel in the exterior panel and the vision glass connect. (Drawing courtesy of Entelechy.)

Figure 14.4c This section through the sill, intermediate horizontal frame, and head of the window system on parts of the west, south, and east walls shows a front-loaded glass with the installation performed from the inside (right side) of the building. Part M573 is a snap bead, which is installed after the glazing has been put in place. (Drawing courtesy of Entelechy.)

Figure 14.4d A section through the adjacent spandrel panel reveals that the glazing will be installed from outside (the left side) of the building. There is, of course, no other way to install it. (Drawing courtesy of Entelechy.)

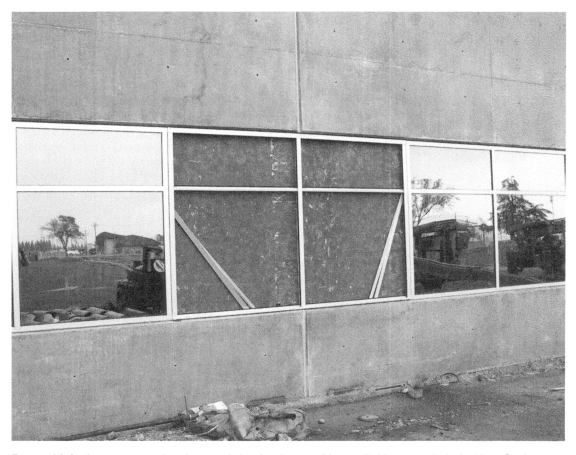

Figure 14.4e It is apparent in this photograph that the glazing will be installed from outside the building. On the upper floor, the installation will require some kind of motorized equipment or scaffolding. (Photo by the author.)

This condition applies to the spandrel panels and vision glass on parts of the west, south, and east sides of the building. The alert reader will notice that in this situation there is only one way to install the glazing in the spandrel panels (from the outside). But such situations are instructive nevertheless: design decisions affect the work, and the builder ought to fully evaluate what is entailed in the work. The glazier on this project will have to rent equipment or scaffolding to install the spandrel panel glass above the ground floor, which is not true of the vision glass adjacent to it.

Summary

Shop drawing production is a critical aspect of the planning invested in the construction effort, and as a part of the submittal process, plays an equally important role in notifying the owner as to the intentions of the builder where critical, detailed systems are concerned. It is in shop drawings that the minute details of the work are clarified and choreographed for an effective installation. The sequence of construction, the specific number and variety of elements, the quality of each, and how they connect with one another and interface with adjacent systems are all memorialized in shop drawings. Without them, too many questions would arise in the course of construction, when options as to solutions are more limited.

CHAPTER 14 EXERCISES

1. Name several purposes of submittals.
2. At which stages of the project are submittals required?
3. How are fabrication, or shop drawings, different from all other construction drawings?
4. How are submittal requirements made known?

15

Project Analysis and Application of Costing Principles

Key Terms

Costing
Geotechnical report
Groundwater
Heat-welded waterstop
Protection board
Permanent slope
Sump
Temporary, or working slope
Temporary structures
Vapor Barrier
Vapor retarder
Waterproof membrane
Waterstop
Water table

Key Concepts

- Problem Identification
- Project Analysis
- Document Review
- Organizing and Planning Work

Objectives

- Apply the concepts introduced and discussed in prior chapters.
- Demonstrate the importance of fully understanding the instructions provided by design professionals and identifying the sources of those instructions.
- Develop an organized program for executing the work depicted in project drawings and specifications.

Construction Graphics: A Practical Guide to Interpreting Working Drawings,
Third Edition. Keith Bisharat.
© 2025 John Wiley & Sons Inc. Published 2025 by John Wiley & Sons Inc.

▨ Introduction

This chapter focuses on the application of the principles of construction document analysis, interpretation, and planning for construction work in the context of a specific project, namely the elevator pit of a two-story education building. While it is seemingly a simple undertaking—what does it take to dig a mostly square hole in the ground with four walls around it, after all? It offers the opportunity to highlight some common challenges in managing and executing construction work.

▨ What Am I Being Asked to Do?

Clearly, the most important question to answer when preparing for construction work is "What am I being asked to do?" What is the nature of the problem? This simple question normally has numerous answers, not surprisingly because of the complexity of even relatively simple construction projects. Once that question has been answered, however, the analysis can follow.

▨ Responsibilities of the Design Professionals

The construction of the elevator pit involves the input of several design professionals, including the geotechnical engineer, the architect, the structural engineer, the mechanical engineer, and the representatives of the building department that must inspect the project. The separation of professional responsibilities is clear in the pit construction project, most notably in the sections of the elevator pit by the architect and structural engineer. Notice that the configuration of the concrete as depicted by the architect is substantially different from that of the structural engineer—it's clear the architect is employing a schematic drawing. The architect is responsible for the drawings, specifications, and other contract documents, but frequently contracts with design consultants for aspects of the work. In the elevator pit, the structural engineer designs the structure, ignoring all other issues, and the architect focuses on nonstructural issues such as rejecting the elements (pit waterproofing, protection of the waterproofing, patching snap-tie indentations, and moisture retardation in the main slab-on-grade). The geotechnical engineer provides the users of its report with information on the soil characteristics of the site, for example its capacity to bear various load types, whether the soil is suitable for backfill, the existence of groundwater, and many other things. The mechanical engineer is responsible for designing heating, ventilation, and air conditioning (HVAC) systems (compressible fluids) as well as plumbing systems (generally noncompressible fluids). In this case the mechanical engineer, in addition to selecting the sump pump, also identified the requirements for the cover of the sump. The representatives of the building department, and this differs from jurisdiction to jurisdiction, confirm that the builder has complied with the design documents and building code requirements. Inspectors, normally very knowledgeable, play a critical role in assuring compliance with applicable regulations. For superintendents and other field personnel, inspectors represent another pair of eyes.

▨ Exploring the Problem

In any investigation of drawings and specifications, including the construction documents generally, it is useful to first determine the nature of a problem, followed by an identification of the professional(s) responsible for its solution, and therefore the location(s) of the information. This is where an understanding of the professional roles of the participants is especially helpful.

As noted repeatedly elsewhere in this text, the answers to problems come from at least two, if not many more, sources: the drawings and specifications. One common way to organize drawings is provided by the National Cad Standard (see Chapter 3). Architects may employ aspects of the Standard, but often enough use an individualized approach. If the design professional employs CSI's MasterFormat for the specifications, the pertinent administrative information is contained in Division 1, and in Part 1 of each specification section. Recall that drawings depict the size, location, and interrelationships of systems and components graphically, while specifications are the written guidelines to primarily the quality of work and materials. Some things are much easier to draw than write about, and others are certainly impractical to draw (the characteristics of drywall finishes for example—how would one depict an "orange peel" or "imperfect smooth" finish graphically, even at full scale?).

The following table describes briefly what the key responsibilities of the common participants in a construction project are:

Professional Responsibilities of the Key Participants in Construction Projects*

Owner	Presents its needs, contracts for design and construction services, provides project funding and an owner's representative for the project.
Civil Engineer	Prime design professional on civil engineering projects (e.g., roads, bridges, waste-water treatment plants); provides demolition plans (see Figure 6.1), site and grading plans for building projects, and underground utility drawings.
Geotechnical Engineer	Responsible for the site investigation report for projects of all types, evaluates the characteristics of a site, and provides the guidelines to be followed by the designers and builders in the design of a wide variety of projects. Also typically provides observation and testing during the execution of site work.
Architect	Prime design professional on buildings, including homes, commercial buildings, schools, hospitals, hotels, and some process facilities. Responsible for the design aspects of a project, focuses on the aesthetics of a project (size, shape, interrelationship of spaces), its resistance to the penetration by the elements, and acts as the owner's representative for the project overall.
Landscape Architect	Designs and describes planting, irrigation, "hardscape" (nonliving elements of a landscape design, e.g., improvements such as patios, trellises, walkways, fountains, gabions).
Structural Engineer	Designs the structure for a project, typically as a design consultant
Mechanical Engineer	Designs the plumbing and HVAC systems for buildings as a consultant, may be the prime design professional for projects such as oil refineries and other process facilities.
Electrical Engineer	Designs the electrical systems for a variety of projects as a design consultant.
Specialty Design Consultants	Act as other design consultants in their area of specialty, e.g., acoustical engineering.
Construction Professional	Translates the instructions of the design professionals responsible for project design into a completed project. A general contractor, much like the architect, takes overall responsibility for construction of the project and contracts with specialty contractors for aspects of the work.
Agency representatives	Includes inspectors representing federal, state, county, and municipal building departments, as well as agencies such as the Department of Fish and Game

*The construction industry is complicated, and the variety of professional relationships is essentially unlimited. Generally, the nature of the project determines who the prime design professional is, so in an oil refinery for example, that might be the mechanical engineering firm. For a project such as a winery, there might be more than one prime design professional for various aspects of the project (an architect for the visitor and tasting room facilities, and a mechanical engineer for the process portion of the project). Further, design-build firms have long been employed for such projects as refineries, computer server farms, and other complex projects. Irrespective of the various roles design and construction professionals may play, the actual work of each professional does not change significantly.

▪ **What to Expect in the Work**

An initial review of the elevator pit drawings (see Figures 15.1–15.6) will reveal that constructing this pit will involve layout (where do things go?), excavation, installation of sub-grade and sub-base material, waterproofing (as distinguished from water resistance media such as slab-on-grade vapor retarders), reinforcing steel, temporary structures (in this case formwork), concrete placement, form stripping, pipe-column installation and grouting, miscellaneous metals (pit ladder, sump grate, embedments in the main slab-on-grade), and clean up.

Building an elevator pit with a concrete floor and four nearly identical walls seems simple enough, but the earthwork alone requires careful consideration. Is the work easily accessed? What is the nature of the material to be excavated? Might it be used on the site as backfill? If not, what is the disposition of the exported material? What is the magnitude of the work? Will special precautions have to be taken, dewatering for example? The answers to most of these questions are normally found in

the site investigation report, often referred to as the geotechnical report, which identifies the characteristics of the earth material on a site as well as the presence and elevation of groundwater and the bearing capacity of the material, among other things. Once the nature of the soil has become familiar, the excavation planning can occur.

Identifying the pit location, depth, and the elements of the system, and where spoil might be temporarily stored, may come from the drawings, but the quality of each element or component of the system as well as the quality of the work are described in the specifications.

Regarding the balance of the work, each operation must be identified, analyzed, and planned in a way that is useful to the builder. Often enough this means organizing the operations into a common and useful sequence and evaluating each operation carefully. Any required exploratory work—what the elevator company might require for example—should be undertaken in the initial review.

The drawings relevant to the example elevator pit follow.

Figure 15.1 This drawing serves to orient the user to the general location of the elevator pit within the building footprint.

ENLARGED PLAN STAIR #1 - FIRST FLOOR

1/4" = 1'-0"

Figure 15.2 Detail 1 on this drawing sheet magnifies the first floor plan of the elevator pit, making its details more apparent.

⑤ STAIR SECTION

1/4" = 1'-0"

Figure 15.3 Detail 5, entitled Stair Section, appears on the same sheet as Detail 1 (the blown-up floor plan), and it provides the ladder and sump locations, in addition to the elevator manufacturer's requirement for a hydraulic ram.

Figure 15.4 Detail H1.2 (1-1) appears on a full sheet of details provided by the structural engineer. Note that issues such as water penetrating the structure are not addressed by the structural engineer. Such concerns are the responsibility of the architect.

ELEVATOR SHAFT WALL SEE
DETAIL C6A

CONCRETE SLAB

LAP WATERPROOFING
OVER VAPOR BARRIER

4" GRAVEL

CONCRETE PIT WALL,
SEE STRUCTURAL

C.P.E. WATERPROOFING
MEMBRANE

PROTECTION BOARD

BACKFILL

FULLY ADHERED LAP

12" MIN LAP

RFI/#PI# O13
Date 12-11-01

WATERSTOP

PROTECTION BOARD

C.P.E. WATERPROOFING
MEMBRANE

2" SAND

4" GRAVEL

TELEPHONE 916 455-1234
FACSIMILE 916 453-1256

DREYFUSS & BLACKFORD   ARCHITECTS

3540 FOLSOM BOULEVARD
SACRAMENTO, CALIFORNIA
95816

**ELEVATOR PIT
WATERPROOFING**
1"= '-0"

REGIONAL & CONTINUING EDUCATION BUILDING
CALIFORNIA STATE UNIVERSITY SACRAMENTO

C9

14 Aug 2001
A0623.00

Figure 15.5 It is quite apparent here that the rendition of the elevator pit manifests the architect's role as the professional concerned about penetration by natural phenomena, NOT structural matters... it is clearly a schematic focused on penetration of the system by the elements. This is a classic example of the clear division of responsibility design professionals share with their consultants.

FROM ABC

TO 55 GALLON DRUM

1"

REMOVABLE LID

CAPTURE TANK RYAN HERCO
MODEL 7174 30"X24"X18"
POLYETHYLENE WITH COVER
& 2" INLET. 55 GAL CAPACITY

U.L. PIPE SEAL "STI" UL-1049 SPEC SEAL SERIES
100 SEALANT. SEAL PER THE LISTING (TYP @ EACH
PENETRATION THRU WALL, 2 PLACES)

WALL ADJACENT TO CAPTURE TANK
SEE FLOOR PLAN

RFI/FI# 64
Date 5-15-02

WALL ADJACENT TO ELEVATOR

FINISHED
FLOOR

1" COPPER TYPE L

18"x18x18" MIN DEEP SUMP

TO
ABC

1"

BOTTOM OF ELEVATOR PIT

EXPANDED METAL REMOVABLE COVER
ABLE TO WITHSTAND FOOT TRAFFIC.

CHECK VALVE

TO 120 V RECEPTACLE

SUMP PUMP "GRUNDFOS" MODEL
SU25, 20 GPM @ 15FT. HEAD 1/4 H.P.,
115v-1∅, FLOAT SWITCH.

FLOAT SWITCH

(A) **SUMP PUMP DETAIL**
NO SCALE

UND TANK
OUT AREA

Figure 15.6 Detail A, a plumbing section, offers information designed primarily for the plumbing contractor, however, the critical information for someone building the pit, namely the size of the sump, is offered here.

Work Analysis

The elevator pit, as noted above, requires temporary structures, in this case concrete formwork. Formwork varies widely, from custom fabricated, complex steel forms to simple boards or even mounded or excavated earth (a clean-cut trench, for example). Thrust blocks for pressurized underground piping (e.g., domestic water and fire protection systems), which minimizes the movement of pipes where they intersect or turn, requires no formwork at all, the clumps of concrete simply act as ballast.

The example elevator pit is a mostly square pit with six-inch thick reinforced concrete walls. At first glance, one might think "OK, I'll excavate the pit with vertical walls (the excavation of the pit would be through engineered fill prepared by the earthwork contractor at the time the rough grading is done), form the inside face of the wall, and cast the concrete against the earth and one form face." Fine, but the architectural detail for the pit requires waterproofing on the earthen side of the pit, as well as at the cold joint between the slabs and pit wall, not to mention that forming the project this way would require penetrating the waterproof membrane (Figure 15.7).

Preliminary Document Review

Layout
Most operations commence with layout, which may require numerous iterations, depending on the nature of the work. The layout of the elevator pit depends in no small part on the nature of the soil. Clearly, an excavation in loose sand, or cobble, is a very different proposition from one performed in fine grained, cohesive soil, such as exists at the Napa Hall elevator pit. The site preparation preceding the excavation of the elevator pit, which calls for three feet of the existing material to be removed and compacted (ppg. 5, 8, geotechnical report*), provides a noteworthy benefit, namely that the material to be excavated is engineered fill and therefore suitable for vertical or near-vertical temporary slopes. This means that a minimum amount of material will have to be removed and either stockpiled or exported. The next question becomes how much room around the pit do the skilled trades need to execute their work efficiently, because it is clear that the concrete cannot be placed against the earthen sides of the pit. What about access to the pit, how do workers get in and out of the fully excavated pit while formwork and steel are being installed? Worker access should always be considered in the analysis of construction operations.

(*The geotechnical report identifies a "minimum of 2 feet of Engineered Fill" for shallow foundations; this potential discrepancy should be clarified by the geotechnical engineer.)

Likely source(s) of information: site plans (civil and architectural); first floor architectural plan; first floor structural plan; underground utility drawings; site survey including building corners; Division 1 of the specifications (for administrative issues); Specifications Section 02, geotechnical report.

Note: This project was designed prior to the Construction Specifications Institute's creation of its 49 Division format. Currently, earthwork information resides in Division 31.

Excavation
Once the slope of the excavation has been decided, an analysis must focus on access to and from the excavation and the type of equipment used, a decision which in turn revolves around equipment availability (not so much where a very small project like this elevator pit is concerned, but certainly on large excavation projects such as the one depicted in Figure 6.8b), productivity, and the disposition of the spoil. Can the excavated material be used for backfill? If yes, where is the best place to store it until the backfill operation commences? If no, it would have to be exported, which changes the nature of the operation. Another question to ask and answer pertains to the existence of groundwater, which can dramatically affect an excavation if any part of it falls into or below the water table. Dewatering might be required, and that can add dramatically to costs and the schedule. The geotechnical report makes clear that even in wet seasons the water table doesn't rise to an elevation that would affect this excavation. The geotechnical report also allows for Class II aggregate base (AB) as subgrade material. Why is this important? AB provides a very stable, relatively easily shaped, very firm, smooth, and durable base layer on which the waterproof membrane and cushion board may be placed. This is a benefit: the ironworkers, unless it is decided that the entire steel cage should be prefabricated, will spend hours on that protective board tying the steel in the slab-on-grade and walls, and a stable subgrade material will lead to a better final product.

Likely source(s) of information: geotechnical report; site plan, first floor plan (architectural and structural), project location map (to identify access to and from project).

Waterproofing
It is critical to understand the difference between moisture retarders, like the fabric placed under slabs-on-grade, and moisture barriers such as the one separating the concrete of a substructure from the earth. There are various products used to provide waterproofing for sub-structures, including sprayed-on materials, sheet goods, and rubber or plastic gaskets called waterstops, or bentonite extruded over cold joints in the concrete, among others. The elevator pit drawings show both a waterstop between the foundation and the walls of the elevator pit (interestingly, the waterstop is not listed in the specifications*) and a chlorinated polyethylene (C.P.E) membrane on the earth side of the walls and footings (see Detail C9, Figure 15.5). The installation of the C.P.E. membrane, given the nature of the foundation, is one of the more challenging aspects of the elevator pit construction. Why? Because the drawings call for the footings to be completely wrapped in the C.P.E. membrane. A quick review of the elevator pit plan reveals two spread footings 3′ × 3′ × 1.5′. . . think about how those footings, with a wall on top of them, which also must be waterproofed, might be wrapped in CPE fabric (see Figure 15.8). It is clearly an involved undertaking, considering the number of seams that would have to be created and properly sealed. Detail C9 calls for a protection board "as recommended by the membrane manufacturer," according to specifications section 07 11 80. This stipulation leads to an important point: *in addition to the drawings and specifications, builders must often explore many other documents, manufacturers' instructions* being just one example.

Additional to the difficulties associated with wrapping the footings in CPE membrane is the challenge of avoiding puncturing the membrane when supporting and bracing the formwork and reinforcing steel.

*This oversight provides an opportunity. . . the builder might suggest to the architect a manufacturer and product with which the builder has had experience.

Likely source(s) of Information: Detail C9; Division 07 of the specifications

Formwork	Considerations in selecting formwork include the size and configuration of the component being formed, access to the component being formed, availability of the formwork system, ease of assembly and disassembly (remember, forms are normally temporary structures), the number of times the forms can be used, finish requirements, and of course cost. The elevator pit is a simple component, with just four 6″ thick walls and a slab-on-grade, rectilinear in shape, and only a few feet tall. It's worth pointing out, however, that there are a number of choices one might make for the pit formwork. How would one select one system over another? Would one first form the slab using edge forms, and then place wall forms on the slab (which would require the slab-on-grade to increase in size by at least half the width of the form), or is there a way to configure a single set of forms that would contain the concrete in the footings and walls? For wall concrete, one available system consists of plywood studs, dual wales, and the hardware to support them; another consists of single wales and the hardware to support them, with secondary vertical members placed periodically against the wales (see Figure 7.13). The choice here is simple; which system is appropriate to the task but involves less work? Certainly, a single wale system is more easily assembled than a dual wale system, not to mention less material. Another issue to resolve is the formwork for the sump. The reinforcing steel around the sump has to be installed per drawings. . . how will it affect the footing? The engineer depicts the sump residing outside of the footing, as shown in the plan of the pit; how does that affect the formwork and the reinforcing steel? (Figure 15.12) As noted above, installing and bracing the formwork and keyways without puncturing the CPE membrane creates special challenges.

Likely source(s) of information: structural drawings, section H-1.2 (1.1); Specifications Section 03 10 00; a company's historical data and preferences for formwork; various internet sites for formwork systems |
| Reinforcing Steel | Reinforcing steel on large commercial projects or bridges can be complicated, necessitating an involved plan for installation, but in this case the installation of the reinforcing steel is straightforward and very simple. Just two bar sizes are required, both of the quality described in Division 03 20 0 of the specifications, and the shape into which the bars will be bent is very simple (U-shaped bars with "ears," continuous from one side to the other in both directions -- see Detail H1.2, Figure 15.4).

Likely source(s) of information: structural drawings, section H-1.2; Specifications Section 03 20 00 |
| Concrete | The concrete mix is determined by the structural engineer and is described in Part II (Products) of Division 03 30 00 of the specifications. One of the challenges concrete subcontractors face is placing concrete, and in a set of forms with walls 6″ apart and about three-and-a-half feet tall, as in this case, the flowability of the concrete is essential. Adding water to concrete makes it flow more easily; but it also weakens the concrete and is normally precluded by the engineer. One can add chemical plasticizers to address this issue, but of course that drives the cost of the concrete up. There are basically two ways to consolidate concrete, vibrate the material internally, or vibrate the forms themselves. A small immersion vibrator is a good choice here, in fact the specifications require it; vibrating concrete externally requires special forms, and normally external vibration works best with elements such as precast concrete cladding and prestressed-precast concrete structural members, and it is more expensive. Another challenge with concreting is the conveyance system used; the "target" is the pit slab-on-grade in the initial place and finish operation, followed by the tops of the formed walls, then the main slab-on-grade. A concrete pump truck, which depending on its size, can convey concrete accurately and quickly from a considerable distance, is a good choice; it saves the concrete finishers the effort involved in using a line pump. Other methods, for example "spooning" the concrete into the pit walls (placing it manually with shovels), which is approved in Part III of the specification section, would be tiring, very time consuming, and would no doubt contribute to unnecessary waste.

Likely source(s) of information: structural drawings, section H-1.2 (1.1); Specifications Section 03 30 00 |
| Column installation | The installation of the column called for in the pits, the elevator cab guide, is very simple: all it requires is layout, column placement, and plumbing up (being made vertical), and the installation of two Hilti KwikBolts (dry anchors) through the base plate into the concrete. Once the column has been placed, the machine bolt called for in the connection at the floor level may be installed (See Figures 15.4 and 15.8). Following the column placement and anchoring, the base plate can be packed with non-shrink grout, normally a manual task.

Likely source(s) of information: structural drawings, section H-1.2 (1.1) |
| Other: Plumbing, Metal Fabrication, Final Clean-Up | The plumbing subcontractor has the responsibility of placing the sump pump and associated piping for it. As for the access ladder and the sump grate and angle iron supporting it, they're both simple operations. . . where connections are required, the installer would lay out the location of the anchors and use a hammer drill and dry anchors to connect the ladder where required. Similarly, the angle iron supporting the grate would be anchored to the wall of the sump with dry anchors. The requirements for the ladder are described in Division 05 Metals.

Most operations include clean-up, which for some, painting for example, can be involved. A thorough project manager will include clean-up in the work scopes of the subcontractors; contractors might also choose to have a clean-up crew.

Likely source(s:) of information: plumbing drawings, Detail A; Specifications Section ; Specifications Section 05 50 00 (elevator guide rail supports, sump grate), 05 51 50 (pit access ladder) |
| Elevator Manufacturer | It goes without saying that the elevator pit work should be coordinated with the elevator manufacturer's requirements, in fact a note to that effect exists in the drawings. A hydraulic elevator system may require embedments, say a piston sleeve, in the floor of the pit (Figure 15.3 and Appendix H.1).

Likely information source(s): architectural drawings, section 5 A7.1; elevator manufacturer's information ("cut sheets," AKA data sheets) |

Discussion

While most of the work in the elevator pit is straightforward, there are two principal difficulties, both of which need to be resolved prior to placement of the concrete. The first involves the dilemma of the keyway and waterstop both being required in the concrete wall; how does one form a keyway as described in the structural section (H 1.2, Figure 15.4) in a six-inch wall, and at the same time install a waterstop in that keyway? Adding yet more confusion, the keyway was not listed in the specifications. The architectural detail C 9—more appropriately described as a schematic drawing—does not show a keyway at all, it merely shows a waterstop in a wall that hardly resembles the outcome described in the structural section (see Figure 15.5). The separation of professional responsibility is evident here; the structural engineer is focused on the wall sliding off its foundation, and consequently addresses the shape and size of the keyway, and the architect, who has delegated the structural design, is focused on preventing water from intruding through the cold joint in the concrete. How should these two demands be reconciled?

The other issue involves installing the formwork, formwork braces, steel, and the waterstop and keyway without puncturing the waterproof membrane.

It is doubtful the structural engineer would allow the waterstop to be tied directly to the vertical bars, because the success of a reinforced concrete system is based on adequate coverage of the bars and concrete. The engineer shows a keyway with about a three-inch trough, a pair of inclined faces that will absorb another inch of wall width, leaving a pair of one inch "shoulders." Unreinforced, those shoulders are vulnerable to spalling.

One solution is worth brief consideration, namely widening the wall to make more room for the keyway and waterstop. It would add to the cost of the concrete, but not substantially, and would add another couple inches to the wall width and slab dimensions. This would allow for adequate separation between the waterstop and reinforcing steel while still providing substantial enough shoulders in the keyway, but the waterstop and keyway still need to be installed. The structural engineer is not likely to refuse adding material to the wall, and with a more easily executed wall assembly, the labor costs might offset the additional quantity of concrete, which amounts to just over one cubic yard.

How might this work actually be accomplished? There is no end to the ingenuity of the skilled trades workers, who are often great sources of information, and seeking input from them is always worthwhile.

There are at least a couple of typical ways to set up temporary support for the reinforcing steel here, one using the bottom edge of a 2 × 4, 1½" or so below the top of the pit slab, and one that includes a keyway through which the bars will pass (see Figures 15.9 and 15.10). There are challenges with both these approaches; the first solution merely requires that the ironworkers tie each bar to the 2 × 4 on edge. . . it involves more tying, but ironworkers are adept at making those connections. The second involves a more complicated keyway (wood or plastic) through which the bars must pass. Threading the bars through the keyway or setting the keyway over the bars creates a stable connection but is time consuming. Keeping the vertical wall bars straight, that is, in a U shape as shown in the drawings, rather than having them pre-bent to tie into the main slab-on-grade, would facilitate the installation of the keyway. In this approach one would install probably both keyways (upper and lower), and after the wall concrete has cured sufficiently and the upper keyway removed from the concrete, workers would bend the bars down such that they tie into the main slab steel. Getting permission from the structural engineer to bend the steel on the job would be wise; field bending and shop bending are two different propositions. What are the drawbacks to these two approaches? Certainly, one of the issues with temporary structures is they must be removed. In Figure 15.9, the waterstop is connected at the top only, which means that during the concreting process, the bottom of the waterstop could wander, since it is pinned only at the top (pinning it at the bottom as well would mean the wood spacer would be very difficult to remove without serious damage to the waterstop after the concrete has set, and workers have only an inch-and-a-half of workspace to remove the fasteners at the top of the waterstop. . . tough to do. In both cases, removing the key would be very difficult and unnecessarily time-consuming. The solution described in Figure 15.11 solves two problems: the waterstop is amply supported by the vertical steel, and nothing needs to be removed. The keyway can easily be created in wet concrete with simple tools. Again, seeking and acquiring permission of the structural engineer for such a modest departure from the drawings is wise.

Organization of the Work

Although the means by which portions of the work might be done are often born during a drawing review, ideally such decisions follow a complete understanding of the design professionals' requirements and a keen understanding of how work is done in the field. Once the work sequence has been determined, the costing of the work can begin. Table 15.1 depicts one approach a builder might take.

Table 15.1 Napa Hall Elevator Pit Construction Sequence

Operation	Reference	Operation	Equipment, Materials	Notes
1.	Site survey, building plan	Set line and grade		*Need ram construction info from elevator supplier*
2. Excavate elevator pit	Vol. 2, 02300-4 (p. 10 pdf);	Excavate elevator pit and spread footings; 17' × 18' at grade, 14' × 15.3' at pit floor-includes 3' work space and 1:2 layback). Pit floor 5.43' BFF; remaining area 6.43' BFF; 1:2 slope. Scarify 8″ of in situ soil under slab and .5' outside pit wall; see excavation plan	Backhoe; ¼ yard bucket.	Use existing earth materials; hand-grade as required and compact in 6″ lifts with a whacker. Specs Vol. 2, p. 10; (02300-4, 3.05 E.3): "Excavate (footings) by hand as required to establish final lines"
3. Compact subgrade	p. 9 geotech rpt; p.124 pdf	Compact subgrade	Whacker to compact elevator pit floor	Surface compaction equipment restricted to vibratory plate compactors, whackers, and pneumatic compactors (p.9 geotech rpt.); geotech must approve compaction
		Excavate for pit wall footings (continuous), spread footings, and sump; compact. Clean trenches and surrounding workspace	Backhoe, 24″ bucket; spade as required, jumping jack @ spread ftgs and sump	Install worker access ladder; overex footing trench 2″ for geotextile membrane and CPE waterproofing
	Note 2, detail H-1.2 (1.1)	*Inspect excavation and backfill*		As required by the geotech
	p.7 geotech rpt.	Wrap footing trench and sump with burlap; stake	Hand tools; metal staples (drip irrigation line staples) or 20d commons to secure burlap trench retainer	Per geotech report, footings must be kept damp until concrete is placed and finished
4. Install Forms 03 10 00	Vol. 2 Specs, Section 2.04.A.4; p. 52 pdf	Install outside wall forms to 10″ BFF (main slab), plumb and align; use 18″ plywood panels for spread footings	¾″ Plyform and single waler form system; 18″ high stud and plywood panels @ spread footings	Plywood forms OK; must use "pressure-sensitive non-absorbent vinyl foam tape" in panel joints per Vol. 2 Specs, Section 2.04.A.4; p. 52 pdf
	Geotech report, p.123 pdf	Install sub-base	Class II AB; vibratory plate compactor	
5.	Vol.4 (detail book) p.72, detail C9; Specs p.129 pdf; Section 07118	Install waterproofing, floor, and wall (see waterproofing detail)	Chlorinated Polyethylene (CPE) membrane	to 30″ above footing; tape; min 3″ overlap, or butt joints and cover w/6″ membrane seam
		Install geotextile cushion		Install in footing trench, slope at interior footing, and over AB
6.	p. S1.2 , detail H1.2, specs 03 20 00	Install footing, slab, wall steel, except for horizontal wall bars	Grade 60 bars	Slab-on-grade calls for #4's 12″ o.c. each way, with #5's longitudinally bottom of wall. Install verts with bend in pit slab only
7.		Install waterstop		See graphic of waterstop and key, Figure 15.11
8. Prep for conc		Install sump form	plywood	Formply box, Figure 15.12
9.		Clean trenches		
		Inspect pit slab steel		

(continued)

Table 15.1(Continued)

Operation	Reference	Operation	Equipment, Materials	Notes
10.		Place and finish pit slab-on-grade; cure; remove sump form 12 hours after pour	3,000 psi concrete	Trowel finish, spray membrane cure
11.		Install horizontal bars in wall		#5's above and below each keyway, #4's elsewhere
		Inspect wall steel		
12.		Install inside forms to FF main s.o.g.; align and brace to pit floor		
13.	p.63, 03310-3, Section 2.03.D	Place and consolidate pit wall concrete to 10" BFF; create upper key in wet conc.	3,000 psi concrete	
14.	Section 03100-2, 2.01.b (p. 52 pdf)	Strip outside wall forms		Minimum 12 hours before forms may be removed;
15.		Break snap ties in outside forms; patch snap-tie holes		
16.	Section 07118; (p. 129 pdf)	Install CPE waterproofing and geotextile cushion, pit wall; allow for 12" lap over vapor retarder in main slab		Minimum lap 12" (Detail C9)
	Note 2, detail H-1.2 (1.1)	Place and compact backfill to FF less 10"; 6" lifts; *Inspect backfill*	Backhoe, jumping jack	Backfill: existing materials; sandy fine silt if import required
17. Prep for slab		Install under-slab materials, main slab-on-grade; seal lap @ pit wall waterproofing		
		Install 6" × 6" × 3/8" angle, corner column base plate templates and 12" anchor bolts @ two corners, and slab-on-grade steel at main slab edge		
		Inspect main slab-on-grade steel		
18. Place main slab		Place and finish main slab-on-grade, cure		
19. Strip forms		Strip inside wall forms, elevator pit		
20. Install columns	S4.1, H1.2; Section 03310-2.02. A and Section 3.03.M, (p. 67 pdf)	Layout; Install column anchors and ¼" × 6" × 5" shear plates at FF; weld; set nut elevation @ baseplate		
		Install elevator cab guide columns; plumb and bolt; pack column bases with nonshrink grout	Gradall to set columns; manually pack grout	
21.	Detail L2	Install ladder, metal angle and galvanized grate, sump.		1" × 1" × 1/8" galvanized steel angle; ¼" × 2" expansion anchors; ¾" × 1/8" galvanized steel bar grate 2 anchors and angle, four angles
22. Clean up				

The sequence having been established, the builder can supplement the written text in the table above with any graphics that describe the work (see Figure 15.7). Once the problem is fully understood, the actual work anticipated in each of the operations listed above can be "costed," (costing here is a verb that means determining the cost of the work) which would include the required equipment and materials and any aspect of the work delegated to specialty contractors (reinforcing steel for example). Labor cost determination logically follows a decision first on how the work should be accomplished, followed by crew composition and some sort of productivity rate that applies to the work at hand.

Quantity Survey

The table above provides a sound basis for quantifying the work involved in the elevator pit. The builder needs to know how much of a given material is required, and normally will summarize the quantities in "buy-quantities" such as cubic yards for earthwork and concrete, tons of aggregate base, tons for reinforcing steel, square-foot-of-contact-area (SFCA) for formwork, and so on.

The order in which one performs a take-off is important. The figures developed in the earth quantity take-off (specifically the dimensions of the base of the pit) lend themselves to other take-offs, such as formwork and concrete.

Earth Quantity

As noted in the beginning of Chapter 6 under Key Concepts, earth material is normally on site, and merely needs to be removed, perhaps treated, and if acceptable,

replaced under conditions prescribed by the geotechnical engineer. Developing the costs for earthwork requires knowing the quantity of earth to be handled, after which equipment may be selected and an estimate of the time to perform the work made. The average end-area method of quantifying earth is a practical and, in this case, quick method for determining the amount of earth to be removed for the elevator pit. For a space this small, cutting vertical sections through the pit, determining their area and averaging adjacent sections, as described on pages 94–100 and Appendix G, is rather involved. However one may still use the average end-area method; by calculating the area of the work at the bottom of the pit first (area of the pit plus the work space around it), then the area at the surface (pit + work space + layback, if any), one can average the area of those two surfaces, then multiply that average by the depth of the excavation.

Formwork

The formwork for the elevator pit is simple and straightforward (see Figure 15.7). The outside forms are installed and braced first, followed by the installation of the subgrade materials, the reinforcing steel, and after those operations are completed, the pit slab-on-grade is placed and finished. That having been accomplished, the carpenter crew (probably no more than two or three workers) can then set and brace the inside wall forms on the pit slab. The inside wall forms rise to the level of the main slab-on-grade. Once the wall and main slab-on-grade concrete have been placed and finished, the contractor can install the tube steel cab guides, access ladder, and sump grate and the angle-iron

Following are graphic depictions of aspects of the elevator pit work.

PIT SEQUENCE
(following excavation)

1 clean trenches

2. install outside forms

3. clean trenches

4. place burlap trench
   retainer (——— ~ ———)

5. install Class II AB,
   compact

6. install CPE mem-
   brane over AB, in
   trenches, & 30" up
   form (——— . ———)

7. install geotextile
   cushion over CPE
   membrane, except
   @ form wall

8. place & tie steel,
   waterstop

3/4" = 1'-0"   steel not shown

Figure 15.7 This sketch lays out a sequence of construction for the elevator pit after excavation to the point of the initial placement of concrete.

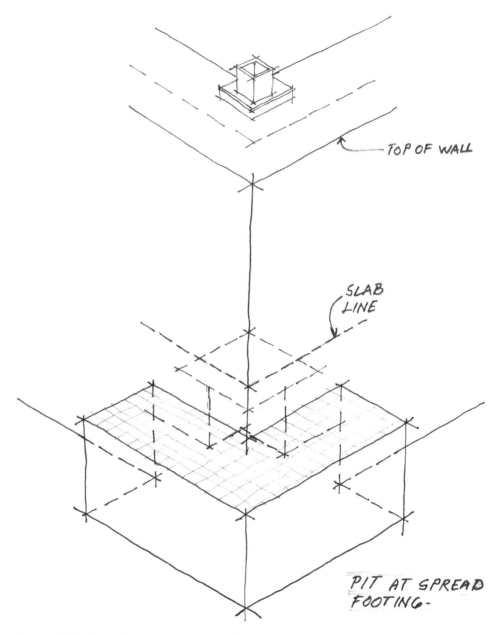

TOP OF WALL

SLAB LINE

PIT AT SPREAD FOOTING-

Figure 15.8 This study of the completed wall shows how the forms for the wall must be cut to accommodate the 3' spread footing. The work involved in covering the footing and wall in waterproof membrane is clear here; there are lots of seams to create and then seal. Bear in mind, in reality, nothing looks quite this neat, especially where excavation of earth is concerned.

Figure 15.9 This quick study of the waterstop-keyway issue reveals a couple issues, namely the fastening of the waterstop to the 2 × 4 leaves the bottom edge vulnerable to movement during the concreting process, not to mention the difficulty of removing the 2 × 4 without damaging the waterstop.

Figure 15.10 While this solution has some advantages (the waterstop and keyway are secure, and the whole component is suspended from the form wall) there is a considerable amount of work involved (milling the keyway and drilling holes for the bars) and removing the keyway would take at least two steps.

Figure 15.11 This solution is a practical one; the waterstop is secure, the steel is adequately covered, the keyway is easily created in wet concrete, and nothing needs to be removed after the concrete is placed and finished.

SUMP FORM
1" = 1'-0"

Figure 15.12 The drawings call for the sump to be tucked into the very corner of the elevator pit, which creates complications with the footing and slab steel (see Detail H1.2), and the concrete in the slab can be sloped such that any water in the pit can be directed into the sump. The form is easily constructed and assembled, it avoids the footing entirely, and it could be easily modified were the architect or engineer to insist that the sump reside in the corner. Since the sump form is suspended from the wall form, no stakes are required to support it, and placing the concrete is straightforward. A cut sheet is easily developed from this sketch, which does not show dimensions (for clarity).

CHAPTER 15 EXERCISES

1. Perform a material quantity take-off for the earth (cy), formwork (sfca), and concrete (cy) for the elevator pit. Example take-offs are available in Appendix D.

2. The potential for a settlement problem exists in this elevator pit, as the drawings make clear. Can you identify it?

3. The elevator pit steel presents some challenges to the reinforcing steel contractor, much like the challenges presented by a continuous chord bar in the walls of the project shown in Figure 8.13. What might you propose to make the reinforcing steel installation more efficient?

4. As noted in the chapter, ambiguities, errors, and omissions occur fairly frequently in construction drawings. These often become opportunities for the builder; design professionals are often receptive to input from results-oriented contractors who respect their roles and are focused on solving problems. In the example project, can you identify a couple such issues?

APPENDIX A

Estimating and the Takeoff—A Brief Overview

▪ Brief Overview

Estimating is analyzing the information contained in a construction contract (primarily the agreement, the general and supplementary conditions, the drawings, and the specifications) and determining a cost and price[1] for a construction project through a careful process of informed guessing. Design professionals conceive the parts and their interrelationships and assemble them in drawings and specifications. Their drawings are the representation of the whole, when it is complete. Contractors reverse this process and deconstruct the whole to understand and evaluate the parts and plan how the work will be accomplished. An essential aspect of this effort is the ability to recognize the correlation between the static images portrayed in construction drawings and the hundreds of dynamic activities involved in constructing a building or other project. Construction drawings provide some clues to the activities required to assemble the parts, but much of the information about their construction comes from know-

ing, by experience or research or both, what is entailed in the work. Indeed, the labor required to construct a building component or system is absent in the drawings. Visualizing a project under construction and understanding the required sequence are, therefore, two critical abilities of estimators, superintendents, and the others who plan and manage construction work.

Once the parts are fully understood and the sequence of assembly is determined, the contractor reassembles the parts in a document that performs several critical functions: it establishes a baseline work scope against which changes can be contrasted, and the baseline compensation for the work. It provides the foundation for accounting and billing for all the described parts; and results in a tool for monitoring progress during the course of the work. That document is the estimate.

In the estimate, work and material costs are segregated into five categories: labor, equipment, materials, subcontracts, and miscellaneous costs. (Subcontracting is a way of packaging equipment and labor costs in specialty areas such as electrical or mechanical work.) As a review of any commercially available database (Means, Lee Saylor, Walker's) demonstrates, it all boils down to units of time and money.

---

[1] Cost is the amount paid for something; price is the amount charged for something.

Quantities of material and work form the basis for construction project costs, and a subprocess of estimating known as the *takeoff* determines them. Types of materials are first discerned (distinguished from one another) and then quantified;[2] costs for the material are quoted, and the costs extended (forecast). Labor units are segregated into various operations,[3] then quantified in a labor take-off (with material quantities as the basis), then related to a company's historical records or some other cost database (or even by prototype analysis for new materials or methods), and extended. Determining material quantities is fairly straightforward; the materials are described in drawings and specifications.[4] As noted, most material costs are determined by multiplying a number of units by the cost per unit, a calculation that requires a vendor to do two things: identify the unit in which it sells its products (cubic yards or meters, square or lineal feet or meters, tons or kilograms, and the like) and specify the cost per unit.

Determining values for labor is much more difficult; in fact, labor costs are what usually explain the disparity from contractor to contractor in estimates of the same project. It is in the determination of labor costs that knowledge of construction processes and the ability to visualize the work are critical. The evaluation of labor costs and productivity warrant regular, close scrutiny by managers and estimators alike. In fact, it is safe to say that in the highly competitive contemporary construction climate, labor efficiency, whether achieved through new technology or methods, is the last bastion of profitability.

In the labor takeoff, operations that were first segregated into equipment or labor-paced operations[5] in the shopping list are analyzed, and labor quantities are determined using the material quantities. Two important productivity equations are used, one that expresses productivity based on total units per worker hour, and one that expresses productivity in terms of worker hours per unit (the reciprocal of the first equation). The labor quantities, once determined, are correlated to a company's historical cost records[6] or commercially available databases, and a production rate[7] and (ultimately) the labor cost are established.

Drawings and the Takeoff

As noted repeatedly in this text, there are limitations to construction drawings as currently produced (two-dimensional multiview drawings depicted on large sheets of paper). Describing in one view only two of the three dimensions that a project lives in is explaining less than 70 percent of the reality. The compensation afforded design professionals by owners is often less than the designers actually need to produce drawings with minimal errors. The architect or engineer may leave a certain amount of design work to the contractor (drawings for electrical or cladding systems are examples). In some situations, the work is better left unexplained, to give the people with the expertise (the specialty contractor, for example) some flexibility as to how to accomplish a given task or build a system.

When the architect and design professional consultants each make their contributions to a single system—the HVAC system, for example—it means the individual performing a quantity takeoff of the system will have to investigate the component in several disciplines, identify the relevant information, prioritize it, and assure that quantities for the complete work are determined.

Imagination, an understanding of what the design professional is trying to accomplish, knowledge of the various roles the design professionals play, familiarity with the basic construction systems, and a willingness to investigate go a long way toward producing an effective quantity takeoff. These realities conspire to make the takeoff, under the best of circumstances, an exercise in informed judgment.

---

[2]The use of these particular terms—discern and quantify—in the context of the quantity take-off comes from *Construction Cost Estimating: A Managerial Approach*, an unpublished manuscript by Richard A. Nickles.

[3]The relevant definitions of *operation* are a process or series of acts performed to effect a certain purpose or result; a process or method of productive activity. ("Frame walls, place concrete; finish, and cure concrete," are examples of operations.)

[4]The one exception to this statement is probably the determination of earthwork quantities. Although total stations and computer software have improved the accuracy of earthwork estimates considerably, there are numerous variables that affect the quantities themselves that come to light only when the work takes place.

[5]Equipment-paced operations are operations performed by machines and operators, and are characterized by relatively few, repetitive steps. Labor-paced operations are performed by workers with hand tools or small pieces of equipment and, by contrast with equipment-paced operations, are characterized by complex, often discontinuous steps. Watch someone performing a task and you will get the idea.

[6]The "reverse" estimating process—my term—consists of evaluating cost records after a project is complete and determining the actual unit costs for operations in a variety of work categories.

[7]Rate of production (or productivity) is a coefficient of work accomplished over time.

APPENDIX B
Answers to the Exercises

Chapter 1

1. Design professionals are responsible for describing and recording what is to be built, which they do in the construction documents. A significant part of their effort is dedicated to producing the drawings and specifications of a project. The contractor is responsible for the means and methods of constructing what the design professional describes. The owner's role is to provide the financing and site, give timely feedback, and operate the project.

2. Design professionals are, professionally and by training, product-oriented, whereas contractors are process-oriented, which is a reflection of their professional responsibility and training.

3. Contractors, who hold the overall responsibility for project construction, delegate some work to specialty contractors such as mechanical and plumbing contractors. Design professionals hire design consultants such as structural and electrical engineers for specialized parts of the work.

4. Architect, landscape architect, geotechnical, civil, structural, mechanical, electrical engineers.

Chapter 2

1. Project uniqueness, project complexity, the number and variety of participants, the differing values and professional responsibilities of the architect or engineer and the constructor, A/E's experience with the type of work being designed, level of activity in the design professional's office.

2. Delivery systems are the legal arrangements through which construction projects are designed and constructed. The seminal delivery systems are: (1) **Design-Bid-Build**, in which the architect prepares complete construction and bidding documents under contract with the owner, contractors compete for the project, and the lowest responsive, responsible bidder constructs the project; it is a cost-oriented system. (2) **CM at Risk** is a delivery system wherein the owner signs a contract with a design professional for design services and a separate **preconstruction services agreement** with a builder. Both providers are normally selected by the owner using Qualifications Based Selection (criteria that weight experience, personnel, project type, and cost rather than simply cost). The design professional and architect collaborate during the design development stage. (Figure 2.1 describes the evolution of design and construction contracts.) The design professional performs the same tasks as under other design contracts; the builder offers estimating, constructability review, scheduling, and value engineering services, among other things; and a set of contract documents is created. The project is then competitively bid or, more frequently, a separate construction contract (sometimes with a Guaranteed Maximum Price) is signed by the builder that offered

Construction Graphics: A Practical Guide to Interpreting Working Drawings, Third Edition. Keith Bisharat.
© 2025 John Wiley & Sons Inc. Published 2025 by John Wiley & Sons Inc.

its preconstruction services. (3) In the **Design-Build** arrangement, the owner signs a contract with a single entity (e.g., a joint venture between a design professional and contractor), which takes responsibility for designing and constructing the owner's project. The selection of the design-build entity may be based on the lowest price, Qualifications Based Selection, or Best Value (criteria reflecting the values of the owner as described in the Request for Proposals).

3. Residential, commercial building, engineering (heavy/highway), industrial.

4. Half-scale reproductions (problems with scale), drawings in metric (with most measurements given in millimeters, there is the risk of significant numerical errors in quantity surveying), evaluating the magnitude of work based on drawings.

5. Programming phase (problem discovery), schematic design (preliminary design work), design (preliminary work is refined), construction drawings (the drawings used in construction that have been approved by the applicable authorities).

6. The five components are:
   - Agreement: Contract type, brief project description, contract document list, start and substantial completion dates, payment terms, consideration, parties' signatures.
   - General and supplementary conditions: Guidelines to the administration of the contract. General are generic, or boilerplate, conditions; supplementary conditions are modifications to general conditions (to reflect the particular circumstances of a project and its participants).
   - Drawings: The graphic depiction of the work.
   - Specifications: Guidelines to the quality of work and materials.
   - Other: Addenda, change orders, and certificates of compliance.

7. Estimate its probable costs and determine its price, a function performed by the estimating department, which breaks a project down into discrete parts and assigns values in time and money to the numerous operations required to construct a project. Basic categories include labor, equipment, materials, subcontracts, and other costs.

Chapter 3

1. The *Project Manual* contains *bidding requirements* (advertisement/invitations to bid; instructions to bidders; bidders' information; preliminary schedules; geotechnical data; existing conditions description, which includes a property survey, description of existing buildings, and a "hazmat" [hazardous materials] report; bid forms, bid form supplements,

representations and certifications; and the *contract documents* (agreement, bonds, certificates of insurance and compliance, general and supplementary conditions, and the specifications).

2. Within each division (civil, architectural, structural, etc.), the information commonly goes from general to specific. In the NCS, general information (legends, notes) are followed by plans (horizontal sections or cuts), elevations (vertical views that show surfaces) sections (vertical cuts through buildings or major systems such as the walls), large-scale views (plans, elevations, or sections that show smaller portions of the project, but are not details), details (large-scale plans, elevations, and sections of very small portions of a project), schedules and diagrams (door, window, and finish schedules or matrices), two user-defined categories that serve miscellaneous needs, and, finally, pictorial drawings such as perspectives and isometric projections. Other drawing sets follow a similar sequence.

3. 4 inches; 100 mm.

4. Soft conversion refers to relabeling of products from U.S. Customary units to SI without a physical change; hard conversion refers to a change in physical size made by a manufacturer that is converting to SI. Concrete block and plywood are examples of products that will undergo a hard conversion.

5. Modular construction generally refers to construction in which specific units—4″, 8″, 16″, 48″, for example, or 100, 200, 400, 1200 mm—are used repeatedly to create a whole system or project. Concrete block and brick structures are examples of units from which whole structures are constructed.

6. Millimeters and meters only.

Chapter 4

1. 1″ = 8′; 1″ = 4′; 1″ = 1′-4″; 1″ = 8″

2. 650 millimeters, or just over 25 ½″ long; 1 meter, or 1000 mm, or about 39.37″ by 600 mm, or 23 ⅝″.

3. Highway sections, utility profiles, site sections, and the like. The ratio of horizontal to vertical is about 10:1.

4. The station point (the observer's eye), the projectors (lines of sight from the object to the station point), the plane of projection (the drawing), and the object.

5. Parallel projections include orthographic drawings such as: (1) Multiview: These drawings are very easy to produce, compared to perspective and axonometric drawings. Their principal shortcoming is that in every view, the third dimension is missing, which makes it impossible to figure out some shapes without one, two, or more views from a different point. (2) Axonometric: Consisting of isometric, dimetric, and

trimetric projections, these drawings are reasonably realistic (they show height, width, and depth simultaneously) and are fairly easy to produce. Their disadvantage is that they are more difficult to produce than multiview projects (round holes become ellipses in each of the faces of the drawings, for example) and unless the scale is adjusted to reflect the reality (the axes in an isometric projection, for example, are about 80 percent of their true length), the dimensions of an object are more difficult to determine. (3) Oblique: Drawings that are parallel projections with the projectors oblique to the plane of projection. Two of three axes are always 90 degrees to one another, with the third axis at any angle, but typically 15, 30, 45, or 60 degrees. They are pictorial in nature; and with two axes being perpendicular to one another, complicated geometry is very easy to produce. Their principal disadvantage is that they are the least realistic of the pictorial projection set.

6. Containment of critical information only, compliance with drafting standards, consistency in how the dimensioning is done, clarity, and closure (dimension strings should reconcile).

7. Axonometric drawings are drawings of objects whose principal edges are inclined relative to the plane of projection. The axes in isometric drawings are 120 degrees apart. In dimetric drawings, two of the three main axes are the same angle, and in trimetric drawings, none of the three is the same. These drawings are useful where a pictorial representation is more effective, since they show height, depth, and width simultaneously. Drawings of decorative concrete block, certain flashing conditions, or even ADA-compliant ramps are examples of objects that are effectively described by oblique or axonometric drawings. Plumbing schematics are frequently shown in isometric, because all the components of the system—piping, connections, as well as the direction of wastewater flow—can normally be shown simultaneously.

8. Architect's scales use a ratio of inches or fractions of an inch to a foot; engineer's scales use whole inches to feet, and SI scales are straight ratios.

9. The plan of projection is placed between the viewer and the object.

10. To expose the anatomy of a structure.

11. Clearly presented, unambiguous, relevant, and complete instructions as to what should be built.

Chapter 6

1. Geotechnical engineers generally focus on site investigations, the classification of soil, and hydrology of a site. Their work culminates in a written report and summary of tests and recommendations to the owner. Civil engineers are hired to evaluate a site, describe its topography, and produce grading and utility drawings. Landscape architects conceive the program for ornamental landscaping and the irrigation systems that support the grasses, ground cover, shrubs, trees, and hardscape that make it up. Their work culminates in drawings and specifications pertaining to the landscape and irrigation drawings.

2. Graphic displays of these systems, primarily plans and sections, would be found in utility or improvement drawings. Textual descriptions would be found in the specifications.

3. Compliance with the Federal Clean Water Act. Their primary focus is erosion control, sediment management, and containment of chemical pollutants.

4. The contour line.

5. Contour lines do not intersect one another; they do not simply end (except at the map's edges); they close upon themselves; they do not split, except in the rare circumstance that the base of a cliff and a particular elevation coincide; every fifth contour line is emboldened; contour lines at streams form V's that point upstream.

6. Characteristics of the soil; groundwater; magnitude of the work; access to, size, and location of the site; and the availability of equipment.

7. (1) Identify specifically what it is you are being asked to do. (2) Take an inventory of the information you have and determine whether you have to generate any more. (3) Define the graphic parameters of the work, in all three dimensions. (4) Select the appropriate vertical and horizontal scales. (5) Determine the number and frequency of sections to be taken and identify them by station point. (6) Plot the existing and design profiles, bearing in mind such requirements as over-excavation. (7) Calculate the difference between the design and existing contours, at each station point, in square feet or meters. (8) Post the area calculations in the proper takeoff format, determine the average end areas, extend the numbers, summarize them, and convert the extended numbers to the proper units.

Chapter 7

4. Deep and shallow foundations. Deep foundations consist of various pile foundations, caissons, and mats. They are used for structures whose geometry results in long, more or less slender lever arms perpendicular to the earth, such as skyscrapers, or in situations where incompetent earth is underlain by a good bearing layer. Shallow foundations employ continuous and

spread footings, and can be used on large projects, provided that the structure is horizontal in its geometry (the principal axis is parallel to the earth).

5. Vertical loads (e.g., gravitational loads, uplift due to hydrostatic loading or seismic events), multidirectional lateral loads (hydrostatic pressure on the substructure, seismic events), and various reactions to the loading (e.g., uplift due to the lateral loading of a tall structure, bending in vertical and horizontal structural members, and shear forces).

6. Drainage, water, and dampproofing and insulation systems.

Chapter 8

1. Cast-in-place reinforced concrete, cast-in-place post-tensioned concrete, precast prestressed concrete, structural steel, masonry, heavy timber, wood light framing.

3. For building construction projects, the architect makes the initial selection based on programming requirements, sometimes in consultation with a structural engineer. The details of the system are left to the structural engineer. For many other projects, engineers decide which system to use.

4. The basic system, of course, but connections between walls and foundations, walls and floors, walls and roof, details around openings through walls, floors, and structural members.

Chapter 9

1. Gravity, momentum, surface tension, capillary action, sound waves, air pressure, water vapor diffusion.

2. A wash will resist the effects of gravity; an internal joint cover or labyrinth joint design will neutralize momentum; a drip will discourage water from intruding by surface tension; a capillary break will address capillary action; mass, a labyrinth, or resilient frame or material will resist sound transmission; air pressure can be neutralized by a pressure equalization chamber; and water vapor diffusion can be resisted by a vapor retarder and internal ventilation of a cavity.

3. Move the aluminum flashing out from the interior wall to the inside face of the panel on the right. It wouldn't be any more visible than where it is now shown and it will prevent water from staining the underside of the soffit.

Chapter 10

1. Horizontal, (or sloped) enclosure.
2. Gravity.

3. Low-slope roofs cover large areas economically, but they are less forgiving of carelessly performed work and complex geometry. Pitched roofs are forgiving of carelessly performed work, and their coverings are easy to install with simple tools, but they require more materials for the area they protect and have a greater falling hazard potential.

4. To protect the roof membrane from destructive light and erosion and to prevent the membrane from being lifted up in a wind.

5. (1) Penetrations in the roof and the required flashing: Minimize the penetrations, use conventional flashing systems to flash and counterflash the penetrations. (2) Complex geometry requiring complex flashing systems: Keep roof simple, reduce the number of twists and turns, and, therefore, flashing. (3) Temperature fluctuations: Design details that use thermally compatible materials or integrate them so that differential thermal movement is minimized (use cast-in reglets rather than surface-mounted reglets); use expansion joints and light-colored protective membranes or ballast; protect the membrane from heat by using insulation.

Chapter 11

1. As nonload bearing elements in a structure, they generally serve the very limited purpose of partitioning space. They are consequently made of a minimal number of light structural elements and are supported by the structural elements of the building.

2. They provide resistance, in varying degrees, to the spread of fire and sound in a building.

3. Any UL-approved wall or floor system of varying fire-resistance capability (typically one to four hours) that segregates a building into zones. The idea is that a fire starting anywhere in a structure will be contained in that area for the rated period.

4. Containment of insulation when the drywall does not extend from floor to ceiling; framing around structural members; installation of draft stops.

Chapter 12

1. Supply lines are pressurized, and drain lines are not, which allows for different materials, methods, and manner of connecting the pipe.

2. Electricity, steam, or hot water or combustible fuels may generate heat in HVAC systems. One common method of distributing heat in electrically generated heating systems is to blow air across resistance strips.

Hot water systems typically propel heated water through pipes to radiators, convectors, coil units, heat exchangers, and the like, using electric water pumps. Heat is distributed either by radiation or convection. Systems that use combustible fuels such as natural gas use furnaces and heat exchangers—air is blown across the heat exchanger.

3. The compressor, the condenser, the evaporator, and the air-handling unit.

4. The design drawings consist largely of plan views and details. Elevations and sections, which in most other drawings of systems reveal the third dimension, are largely absent in mechanical drawings; consequently, visualizing the interstitial space in buildings is critical. Shop drawings generated by mechanical contractors may list the elevations of the top and bottoms of ducts in the interstitial space.

5. Both are pressurized systems, and the area in section of the distribution system conduit (ducts in the former, pipes in the latter) reduce in size as the conduit gets further and further away from the source.

Chapter 13

1. Both consist of loops of sorts; electrons flow through wiring from a source through electrical controls and devices back to the source (ultimately, the electrical grid operated by the utility company). HVAC systems distribute conditioned air through ducts to remote locations, where it is enjoyed by a building's occupants, after which it is drawn back to the source through some form of return air system. Wiring, like ductwork, diminishes in size as it gets further and further away from the source.

2. Service equipment (transformers, distribution panelboards, switchgear, feeders and raceways) branch circuitry, loads and devices.

3. Any channel or pipe for conveying fluids, or (in electrical work) a tube or duct for enclosing electric wiring.

4. Electrical systems are generally more schematic in nature, large portions of them are not drawn to scale, and very little of the vertical component is ever addressed; that information is included in architectural, structural, and other sheets. In general, much more is left to the user's imagination.

Chapter 14

1. To inform the design professional how the contractor intends to execute portions of the construction project and how those portions correlate to other portions of

the project; to provide a standard against which proposed substitutions can be evaluated; to establish and memorialize quality standards for the construction project; and to provide the owner with critical operations and maintenance documentation. Submittals manifest the contractor's interpretation of the project requirements.

2. Depending on the purpose of the submittal, prior to construction, during construction, and upon project closeout. Design professionals frequently stipulate that approval of the submittal must precede construction of the parts of the project requiring submittals.

3. The principal distinction is that shop drawings are produced not by the design professional, but by the contractor or subcontractor, and they typically depict systems and components in greater detail than do other drawings.

4. For projects organized according to the Construction Specifications Institute's MasterFormat, the general requirements section of the specifications (Division 1) make the submittal requirements known. The general and supplementary conditions of the construction contract set forth the procedural requirements for submittals.

Chapter 15

1. Assuming excavation of the main pit to 5.4' below the main floor elevation, a 1:2 layback, and excavation of the continuous and spread footings to −6.4', there are some 65–70 cy of earth.

   Formwork for this project will amount to approximately 425 SFCA.

   Concrete for the elevator pit, including footings, slab, and wall concrete from the pit slab to 10″ below the main slab-on-grade should be about 10 cy.

2. The architect calls for protection board for the waterproof membrane, which means the bottom of the continuous and spread footings will rest on what might be a compressible barrier. Depending on the CPE membrane supplier, this could mean settlement in the wall foundation. Protection board as thin as 3.5 millimeters to about 6 millimeters (about 1/8″ to ¼″) is available, which should assuage an engineer's concerns about settlement, but assuring that it is not a problem is a worthwhile undertaking.

3. The drawings call for U-shaped bars with "ears," which certainly could be installed as is, however they are an awkward shape for the ironworkers; the bars would be continuous and about 24' long with four bends. This configuration increases the difficulty of setting the two required keys as well,

if one chooses to install a wood or plastic keyway. Getting permission from the engineer to overlap the bars in the bottom of the wall is one option that might make the pit slab steel as well as the wall steel easier to install. . . the slab-to-wall steel would be installed and tied first, followed by the vertical wall-to-main-slab bars and the horizontal steel.

4. The waterstop and keyway issue present such an opportunity; the waterstop manufacturer is not listed in the specifications, nor is it clear just how this issue should be resolved, which provides the construction professional with an opportunity to select products and installation methods. True of the reinforcing steel as well, as noted in #3 above.

APPENDIX C
Residential Construction Drawings

Part 1: Residential and Commercial Building Drawings: A Comparison
Part 2: Sample Drawings[1]
Part 3: Inventory of Information in a Residential Construction Drawing Set

■ Part 1: Residential and Commercial Building Construction Drawings: A Comparison

There are four basic categories of project: residential and commercial building construction, engineering construction, and industrial construction (refer to Chapter 2). Two of those four types, residential and commercial building construction, constitute approximately three-fourths of the overall construction market. That proportion may change slightly as funds shift from market to market and region to region, but it is likely that these two segments of the industry will continue to dominate the construction market place. A comparison of the drawings produced in both sectors of the market is therefore worth making.

Residential Design

As noted in Chapter 2, some states do not require designers of detached single family and duplex homes to be licensed, and consequently architects, individuals, builders, and building designers all produce the documents used in construction. Architects and the engineers in their employ design multiple-family projects (three- and four-family structures, townhouses, apartments, condominiums, and multistory residential projects) usually as a matter of state law.

Nature of the Work

While there are many variations in the documentation of residential projects,[2] the drawings for homes and duplexes bear numerous similarities to commercial building drawings but generally are less extensive, have fewer details, and are less prescriptive. Designers of buildings used by the public, particularly by large numbers of people, are subject to myriad requirements having to do with minimum standards of health and

---

[1]Building design drawings courtesy of Tamano and Chaw Residential Design and Monarch Homes. Structural drawings courtesy of PFS Consulting, Inc.

[2]The influences on documentation include applicable state and local law, housing type (the two primary types being production housing and custom homes), developer or owner requirements, the individual preferences of the producers of the documentation, indigenous building practices, and cost.

Construction Graphics: A Practical Guide to Interpreting Working Drawings, Third Edition. Keith Bisharat.
© 2025 John Wiley & Sons Inc. Published 2025 by John Wiley & Sons Inc.

safety, many of which do not apply to buildings occupied by a single family. Significant access and egress requirements (including by people with mobility challenges), panic hardware, signage requirements, and fire suppression systems are examples of constraints that do not commonly apply to single family and duplex projects. As in comparable creative undertakings involving construction, architects or building designers conceive of solutions to their clients' design problems and then describe those solutions using drawings and text. Site and grading plans; architectural floor plans, roof plans, exterior elevations, building sections, wall sections, interior elevations, and architectural details; as well as foundation and floor framing plans and structural details are all used (see Chapter 4 for an explanation of the various views used in construction drawings). The simpler the project, the simpler the task of describing it. There are notable differences, however. For all their simplicity, small residential projects can be much more complicated than the typical commercial building project, particularly where building site issues and the geometry of the roof structures are concerned. A far greater number of homes are built in areas with severely sloping sites (the San Francisco Bay Area, the Hollywood Hills, Seattle, Portland, Oregon, and Boston come to mind) than are commercial building projects. The cost of creating a level building pad for a home on a steeply sloping site, if it can be done at all, is often prohibitive. Consequently, designers resolve the issue by requiring complex foundations comprising, for instance, stepped foundations over reinforced concrete piers, grade beams, and tie beams.

Simple, low-slope roofs hidden behind parapets or mansard-style roofs predominate in commercial building. Roof structures play a critical role in the esthetic appearance of residential buildings and consequently pitched, ornamental structures are preferred. When those roof structures intersect at odd angles particularly, or when numerous load-bearing elements intersect in a single location, designing and building them can be challenging. Certain interior assemblies—curved or circular staircases, for example—can be elaborate in a residential building, particularly in more expensive homes. This complexity, combined with the tendency of building designers to be less prescriptive than architects who describe commercial buildings, can make residential building construction a frustrating exercise.

Documentation

The drawings for most small projects generally follow the sequence used in commercial building projects: In the drawing set, site drawings (which may be produced by the architect or building designer rather than by a civil engineer) precede the architectural drawings. Architectural drawings may consist of floor plans followed by a roof plan, exterior elevations, building sections (which may include interior and exterior elevations as a part of the section), and architectural details. Specifications in residential drawings, rather than existing in the separate book or binder form common in commercial construction, may be recorded on the drawings themselves. Some designers follow the same approach that structural drawings in commercial construction do: the first page or two is devoted to text and some drawings of generic details (standard details the designer includes on every project), followed by the building drawings (refer to the appendix for an inventory of information in a sample residential construction project). For all practical purposes, mechanical drawings in a residential construction set are nonexistent, unless the project approaches commercial building size (10,000 or more square feet). Electrical drawings simply show the floor plan, the location and type of fixtures, and the location and the characteristics of the devices that control them (e.g., three-way switch, rheostat, number of switches in a "gang"). Specific information regarding key aspects of the drawings follows.

Site, Grading, and Utility Drawings

Every government entity has its idiosyncrasies, but many require little in the way of grading or utility drawings for a home that is constructed in a medium- to high-density subdivision—most of that work is produced by the civil engineer responsible for designing the subdivision and put in place by the earthwork contractor who creates the building pads and streets in the subdivision according to the engineer's design. Electrical power, the domestic water supply, and the sanitary sewer are installed in the streets or adjacent to them and are brought just inside the property line of each building lot. The constructor of the home on each lot connects the plumbing in the home to those utilities. Storm runoff on the lot is directed to gutters and storm drain inlets in the street or detention basins and bodies of water such as ponds or lakes, in the subdivision via simple surface drainage mechanisms (slopes and swales). When other methods such as underground drain lines are desired, it is usually a landscape contractor who installs them as part of a landscape plan. Fire service underground is required in most commercial building projects because owners want to take advantage of the building square footage bonuses offered by the building code and lower insurance premiums that result when an automatic fire suppression system is used. Automatic fire suppression systems are used in residential construction (mostly for insurance reasons), but a separate underground system is not required: these systems use the domestic water supply. In some subdivisions, the improvements (streets; maybe curb, gutter, and sidewalks; and utilities) are installed but the home sites are not graded. In this circumstance site and grading plans, particularly on hilly sites, are required.

Architectural Drawings

The architectural drawings set for a residential building is very similar to that of a commercial building with one common exception: the walls frequently play a more integral role in the performance of the structural system in a residence. Businesses start, they grow, they move, and they disappear as well, and consequently commercial building owners require considerable flexibility within the building envelope to accommodate a changing clientele. The design of the structural system, then, anticipates those changes and carries out its role independently of the partitions that divide multiple businesses in the shell. Many commercial building contractors construct a shell for an owner under one contract and install tenant improvements (TIs) under separate contract with the owner or lessee after the space has been designated (see Chapter 11 for a discussion of tenant improvements). While remodeling certainly takes place in residential projects, changes are not anticipated during the design development process as they are in a commercial building. In the drawings themselves, wall types, width, length, and orientation to each other are shown in floor plans. As in drawings for commercial buildings, architectural floor plans show room names, uses, or numbers; features such as sinks, counters, appliances, storage space, and plumbing fixtures; dimensions; and the numerous references to relevant drawings (major and minor building sections, wall sections) and legends for the various symbols used. As an alternative to employing a room finish schedule, reflected ceiling plan, or wall sections to identify ceiling heights, as commonly occurs in commercial buildings, the designer may list them directly on the floor plan, elevations, and/or sections.

Foundation Plans

The instructions to builders regarding foundations are generally the same in both residential and commercial building, although it is fair to say that commercial buildings exhibit a wider variety of foundation type and complexity than residences, which weigh less and typically employ slabs-on-grade or conventional foundations (footing-stemwall systems with dimensioned or manufactured lumber floor structures).[3] The biggest difference is in the detailing of the foundation system, which for a commercial building project of any consequence can represent hundreds of drawings.

Roof Structure Framing

As noted above, roof structures in residential projects are frequently more complicated than commercial building roof structures, due in no small part to their complex

geometry. Pitched roofs (roofs with slopes greater than 3 in 12—three vertical inches to 12 horizontal) typically require the installation of many small pieces (e.g., shingles, shakes, slate, concrete roof tiles), whereas low-sloped roofs employ large sheet goods (built-up or single ply roof coverings). From the builder's standpoint a pitched roof can be much more of a challenge, when access to the work, safety, and productivity are considered.

Exterior Elevations

Exterior elevations in both residential and commercial projects are very similar. They are pictures of the structure as it would appear to a bystander—views of the surface features of the structure from all sides, written descriptions of its physical features, with leaders pointing to specific elements, components, or assemblies.

Building and Wall Sections

Major building sections in both commercial and residential drawings are cuts through the entire project that expose the anatomy of main elements in the building. These drawings, depending on where in the structure they are taken, combine views of inhabitable spaces, the structure itself, and exterior surfaces (see Part II for example graphics). Wall sections, on the other hand, take smaller segments of the building and look at them in greater detail. Both architects and engineers make extensive use of wall sections in commercial building projects, but to a lesser extent in residential construction drawings. In the architect's case, the focus is on the general size, shape, location, and esthetics of the system; its fire resistance capability; ventilation characteristics; and resistance to an invasion of the elements. In the case of structural engineers, wall sections are popular for identifying the structural elements required, the manner in which they are connected to one another, and the relationship of the wall system to larger structural elements. Considering that their professional responsibility is to make sure a structure performs as intended under a variety of static and dynamic loads, it is not surprising that structural engineers frequently ignore other aspects of the wall system (e.g., wall flashing).

Interior Elevations

Design professionals who produce drawings for commercial buildings frequently make use of interior elevations to describe the surface features of rooms in the structure. Particular attention is paid to public spaces such as lobbies, waiting rooms, restrooms, and specialty rooms such as locker rooms. In a hospital or health care project, for example, interior elevations of the typical hospital room and special rooms such as imaging or examination rooms are

---

[3]See Chapter 7 for a discussion of foundation types.

widely used (see Figure 11.5o). Interior elevations are used in residences as well, but not to the same extent.

Detail Drawings

The focus of detailing—describing the nature of smaller components and assemblies in the project—is the same in both residential and commercial building projects, although the specific subject matter may differ. Specifying cladding and roof system components (the building envelope), a task that falls on the shoulders of architects, focuses on resistance to the penetration of the elements and sound. For the engineer, how the cladding will perform with various loads applied to it (including thermal expansion and contraction), and the response of the structure to bearing the load of the cladding are the major concerns. Consequently, connections are of great interest to both design professionals. For projects that employ a system other than curtain walls—for example, a concrete tilt-up building (site precast concrete panels)—the detailing of the panels is shared: architects focus on describing the esthetic features of the panel, its resistance to the elements, floor and roof connections and their resistance to fire, and compliance with zoning codes, among other things (refer to the figures in Chapter 9, in particular Figures 9.4i–p, for examples of the architect's work pertaining to cladding). Engineers focus on structural concerns involving the panel: how they are connected to one another, how floor and roof structures connect to the panels, and the structural elements all these assemblies are composed of.

Mechanical Drawings

Mechanical systems, consisting of piping (hot and cold water supply; drain, waste, and vent piping; hydronic systems; and fire suppression); pumps, valves, and other equipment; and heating, ventilating, and air conditioning (HVAC) systems and the equipment that controls them can be extremely complex in a large office building or hospital, due to the public nature of the buildings and the number and physical condition of the people occupying them. These systems are designed by mechanical engineering firms that operate as subconsultants to the prime design professional. In contrast, mechanical systems in residences serve a handful of occupants in a far simpler interior environment. Conditioned space in commercial buildings is subdivided into zones that various types of equipment serve, while in residential construction the entire home represents a single primary zone or perhaps two zones, with individual rooms considered as sub-zones. Consequently, competent mechanical specialty contractors design and install HVAC systems on a design-build basis with few instructions from the design professional, with the possible exception of identifying the system type and location of large equipment (condensers and air handlers) and perhaps the key access points. The air distribution system—the routing of ducts and location of registers—is frequently left to the contractor.

The design-build nature of HVAC systems applies to plumbing systems in residential projects as well. Beyond identifying where the fixtures will be, the designer does little to describe the plumbing system. In commercial building projects, mechanical engineers produce complete plumbing drawings, including isometric drawings of drain, waste, and vent piping and hot and cold water supply (as well as other systems). Plumbing systems may consist simply of a ghosted floor plan with fixtures emboldened on them.

Electrical Drawings

Electrical drawings in residential construction are similar in their character to the plumbing systems in that the designer identifies the location, number, and type of light or appliance and the switches that control them, the location and number of power outlets desired, the location of the service panel, and perhaps the type and color of switches, outlets, and covers. Beyond that, the installation is up to electrical specialty contractors who identify the number, type, and size and routing of the conductors. In contrast, electrical drawings in commercial buildings, though not as explicit as other drawings in the set, are far more prescriptive. (Nearly all information in electrical drawings is given in plan form and equipment lists and schedules. The third dimension—the vertical element between floor and ceiling—is ignored, leaving the mapping of the distribution system up to the specialty contractor.)

Summary Comment

What gets described in any drawing set is really the province of the individual or entity employing the design professional or building designer who resolves complex design issues on behalf of the client. A company that constructs projects for its own account using in-house personnel who know their company's preferences well (some apartment builders who develop, own, and operate their own housing complexes function this way) probably require less instruction from an architect than a construction company building an office building or apartment complex for an investment group that has employed an entity to develop, design, and describe that project. It should come as no surprise that commonalities in residential and commercial building drawings exist, but variety in the information abounds. Part II is a visual comparison of some of the drawings used in a residential and commercial building projects in California.

---

[4]All building design drawings in Appendix C courtesy of Tamano and Chaw Residential Design and Monarch Homes. Structural drawings courtesy of PFS Consulting, Inc.

▪ Part II – A Visual Comparison: Residential and Commercial Building Graphics

The following drawings are excerpted from a 20-page drawing set for a single-family residential project in northern California.[4] The following table is the index for the set:

G.1	Title Sheet, General Notes	S1.1	Foundation Plan
1	First Floor Plan	S1.2	Second Floor Framing Plan
2	Second Floor Plan	S1.3	Roof Framing Plan
3	Exterior Elevations	SN 1	Standard Notes and Details
4	Exterior Elevations	SN 2	Standard Notes and Details
5	Sections	SD 1	Structural Details
6	First Floor Electrical Plans	SD 2	Structural Details
7	Second Floor Electrical Plans	SD 3	Structural Details
AD 1	Architectural Details	SD 4	Structural Details
T-24	Energy Compliance	SD 5	Structural Details
		C 1	Site/Grading Plan

FRONT ELEVATION
SCALE: 1/4"=1'-0"

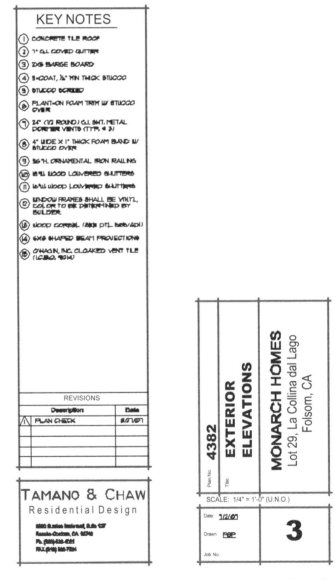

KEY NOTES

1. CONCRETE TILE ROOF
2. 1" G.I. COVED GUTTER
3. 2X8 BARGE BOARD
4. 3-COAT, 7/8" MIN THICK STUCCO
5. STUCCO SCREED
6. PLANT-ON FOAM TRIM W/ STUCCO OVER
7. 24" (1/2 ROUND) G.I. SHT. METAL DORMER VENTS (TYP. 4 3)
8. 4" WIDE X 1" THICK FOAM BAND W/ STUCCO OVER
9. 36"H. ORNAMENTAL IRON RAILING
10. 18"H. WOOD LOUVERED SHUTTERS
11. 15"H. WOOD LOUVERED SHUTTERS
12. WINDOW FRAMES SHALL BE VINYL, COLOR TO BE DETERMINED BY BUILDER
13. WOOD CORBEL (SEE DTL 5,6,8/A24)
14. 6X8 SHAPED BEAM PROJECTIONS
15. O'HAGIN, INC. CLOAKED VENT TILE (ICBO. 4214)

REVISIONS

	Description	Date
△	PLAN CHECK	8/15/07

TAMANO & CHAW
Residential Design

9980 Sunrise Boulevard, Suite 137
Rancho Cordova, CA 95742
Ph. (888)-538-6211
FAX (916) 330-7354

Plan No. 4382

Title: EXTERIOR ELEVATIONS

MONARCH HOMES
Lot 29, La Collina dal Lago
Folsom, CA

SCALE: 1/4" = 1'-0" (U.N.O.)

Date: 7/2/07

Drawn: PBP

Job No:

3

Figure C.1 This is an exterior elevation of the home, showing its key features. Notes and pertinent dimensions are listed within or alongside the drawing itself.

FIRST FLOOR PLAN
SCALE: 1/4"=1'-0"

Figure C.2 This is the first floor plan, what would be known as the architectural floor plan. The key purpose of this plan is to define the spaces by name and by dimensions in all three directions, some of the elements within spaces (e.g., a kitchen island doors, shelving; the range, fireplace, and hot water heater; floor finishes) and how they correlate. The designer refers the user to other pertinent drawings (e.g., Sections A and B). Walls types are differentiated by poché. Notice how the designer identifies areas with thickened walls by cross-hatching the entire width, saving considerable drafting time and giving the framer some flexibility as to how to frame the walls outlining the space. Make special note of the wall between the kitchen and bathroom, against which the range is placed. It is the subject of several drawings to follow.

393

ELECTRICAL PLAN - FIRST FLOOR
SCALE: 1/4"=1'-0"

Figure C.3 Here is the electrical plan for the same floor, with extraneous information removed for greater legibility. The locations of the service panel, equipment disconnects, power, fixtures, and devices that control them are described in notes or by using distinct symbols. Dimensions are not given: it is assumed that the electrical contractor will locate them according to the layout shown. Electrical devices and wiring are "roughed in" following the completion of the rough framing, so the electrical contractor uses the existing framework to measure from.

SECOND FLOOR PLAN
SCALE: 1/4"=1'-0"

Figure C.4 The second floor plan duplicates the nature of the information described on the first floor plan, but of course it applies to the second floor. In addition to the longitudinal and transverse sections (A and B) first introduced on the first floor plan, the designer has added a wall section for a portion of the building of particular interest. The wall between the master bedroom walk-in closet and the bathroom at the top of the drawing sits directly over the kitchen/laundry wall below.

ELECTRICAL PLAN - SECOND FLOOR
SCALE: 1/4" = 1'-0"

Figure C.5 As in the first floor electrical plan, the fixture types and the devices controlling them are shown in their desired locations. The different fixture types, the power, and communications outlets (TV and telephone) are identified by symbols, correlated to a legend on the drawing sheet.

ROOF PLAN
SCALE: 1/8"=1'-0"

Figure C.6 It is not unusual to see roof plans like this one, where the configuration and slope are described, but not much else. Truss manufacturers produce detailed drawings as a part of their contracts. They have their own ways of organizing the work, so all they really need is an understanding of what the designer wants. The addition of ghost lines to orient the user to the positions of certain parts of the roof is helpful.

Figure C.7 This detail and Detail 502 (Figure c.8) address the same issue, a code-required one-hour separation wall between the garage and living space. The required drywall installation in the two situations (on walls where joists are perpendicular to the wall and where they are parallel) is different. Which one presents the greater challenge to the builder?[5]

(PARALLEL JOISTS)   (PERPENDICULAR JOISTS)
**ONE-HOUR SEPARATION
HOUSE TO GARAGE**
SCALE: 1"=1'-0"

(PARALLEL JOISTS)   (PERPENDICULAR JOISTS)
**ONE-HOUR SEPARATION
EXTERIOR WALL**
SCALE: 1"=1'-0"

Figure C.8 The conditions described here are similar to those in Figure C.7, except that the exterior finishes, rather than interior finishes, are shown.

---

[5] Where the I-joists and wall are perpendicular, the drywall hangers have to cut around every joist. Where they are parallel, the drywall is installed to the underside of the sheathing, and a ledger to carry the ceiling board is installed over the drywall. In the wall between the dining room and the one-car garage, where this detail applies, there are sixteen bays in which to install the gypboard this way. It's worth pointing out that the framer is also affected: s/he has to notch the two LSL blocks on top of the wall to fit them between the trusses (due to the "I" shape of the trusses).

Figure C.9a The transverse section referenced on the first and second floor plans.

Figure C.9b This is the longitudinal section of Figure C.9a. Note how parts of the structure are dissected and the basic structure exposed. Where the section passes through an open space along its line, the drawing turns into an interior elevation (when it is inside the structure) and an exterior elevation when the section line passes through an open space outside the building. The designer addresses details such as insulation, wall plate heights, and interior finishes. The numbered bubbles refer to key notes in the drawing margin. In this case, 1 is "pre-engineered wood trusses," 5 is "¹⁄₂" gyp board (typ.)," and 3, 11, and 7 are batt insulation notes.

Figure C.10 This is the only wall section the designer felt compelled to draw: it is an exterior balcony with specific dimensions and architectural interest (note the foam corbel under the balcony). In a commercial drawing set, this feature would be detailed. In residential construction more is left to the contractor. There are contractors who specialize in this sort of work, and follow generally accepted guidelines established by manufacturers and the building code.

FOUNDATION PLAN
SCALE: 1/4"=1'-0"

Figure C.11 Although it may be the last system the structural engineer designs, the foundation drawings are usually one of the first in the drawing set.[6] Full-footprint plan views in structural drawings (foundation, floor, and roof plans), building sections, wall sections, and structural details all describe the elements of the structural system and the manner in which they are to be connected. The foundation system shown here, a reinforced concrete slab-on-grade with shear walls comprising 2×4 and 3×4 PTDF (pressure-treated Douglas Fir) and 2×4 #2 Douglas Fir studs and plates, 3×4 #2 DF studs and plywood, are common. This structural engineer makes heavy use of a keynoting system, thus leaving the drawing uncluttered. Unlike the dimensions shown in the architectural floor plan, which follow the boundary-plane dimensioning technique, these combine boundary-plane and axial dimensioning (see Chapter 4 for more information on dimensioning protocol). As in the architectural sheets, the various walls are distinguished by poché; however, the engineer is specific as to which walls are load-bearing walls. Notice, for example, the cross-hatched walls in the two-car garage opening, and the underlying continuous footing, shown dashed. Where concentrated loads exist (see footing drawings 1 and 2 in this garage), the footings are enlarged to bear them. The ghosting of floor plan walls is helpful in orienting the user of the drawings. Trace the wall between the kitchen and bathroom (top of the drawing) up through the roof in the following drawings. Using the numbers and symbols shown on the drawing for this wall, and the adjacent Shearwall Schedule, Keynotes, Hold Down Schedule, and Footing Schedule, one can identify all the requirements for the footing in this wall.

[6] The analysis and design of the structural systems frequently begin with the uppermost floors and is carried down through the structure to the foundation.

Figure C.12 The second-floor framing plan describes the floor frame as well as the walls below it. Special posts are shown in some walls—for example, in the one between the kitchen and laundry room (through which detail bubble 906 is drawn, top of the drawing). Post type 5 (see hexagon with the number 5 in it in the Keynotes) is a 6×6 #1 Douglas Fir post. The shearwall schedule calls for a type 5 wall –3×6 PTDF sole plate, 2×6 #2 DF studs 16" o.c., 3×6 #2DF studs at plywood seams, and $^3/_8$" APA rated plywood sheathing from plate to plate, with blocking at all seams, (which means 3×6 blocking at 8 feet or 4'×10' sheets of plywood). The length specified on the plan is 9'-0". (PTDF stands for pressure-treated Douglas Fir–lumber that has been chemically treated so that it maintains its structural integrity when in contact with concrete, which sweats.) The designation #1 or #2 pertains to the quality of the lumber as determined by a testing group like the Western Wood Products Association. APA is the American Plywood Association, which rates the quality of plywood.) The letter P with a number and arrow associated with it, shown primarily around the corners of the structure, is the designation of lines of force, to which a set of engineering calculations is associated. It correlates the calculations to the drawing, so that plan checkers in the building department can ascertain the appropriateness of the calculations.

APA RATED SHEATHING
PER PLAN

P.E.N. 8d @ 6"O.C. U.N.O.

2X BACKING

JOIST PER PLAN

A35 CLIP @ 8" O.C.

SHEAR WALL WHERE
APPLICABLE

Figure C.13 This detail, drawn through the wall between the kitchen and laundry room, instructs the builder to tie the floor diaphragm to the top of the first-floor wall using off-the-shelf hardware, in this case Simpson A35 connectors 8" o.c. (center to center). A35s are light-gauge metal angles used for connecting framing elements. For more information on this and other framing hardware, see Simpson's web page at www.strongtie.com.

Figure C.14 The roof plan shows the roof structure, in this case prefabricated wood trusses, the key elements describing the configuration (thick bold lines, in some cases dashed, showing the ridges, hips, partial hips, jacks, valleys, and overhangs), the cases where one roof supports another (see the shingle-like symbol), and the location of the platforms for the furnace and air handler units, shown as rectangles with lines on a bias (the number 21 in the octagon) and a rectangle with lines from corner to corner (number 20 in an octagon). The second-floor walls that have a special structural connection to the roof diaphragm are also shown cross-hatched. Note how the wall between the Master Bedroom walk-in closet and the Bedroom 2 bath has changed to wall type 2 (see shearwall schedule) and is 10'-6" long–1'-6" longer than the first-floor wall it sits over (refer back to the architectural and structural plans, C.2, C.11, and C.12). This is because the first-floor wall is a collector for loads, primarily wind load, but also reactions to seismic events, from the roof, second-floor walls, and floor diaphragm. The second-floor wall (Type 2) is a collector for the same loads, but only from the roof and second floor walls. As in C.12, the lines of force that the engineer correlates to calculations submitted to the local building department, are shown as Ps with numbers and an arrow. Note that the only indication of the location of the furnace and air handler is in this drawing–the weight of the equipment must be accommodated by the structural engineer.

403

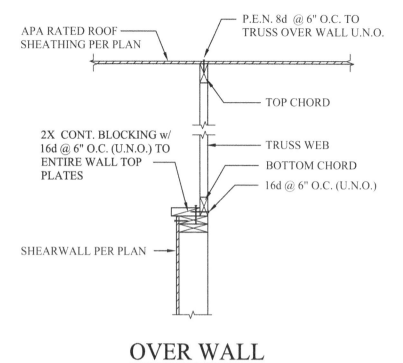

APA RATED ROOF
SHEATHING PER PLAN

P.E.N. 8d @ 6" O.C. TO
TRUSS OVER WALL U.N.O.

TOP CHORD

2X CONT. BLOCKING w/
16d @ 6" O.C. (U.N.O.) TO
ENTIRE WALL TOP
PLATES

TRUSS WEB

BOTTOM CHORD

16d @ 6" O.C. (U.N.O.)

SHEARWALL PER PLAN

OVER WALL

P.E.N. 8d @ 6" O.C.

APA RATED ROOF
SHEATHING PER PLAN

TOP CHORD

WEB

2x BACKING

BOTTOM CHORD

16d @ 6" O.C. (U.N.O.)

VARIES

ALT. OFFSET

Figure C.15 This detail is similar to Detail 906 (see Figure C.13), but rather than connecting a floor diaphragm to the shear wall, it ties the roof diaphragm to the shear wall through the roof truss.

8d @ 6" O.C EDGE NAIL
SHEATHING ON THIS
TRUSS T.C. BOTH SLOPES

EQ.   EQ.

STRAP TO TOP PLATES

TRUSS B.C. & SCAB 8' LONG
w/ (26) 16d NAILS MINIMUM

STRAP PER PLAN. FOR STRAPS
GREATER THAN 1 1/2" WIDE
INSTALL 2X4 SCAB @ BOTTOM
CHORD AS SPECIFIED ABOVE
ON SINGLE MEMBER TRUSSES.

POST PER PLAN

ISSUE DATE:	8/3/2007
PROJECT MANAGER:	PM
DESIGNER:	PM
DRAWN BY:	SRB
CHECKED BY:	RR
REVISIONS:	–

SHEET TITLE:

STRUCTURAL DETAILS

SHEET NUMBER:

SD.3

JOB NUMBER:

413	DRAG TRUSS STRAP

Figure C.16 Detail 413 is a "drag" connection, a connection that transfers a shear load caused by forces acting laterally from a member in one plane to a member in another, in this case, between the bottom of the truss and the top of the wall.

Commercial Building Drawings

The drawings in the following pages illustrate similarities and differences between commercial building construction and residential drawings.

Figure C.17 This drawing typifies the foundation system for a small commercial building. Note the use of a grid line system and the numerous references to details. You'll also notice that this structure employs CMU for walls and structural steel columns for the frame—not common in residential structures.

52'-

8"   14'-0"   8"

6'-0"   3'-4"

A
A-4

SIM.   3
A-4 A-6

WEATHER PROOF
LOUVER WITH
SECURITY BARS

F.E.C.

7

HYDRAULIC ROOM

118

SINGLE

A-4 A-6

A.L.

6

A
A-4 A-4

B
A-4 A-4

E
A-4 A-11

NEW ½" GWB ON 3⅝" METAL STUDS @
16" O.C. W/ BATT INSUL. BETWEEN
THE STUDS

NEW 8" CMU

NEW SEALANT JOINT TYPE MJB

NEW 2"x⅛" STEEL STRAP ANCHOR
ANCHORED W/ ⅜" Ø 4" L EXP.
ANCHORS @ 2'-0" O.C.

NEW ¼" EXP. ANCHOR 4'-0" O.C.

EXISTING CONC. COLUMN

EXISTING 8" CMU WALL

1½"

6"

8"

7"   7"

℄
COLUMN

DETAIL (TYP.)   A
SCALE: 3 INCHES=1 FOOT   A-4 A-4

0   3"   6"   12"

DETAIL (TYP.)   B
SCALE: 3 INCHES=1 FOOT   A-4 A-4

0   3"   6"   12"

Figure C.18 These three drawings and the two on the next page reflect the typical progression of drawings, from general to increasingly specific. Examine them all one by one starting with the plan at the top of this page.

ING
SPOUT

SHT. METAL
SHPAN

ROOF MOUNTED
EXHAUST FAN

EXIST CONC
BEAM

3/4" H.M. INSUL.
ITY DOOR.

NEW SHT METAL
DOWNSPOUT

NEW WEATHER-
PROOF LOUVER
W/SECURITY BARS
(TYP)

3'-4"

4'-0"

ELEVATION
FOOT

A
A-4   A-5

CONC. BEAMS & COLUMNS
OND

NEW MECH. EXHA

STEEL JOISTS

OPEN
ICE BAY

CONC. COLUMNS BEYOND

ONC FLOOR SLAB
PILLARY WATER BARRIER.

HYDRAULIC ROOM
(11B)

5
A-6 A-6

6
A-6 A-6

BUILDING SECTION

1
A-4 A-6

EXHAUST

SCALE: ½ INCH = 1 FOOT
12" 0   5

Figure C.18 (*Continued*)

Figure C.19 These drawings take off from where the drawings referred to in Figure C.18 end. Wall sections and details of the sills and heads of the wall contain significant numbers of notes. The three drawings on the next page describe parts of these two walls in greater detail, as does the drawing on the following page.

JAMB DETAIL (B) A-3 | A-8 ✓
SCALE: 3 INCHES = 1 FOOT

1¾" INSUL. H.M. SECURITY DOOR

SWEEP STRIP

THRESHOLD-BED IN SEALANT-(TYP.)

CONC. FLOOR SLAB

4"x10" PRECAST CONC. SILL W/2 #5 BARS AND NOSING

PAVEMENT

2"

1"

SILL DETAIL (C) A-6 | A-8

TYPICAL DOOR DETAILS

SCALE: 3 INCHES = 1 FOOT

0   3"   6"   12"

½" GWB ON 3⅝" METAL STUDS @ 16" O.C. W BATT INSUL. W/ VAPOR BARRIER BETWEEN THE STUDS

8" CMU WALL

8"x16" H. C.M.U. BOND BEAM LINTEL W/ 3 #5'S

DAMPPROOFING-(TYP.)

8" C.M.U. BOND BEAM LINTEL W/ 2-#5'S

SEALANT- JOINT TYPE MJB BOTH SIDES (TYP.)

5¾" P.S. DOOR FRAME-FILL W/ GROUT

1¾" INSUL. H.M. SECURITY DOOR

SEALANT-JOINT TYPE MJB (TYP.)

3/8"

2"

1'-2"

5¾"   1⅞"

HEAD DETAIL (M) A-8
SCALE: 3 INCHES = 1 FOOT

HEAD DETAIL (L) A-6 | A-8
SCALE: 3 INCHES = 1 FOOT

Figure C.19 (Continued)

BUILT-UP ROOFING
OVER RIGID INSULATION

STRIP FLASHING

SHT. METAL FASCIA

WOOD NAILERS
W/ $\frac{3}{8}$"ø EXP. BOLTS @ 4'-0"
O.C. KERF AND TREATED

TOP OF CMU BOND BEAM.
ELEV. VARIES

CONT. EDGE STRIP (TYP.)

8"X16" HIGH CMU BOND BEAM
W/2-#5 BARS, CUT TO
MATCH ROOF SLOPE

8" CMU W/JOINT REINF. @ 8" O.C.

BATT. INSULATION W/VAPOR BARRIER
$\frac{1}{2}$" GWB ON 3$\frac{5}{8}$" METAL STUDS
@ 16" O.C.

METAL ROOF DECK
OVER STL. JOISTS

PREMOULDED
CLOSURE STRIPS

DAMPPROOFING
(TYP.) @ INT. FACE
OF ALL WALLS FURRED
OUT FOR INSULATION

6"

ROOF DETAIL

SCALE: 3 INCHES = 1 FOOT

F
A-8 | A-10

Figure C.19 *(Continued)*

Figure C.20 This floor plan for a commercial building bears a significant resemblance to the floor plan in Figures C.2 and C.4 in its focus.

Figure C.21 The simplicity of this roof is significant compared to the roof plan shown in Figures C.6 and C.14.

Figure C.22 In stark contrast to the residential drawings, which have no mechanical information in them except for the locations of the condenser unit and furnace and air handler, mechanical drawings in commercial drawings are elaborate. This drawing shows part of the distribution and air recovery portion of the system for a portion of one floor.

Figure C.23 Commercial building drawings often employ multiple plans for power, signal, and lighting, unlike residential, which typically combine them on a single plan.

Figure C.23 (*Continued*)

Figure C.23 (Continued)

Part III: Inventory of Information in a Sample Custom Home Drawing Set

The following table is a description of the information contained in a set of construction drawings produced for a custom home builder. Custom home builder in this example describes individuals or legal entities that purchase multiple finished lots from subdivision developers, employ architects or building designers to produce project drawings (or create their own), and build homes on speculation or sell their projects to consumers on a build-to-suit basis.[7]

---

[7]The drawings that this table is derived from were created by Tamano and Chaw Residential Design of Rancho Cordova, CA, for Monarch Homes, Inc., of Cameron Park, CA.

Page Number and Title	Description of Information	Word Count
G.1-Title Sheet and General Notes	• Graphic information—vicinity map • Index of Drawings • Title Block: consultants list and contact information (structural engineer, Title 24 compliance analyst), change data (revisions made after publication of the construction drawings; e.g., changes required by the building department having jurisdiction over the project), designer contact information, sheet title; project owner, project name, and building model number • Building Code Data: applicable building code, occupancy group (type of construction, occupancy group, number of stories), building and garage areas • General Notes: (e.g., seismic straps for water heaters, glass type for glass in doors and windows used for escaping a room, references to building codes regarding components such as fireplaces); 39 topics, 112 lines, 15 words/line	
		1680
	• General Framing Notes (e.g., bracing requirements, quality of lumber, references to building code); 18 topics, 43 lines, 14 words/line	602
	• Attic Mount FAU Systems (e.g., notes regarding furnaces and air handlers located in the attic of the structure); 3 topics, 13 lines, 14 words/line	182
	• General Electrical Notes (e.g., smoke detector requirements, equipment requiring dedicated circuits, duct sealing requirements [reference to the Uniform Mechanical Code, California Energy Standards], GFCI requirements in wet locations, references to the National Electrical Code); 27 topics, 77 lines, 16 words/line • Symbols—changed data	1232
1-Floor Plan	• Graphic information—Architectural floor plan with dimensions • General Notes (e.g., substrate for shower enclosures, reference to the Uniform Building Code); 10 topics, 30 lines, 15 words/line	450
	• Key Notes (e.g., spring height of arches, fireplace type, appliance description), 24 notes, 6 words/note	144
	• Plan notes (e.g., window type, door size, ceiling height, floor covering type), 165 notes, 2 words/note • Dimensions— 95 on horizontal and 101 on vertical dimension lines[8] • Ceiling height notes (9', 10', 11',12', 13', 14', 15'-6", high ceilings with cornice soffits in the master bedroom and foyer) • Symbols used include poche representing two different wall types (2×4 and 2×6), two major building section symbols, 1 material symbol, 57 applications of 19 different template symbols, change clouding (the method by which changes to a drawing set are brought to the attention of users who may have earlier copies of the drawing set) • Line types: at least 8 different line types including dimension, extension, object, profile, center, title, hidden, leader, and border lines • All the title block information mentioned in G1 above plus the drawing scale, and area calculations for the footprint of the home, the garage, the covered patio, and the porch	330

---

[8]See Figure 4.14 for a description of line types.

Page Number and Title	Description of Information	Word Count
2-Exterior Elevations	• Graphic information: front and left elevations • Elevation notes: specifications regarding exterior finishes, (e.g., conformance of flashing under plaster to building code sections pertaining to moisture barriers), 4 topics, 20 lines, 15 words/line • Attic ventilation notes (e.g., 10 calculations determining, among other things, the area of required attic ventilation) • Key Notes: 12 notes, 4 words/note • Dimensions: finished floor elevation; 4 plate heights (10'-0' $^3/_4$", 11', 13', 15'-6"), the spring points, radius, and rise of 4 arches, elevation of illuminated building numbers, window head heights, elevation of fireplace stack relative to any other portion of the building 10' away • Symbols: 4, including roof pitch indicator, datum points, reference symbols to architectural detail sheet, glass • Line types: object, hidden profile, "ghost," dimension, extension, leader, and border lines • All the title block information mentioned in G1 above plus the drawing scale	300 48
3-Exterior Elevations	• Graphic information: rear and right elevations • Elevation notes: specifications regarding exterior finishes, (e.g., conformance of flashing under plaster to building code sections pertaining to moisture barriers), 4 topics, 20 lines, 15 words/line (repeats above) • Attic ventilation notes (e.g.,10 calculations determining, among other things, the area of required attic ventilation); repeats above • Key Notes: 11 notes, 5 words/note • Dimensions: finished floor elevation; 4 plate heights (10'-0' $^3/_4$", 11', 13', 15'-6"), windowsill heights • Symbols: 4, including roof pitch indicator, datum points, reference symbols to architectural detail sheet, glass • Line types: object, hidden profile, "ghost," dimension, extension, leader, and border lines • All the title block information mentioned in G1 above plus the drawing scale	55
4-Foundation Plan	• Graphic information—foundation plan with dimensions • Key Notes (e.g., slab-on-grade materials, treatment of slab at door openings); 11 notes, 6 words/note • Plan notes (e.g., spread footing dimensions) • Shearwall schedule—matrix of 5 columns, 7 rows, 20 separate instructions regarding shear wall construction • Dimensions—53 on horizontal and 35 on vertical dimension lines,[9] 4 different spread footing sizes in 11 locations • Symbols—2 major sections, 10 different detail references in 22 locations, door threshold (created by designer), shear wall designation (three different symbols), and a symbol for hold-down hardware. • Line types: at least 9 different line types including dimension, extension, grid, object, profile, center, ghost, hidden, shear wall (showing length of wall and type, created by designer, 17 locations, 16 different lengths, 4 different wall types), leader, and border lines[9] • All the title block information mentioned in G1 above plus the drawing scale	66
4S-Shear Wall Plan	• Graphic information—architectural floor plan in ghosted lines • Key Notes—n/a • Shearwall Notes—12 topics, 17 lines, 10 words/line • Plan notes (e.g., hold-down types, wall shear length) • Shearwall schedule—matrix of 5 columns, 7 rows, 20 separate instructions regarding shear wall construction • Dimensions—shear wall length only • Symbols—shear wall designation (3 different symbols), 5 different symbols for hold-down hardware • Line types: at least 7 different line types including grid, object, profile, ghost, hidden, shear wall (showing length of wall and type, created by designer, 17 locations, 16 different lengths, 4 different wall types) and border lines[9] • All the title block information mentioned in G1 above plus the drawing scale	170

---

[9]See Figure 4.14 for a description of line types.

Page Number and Title	Description of Information	Word Count
5-Roof Framing Plan	• Graphic information—roof structure plan with notes referring to wall elements • Key Notes (e.g., roof sheathing material requirements, overhang details, hardware requirements for roof-wall connections); 15 notes, 9 words/note • Roof Framing Notes (e.g., material quality, bracing information), 8 topics, 20 lines, 7 words/line • Plan notes—wall framing materials at beams, beam types and sizes (4 different beam sizes/materials in 46 locations), 110 special notes regarding beam posts, roof truss information • Legend (e.g., reference to beam calculations, beam material requirements), 5 abbreviations, 5 words/abbreviation • Dimensions—overhang (2 different lengths), plate heights (4 different heights and two locations of raked [sloped] walls) • Symbols—2 major sections, 11 different detail references in 20 locations, roof pitch indicator, poché at roof overbuilds,[10] continuous wood member, drag and girder truss (created by designer) • Line types: at least 5 including object, profile, ghost, hidden, leader, and border lines • Other—number of different trusses required (4 locations, all different), 7 girder trusses in 6 different configurations (2 integrated truss/girder locations), location and extent of 5 overbuild roof sections • All the title block information mentioned in G1 above plus the drawing scale	135 140
6-Sections	• Graphic information—major building sections in two locations perpendicular to one another, combined with interior and exterior elevations where applicable • Key Notes (e.g., roof structure elements and components, insulation requirements in roof, truss types), 12 notes, 16 lines, 6 words/line • Dimensions—plate heights (4 different heights) • Symbols—insulation, earth, concrete, continuous wood, roof pitch indicator, glass • Line types: object, profile, dimension, extension, ghost, leader, and border lines • All the title block information mentioned in G1 above	96
7-Electrical Plan	• Graphic information—floor plan (e.g., walls, appliances, HVAC equipment, condenser unit disconnect, exhaust fans, cabinets, doors, windows, plumbing fixtures, locations of lights, switches, outlets, communication system ports including telephone and television, fire safety equipment, service panel type and location) • Symbol legend—30 electrical symbols, many of the designer's creation • Key Notes—n/a this sheet • Plan Notes—(e.g., fixture types, location of condenser unit disconnect, service panel type) • Dimensions—n/a this sheet • Symbols—30 electrical symbols, many of designer's creation, changed data indicator, concrete, template symbols (e.g., range, washer, dryer, lavatories, water closets, tubs, water heater) • Line types: object, profile, hidden, leader, and border lines • All the title block information mentioned in G1 above	
Architectural Details	• Graphic information—5 details (sections through) eave and trim band ornamentation • Key Notes—n/a this sheet • Drawing notes—average of 3 per detail • Dimensions—average of 5 dimensions per detail • Symbols—rigid insulation, batt insulation, large-scale plywood in section, continuous wood • Line types: object, profile, hidden, leader, and border lines • All the title block information mentioned in G1 above plus the drawing scale	

---

[10]A roof overbuild in this context refers to a portion of one roof that supports another roof structurally: rafters or trusses are installed and sheathed, and the "overbuild" portion of the roof is constructed over the initial structure.

Page Number and Title	Description of Information	Word Count

EN-1 Energy Compliance

- The subject here is compliance with energy requirements (in this case Title 24, Parts 1 & 6, California Code of Regulations)
- Graphic information—series of tables and forms duplicated from computer generated energy analysis
  - Project title and address, consultant (analyst) contact information, climate zone the project is located in, analysis method
  - General Information, including area of conditioned space, construction type, polar orientation, ratio of window area to building footprint, window specifications, average wall height
  - Building shell insulation (an analysis of the R-Value[11] of wall assemblies and the roof system)
  - Analysis of the U-Factor[12] of the building fenestration (the arrangement of windows and doors in the elevation of a structure), including a list of the number and type of each window and door, their area, U-factor, Solar Heat Gain Coefficient, orientation to the sun, and any interior or exterior features (of the building or window) that might shade it
  - Water heater performance evaluation and specifications for compliance
  - HVAC System analysis including duct R-Value, efficiency ratings of the furnace, air handling unit, and condenser unit
  - Compliance statement—an affidavit that the drawings list the "building features and performance specifications needed to comply with Title-24, Parts 1 and 6 of the California Code of Regulations, and the administrative regulations to implement them."[13]
  - "Mandatory Measures Checklist"—a checklist signed by the designer and (after the construction is complete) the enforcement agency responsible for compliance. This document consists of some 45 specifications describing **Building Envelope Measures** (insulation requirements of roof and wall assemblies), **Space Conditioning, Water Heating, and Plumbing System Measures** (criteria for heating and cooling loads, thermostat requirements, insulation of storage and distribution systems including tanks, ducts and pipes, and **Lighting Measures,** for example, the minimum lumens/watt[14] required for lighting

SD-1[15] Structural Details

- Graphic information—framing elevation showing the elements comprising a typical wood light frame wall, with nailing requirements between elements (for example stud to sole plate, stud to top plate, king stud to header, trimmer to king stud), minimum lap requirements of top plates, the connection between sole plate and foundation, shear blocking requirements, ratio of header depth to opening width, among other things.
- General Notes—notes referencing specific building code sections pertaining to, e.g., glass requirements for windows in a room used for escape, rise and run ratio of stairs, firewall requirements, material quality requirements, and other specifications; 29 notes, 75 lines, 11 words/line — 825
- Foundation notes—26 specifications covering such topics as minimum concrete quality for foundations and slabs-on-grade, reinforcing steel quality and coverage, compaction requirements for subgrade (engineered earth that underlies and supports paving or a foundation); 26 specs, 62 lines, 11 words/line — 682
- Framing notes—16 building code references and other specifications covering, among other things, lumber quality, minimum size for various frame elements, nail size and type, bolting requirements, manufacture of glued laminated beams; 16 notes, 43 lines, 11 words/line — 473
- Structural steel notes—6 notes, including references to the materials, fabrication, and erection requirements of the American Institute of Steel Construction (AISC), to the American Society of Testing and Materials (ASTM) for bolting requirements and the quality of materials used in pipe and tube steel elements, and to the American Welding Society (AWS) for welding requirements; 6 notes, 10 lines, 11 word/line — 110
- Roof truss notes—11 notes referencing specific building codes sections on truss manufacturing, quality of material used in top and bottom chords and webs, submittal requirements, temporary erection requirements, and others; 11 notes, 20 lines, 11 words/line — 220
- Drawing notes—some 22 notes on the framing elevation averaging 10 words each — 220
- Shearwall schedule and legend—matrix of 5 columns, 7 rows, 20 separate instructions regarding shear wall construction
- Dimensions—n/a this sheet
- Symbols—Legend correlated to shear wall schedule
- Line types: object, leader, break, and border lines
- All the title block information mentioned in G1 above plus the drawing scale

---

[11]R-Value is the resistance of a given barrier such as the exterior sheathing or a wall cavity to the flow of heat across it.

[12]U-Factor is the rate at which heat between two fluids (exterior and interior air, for example) is lost per unit of area.

[13]From the compliance statement shown on EN-1 of the drawing set.

[14]A lumen is an SI (metric) measurement of light intensity and flow.

[15]This drawing is similar to the "S1 Sheets" produced by structural engineers in commercial building projects. They are generic specs pertaining to typical situations (e.g., how the typical masonry wall reinforcement should be installed, how the structural members in a wood light frame should be connected—nailing requirements for myriad connections of wall elements is an example—and other structural details).

Page Number and Title	Description of Information	Word Count
SD-2 Structural Details	• Graphic Information— 13 details (all sections each with its own scale) showing footing requirements in 10 different situations, 1 hold-down detail, 1 section through a concrete slab with a thickened edge, and 1 shear wall • Notes—an average of 11 notes per detail, with 4 words/note • Dimensions—an average of 4 per detail • Symbols—earth, crushed rock, welded wire fabric, reinforcing steel, concrete, continuous wood, insulation, flexible expansion joint, masonry • Line Types— object, profile, hidden, center, dimension, extension, leader, break, and border lines • All the title block information mentioned in G1 above	572
SD-3 Structural Details	• Graphic Information—20 details, all with their own scale, including 1 isometric (showing the connection between the bottom chord of the wood trusses and the top plate of nonbearing walls), 1 plan view of the material requirements of curved walls and the connection between walls intersecting at something other than 90°, 2 framing elevations (one of the overbuild requirements and the other of a truss heel-to-wall plate connection), and 16 sections. The sections show, among other things, the connections between ridge and rafter, truss bottom chord and wall top plate, roof truss and wall; as well as gable end bracing; 6 different truss to wall shear transfer connections; framing requirements where two gable ends of different heights connect; and a ledger-to-rafter connection. • Notes—an average of 9 notes per detail, with 5 words/note • Dimensions—an average of 4 per detail • Symbols—earth, crushed rock, welded wire fabric, reinforcing steel, concrete, continuous wood, insulation, flexible expansion joint, masonry • Line Types— object, profile, hidden, center, dimension, extension, leader, break, and border lines • All the title block information mentioned in G1 above	900
SD-4 Structural Details	• Graphic Information— 9 details, all with their own scale, generally more complex than those on the previous two sheets, including 4 framing elevations (one of a drag connection,[16] one of an alternate top plate splice, to be used when the normal 4' offset of splices between the two top plates cannot be achieved), one of the required connection between walls that align but are of two different heights, and one describing the required connection between the top plates of a wall and a beam supported by the end of the wall or, in the case of two intersecting walls, the corner. The sections show another drag connection, a floor diaphragm shear transfer connection, and three situations in which two roof structures of different heights and the walls associated with them are described. • Notes—an average of 9 notes per detail, with 5 words/note • Dimensions—an average of 4 per detail • Symbols—datum points, small scale plywood, continuous wood • Line Types— object, profile, hidden, center, dimension, extension, leader, break, and border lines • All the title block information mentioned in G1 above	405

Summary		*Word Count*
13 pages	*At least 58 drawings, including 5 full-footprint plans, 4 exterior elevations, 2 building sections, and 47 details*	*10,037*

[16]Drag connection, at least in this context, is a connection between a component or assembly in a roof structure (in one plane) and a shear wall (in another plane). A shear wall resists lateral movement in a building caused by wind or seismic activity.

APPENDIX D

Drawing Review: A Journey through Commercial Building Construction Drawings

The following scenario as an example of a typical drawing review exercise, and it represents just one approach to interpreting the project data.

Project: A single-story, 15,000-square-foot California Highway Patrol office and separate service facility and fueling station in San Francisco, CA.

Description[1] *Office Building*, B2 occupancy,[2] Type V, nonrated, sprinklered

 Service facility: B1 occupancy, Type V, 1 hour fire resistive, sprinklered

---

[1]This project was constructed under the Uniform Building Code. See Chapter 2 for a discussion of building codes.
[2]A "B" occupancy describes office buildings, fire and police stations, professional offices, and post offices among other things. Refer to any recent U.S. building code or the International Building Code for more specific information on occupancy and fire rating.

Task: Determine the board footage of framing lumber, manufactured lumber, shear plywood, and all the hardware and fasteners required in the load-bearing walls of a single-story wood light frame structure. This is a focused task. Some of the exploring that might normally be done (say as a project manager or superintendent preparing to manage the construction of the entire project) can be postponed or ignored completely in the interest of time.

Preliminary Comments

As will be evident after reading the following table, there is a substantial quantity and variety of information to sort through, categorize (if only mentally), take notes on,

Construction Graphics: A Practical Guide to Interpreting Working Drawings, Third Edition. Keith Bisharat.
© 2025 John Wiley & Sons Inc. Published 2025 by John Wiley & Sons Inc.

and prioritize. It should be pointed out that regardless of whether a drawing sheet "informs" the task at hand, all the drawings and notes must be skimmed at least once, if only to identify the information that is irrelevant. Of the information quantified in the last row of the following table, some 6 architectural plans, 6 structural plans, 4 major building sections produced by the architect, 9 exterior elevations, 59 architectural details, 64 structural details, 1 mechanical, and 1 electrical detail, bear to varying extents on a framer's typical responsibilities. Having said that, a framer would devote the most time and energy (particularly with a schedule-driven task like a material quantity survey) to the structural sheets, with periodic visits to the architectural sheets primarily for refreshers as to space, finishes, dimensions, and the correlation of components. Although this is a graphics-oriented textbook, text plays a key role in the description of all projects, and it is worth reiterating that drawings and specifications, which form the bulk of the instructions regarding what is to be built, play different roles but are equally critical and should be reviewed in tandem. For example, in the "S1 sheets" of the subject project (and in many others), the structural engineer calls for predrilling wood members to avoid splitting the wood. This is critical information in ultimately determining labor costs, but how would one draw that? The point here is that the specifications are inseparable from drawings when it comes to fully understanding the requirements of a project, and they should be close at hand when doing a material quantity survey.

The challenge an estimator faces is translating all those written and graphic instructions into discrete processes to which a value in time and money can be attributed. When one sees a drawing of a beam-to-column connection in which the bolts and nuts must be countersunk, say into a beam that is connected to the knife plate[3] in a tube steel column, what does the drawing represent in terms of work? Well, a beam manufacturer has to be contacted and given a set of drawings or at least enough information to estimate the manufacturing and transportation costs and to price the beam. Once both the contractor seeking the quote and the manufacturer or supplier agree to cooperate, they sign a contract that (hopefully) reflects their mutual understanding. The manufacturer produces the shop drawings manifesting how the beam will be manufactured and sends the drawings to the contractor for approval. The contractor reviews the drawings for compliance with the contract documents and sends the shop drawings to the structural engineer, who reviews them and approves them as is, with comments, or rejects them and they return to the contractor. If they are approved, they are returned to the manufacturer, who then produces the beam and prepares it for transportation to the site. Once the beam arrives

at the site, it is off-loaded, usually by project personnel using hoisting equipment on the site, and stored in the installed position (with the long, or y, axis vertical) somewhere on the site. When the time comes to install it, a carpenter will unwrap the beam, mark it, cut it to length, and cut the slot in the ends so the beam can be placed over the knife plate on the haunch (seat) attached to the column, probably with a chain saw. A hoisting device then picks the beam up and places it on the column, guided by the carpenter, who has had to set up a ladder or hoist to provide safe access to the connection. The beam is stabilized or braced as required and the rigging used to hoist it is removed. Now, unless the carpenter has a special drill bit that produces a countersunk hole simultaneously with the hole for the bolt, the drilling work must be done in at least four steps: 1) the location of the holes on both sides of the beam has to be identified and marked, after which 2) the larger of the two holes can be drilled, first on one side of the beam, then on the other. Then, 3) the drill bit has to be extracted from the drill motor and replaced with a bit slightly larger than the shank of the bolt and long enough to penetrate the beam, and the drill operator 4) begins drilling the remainder of the hole, hoping the drill bit actually lines up with the hole in the knife plate hidden inside the beam and doesn't break in the process (bits often bind and snap when they are drilled through wood into a metal hole). Or the hole is drilled part way from one side and finished from the other. For a typical column-to-beam saddle connection, wherein the beam sits in a U-shaped metal column cap that wraps up around the beam, the task of drilling is much simpler: only one bit is required, and no layout is necessary because the drill operator has a hole in the saddle to indicate the location of the hole, and has only to imagine where the corresponding hole on the other side of the beam is, or better yet, the drill operator starts on one side, drills half way through, then moves the drill to the other side to finish the task. How long would this take? How much longer would it take if the drilling were done 30' off the ground, from a scissor lift? Is there room to work? Is there adequate light in which to perform this work? This is the sort of exploration that must take place when a detailed estimate is required.

■ Quantifying the Work

So how is such work quantified? There are a couple reasonable answers to that: One can determine a time value for drilling each hole and multiplying that factor by the number of holes to be drilled. One might also apply a production rate for drilling holes on the basis of inches per unit of time required. The latter may be a more prudent approach in that this unit accounts for differences in beam width, which can be significant. Selecting units

---

[3]A knife plate here is a piece of steel plate with bolt holes that is welded to the center of the tube steel column from which it projects.

organizing the information, and deciding what to monitor and account for is the responsibility of the estimator, or the person managing the estimating process. (Refer to "Interpreting Drawings" in chapter four for more information on unit selection.)

The Take-off Process

As noted in the foregoing paragraph, someone has to develop a plan to approach the estimate and ultimately the construction of the project, but exactly what transpires varies with individuals, project types, time constraints, and workload. One ubiquitous fundamental step to estimating construction projects, however, is executing material quantity take-offs, which forms the basis for much of the "costing" and subsequent pricing of construction projects.

Purposes of the Take-off

The industry uses material quantity take-offs for various purposes, among them to develop the budget for projects in the design development stage; to forecast the labor, equipment, and material costs of a project; to compare alternative designs, develop cost proposals for changed work, or compare subcontractor estimates; and to analyze alternative construction methods. Additionally, preserving completed estimates (in which take-offs play such an important role) for the life of a company is prudent protection against future claims and results in the contractor having a historical database for use in future projects. Performing an effective take-off requires imagination, knowledge of numerous materials and processes of construction, knowledge of construction documents and the law, subtleties in language, familiarity with construction graphics, and good "peripheral vision,"[4] meaning becoming aware of aspects of a project that are outside the focus of one's immediate task. Individuals vary in their approach, however some common steps occur: The project must be "deconstructed" into parts that are recognizable and manageable; its components and assemblies discerned, listed, and quantified, in a format that is useful to the builder, and finally reassembled in a format that is conducive to forecasting its likely cost and helpful in monitoring progress. Commentary on presenting the information gleaned through the take-off process and examples of effectively executed take-offs follows.

Presenting the Information

The following information consists of suggestions as to how to approach a material quantity survey, some thoughts regarding presenting the information, and examples of a variety of material quantity take-offs.[5]

Notes to the take-off process

1. Define the problem clearly, in all it facets.
2. Develop a plan to solve the problem.
   - What is the level of detailed required?
   - Review the drawings and specifications; develop a shopping list as you go (MasterFormat is helpful for this).
   - Look for opportunities to maximize your efforts by identifying patterns (is the structure symmetrical, are there repetitive elements?)
   - How will you sort the project into manageable parts? (MasterFormat, work breakdown structure, "natural" demarcation points; e.g., building by building, room by room, vertical, horizontal, and sloped framing, wall and roof structure, or by elements with equivalent production rates, etc. Remember, if you can't measure it, you can't manage it.)
   - How many different elements are involved (discern)?
   - How are they distinguished from one another (grade, type, species, etc.)?
3. Determine how to organize the information you will record on the spreadsheet, with the following in mind:
   - Develop useful information according to priorities—time is limited.
   - To start, do those tasks involving major, or simple, or well-detailed items. If you're working on an item of minor significance and run out of time, the consequences of incorrectly estimating it are not likely to be grave.
   - Sort the components; list their elements, dimensions, and related notes; and calculate the quantities, using the shopping list (see Figures 7.1 or 11.2 for examples of a shopping list) and priorities as a guide.
   - Make the information relevant; often this means using "buy" quantities (cy [cubic yards], cm [cubic meters], tons, squares [100 square feet], ea [each], MFBM [thousand foot board measure: 1" × 12" × 12" = 1 board foot], sf [square feet], sy [square yards]), lf [lineal feet], SFCA [square foot of contact area], or production data figures, which are not always the same.
   - Make it easy to find, read, and check the information, for you and others.

---

[4]This term, and the phrase "discern and quantify" in the context of quantity take-offs, originated in an unpublished manuscript by Richard A. Nickles entitled *Construction Cost Estimating and Bidding: A Managerial Approach*.

[5]The example take-offs are from *Construction Cost Estimating and Bidding: A Managerial Approach*, an unpublished manuscript of Richard A. Nickles, Emeritus Professor California State University, Sacramento.

- Develop a good audit trail (road map to the decision making that occurs during the course of a project).
- Describe the element, component, or assembly succinctly and accurately, using the proper terms.
- What assumptions have you made?
- Where did you get your information (plan page numbers, reference details, cost information references, etc.)?
- Where is (are) the component(s) located in the project?
- How did you arrive at your figures (units of measure, formulas, and so on)?

4. Determine that the format is the right one by sketching the spreadsheet, using the components you will be quantifying.

5. Adjust as necessary and do the spreadsheet (column and row labeling, "noncalculated" numbers, formulas).

6. Do the numbers.

7. Check your work.
- Inspect your work randomly for obvious errors using a magnitude sensitive test.
- Apply "check" formulas to verify accuracy (ratios help, and are easy to develop).

8. Summarize, or total, your work, clearly identifying the "bottom line" and make a note that you've checked your work.

Notes: The management of the construction process consists largely of effectively managing and communicating lots of information. In doing take-offs, your fundamental goals are to correctly analyze the problem or task, plan your solution to it, "break down" the task into manageable parts, reorganize the information into a pattern that is useful and easy to access for you and others, and properly explain your decisions and assumptions (audit trail).

1. Format and Organization
   a. On each sheet, clearly identify the project, the scope of the take-off, the author and the page numbers.
   b. Use separate sheets of paper for components and elements—it's easier to combine than separate.
   c. If you include different components or elements on the same page, clearly distinguish them from one another.
   d. Organization should reflect a sensible project breakdown structure; e.g, components with similar production rates should be taken off together.
   e. Include essential information only; e.g., 2.0', especially if entered in a column in which the unit (feet) is already designated, adds to clutter, and increases the potential for error.
   f. Columns should line up from the top to the bottom of the page.

g. The format should facilitate a simple summary of the quantities.
h. Allow adequate space for readability and for adding quantities.
i. Include estimates of other materials or operations where possible, if the dimensions are identical (maximize your efforts).
j. Total quantities, in the correct units, should be visible at a glance.

2. Discern
   a. When listing, describe the element or component accurately and concisely.
   b. Identify the required materials, including type, quality, any specific lengths required, etc.
   c. Columns should include an abbreviated description of the material (grade, species, strength) or operation, units of measure (ft, sf. sfca, cf, ea), and applicable formulas.
   d. When listing, identify where in the project the component or element is (n, s, e, w, by room number, etc.) using words, detail reference numbers, or other identifiers that anyone possessing a set of the same drawings can use to locate them.
   e. Distinguish between like and unlike elements of the same sort; for example, concrete benches might be divisible by size, configuration, length, etc.
   f. Sketches help, but avoid duplicating what is shown in the drawings.
   g. If in doubt about something, take it off more than one way.
   h. Avoid duplicating listings and calculations.
   i. When listing, combine like elements into one calculation where possible, using a "notes" section to show your tally.

3. Quantify
   a. Develop some sort of system or method of determining where you are and where you have been, so that nothing is overlooked.
   b. Use the proper units—convert inches to tenths of feet, etc.
   c. Use the proper quantities—buy quantities, production figures, etc. For example, concrete is usually taken off and purchased in cubic yards, but labor to finish it is often calculated in square feet.
   d. Use written dimensions; when you have confirmed that a drawing is done to scale, and you have no other choice, use a scale.
   e. Round up at .5 or more, round down for .4 or less—two points past the decimal is adequate for most calculations, except board foot calculations

for large quantities of dimensioned lumber, which carry five points past the decimal.

    f. If one element (blocking in a wall for example) is accounted for in another element (in the stud count for the wall), account for the element by listing it separately, and make a note to that effect.

    g. Clearly identify deductions; brackets < > work better than parentheses ( ).

    h. Line out spaces where no calculations are intended.

    i. Rough calculations are helpful in checking work. Apply a magnitude-sensitive check of the numbers early in the take-off.

    j. Develop and use simple ratios to check work; e.g., 81 sf of slab-on-grade 4 inches thick is equivalent to 1 cubic yard (cy) of concrete; 54 sf of 6" slab-on-grade would consume 1 cy of concrete.

    k. Stick with net quantities; adjustments for waste can be made in an estimate summary sheet.

    l. Clearly identify subtotals, cumulative totals, and totals.

4. Audit Trail

    a. When recording dimensions and other data required to calculate quantities, use plan page numbers, detail numbers or other notes to relate the take-off to the documents.

    b. Listing multipliers (numbers of identical elements) is in itself a way to develop an audit trail; e.g., 5 concrete benches each with 9 #4 longitudinal bars 12 feet long is more descriptive than 540 LF of #4 bars.

    c. Record your assumptions and formulas.

    d. Retain pertinent notes and calculations (area calculations, for example) and reference them on your take-off.

Figures D.1–D.9 exemplify all the characteristics of a good quantity take-off: legibility, orderliness, proper formatting, accuracy, and effective mapping (a good audit trail). Their only shortcoming is the boldness of the grid. Taking a cue from Edward Tufte,[6] one could easily improve these displays by ghosting the gridlines and creating more space between the different components.

---

[6]Author of several wonderful books on data communication (*Beautiful Evidence, Envisioning Information, The Visual Display of Quantitative Information, Visual Explanations*).

Summary of the Pertinence of Drawing Sheets to a Framer's Typical Work Scope

Discipline	Title	Subject Matter	Comments
Title Page	No title block on this sheet: project name and address listed in a large frame on the front page of the set. (1)	List of abbreviations; symbols; building data including the applicable building, mechanical, plumbing, and electrical codes as required (in this case) by Title 24, Part 3, of the California Code of Regulations; and a sheet index. Brief project description including building area, occupancy group, construction type, and the deferred approval[7] items.	The title sheet often lists the contact information of the key participants (owner, architect, design consultants). In addition to the architect's contact information, the structural engineer's is critical for a framer to have. Despite the attempts to standardize graphic conventions, no single standard yet exists, so it is always useful to have a legend and list of abbreviations used by the particular design professionals involved. It is not unusual to find legends of this sort in the first page or two of every discipline. By listing their abbreviations and symbols, the design professionals are declaring the "dialect" they use. For a framer the timing of roof trusses (that is, when they are available for installation) is of interest from the scheduling standpoint, so a quick review of the deferred approval items is useful.

(Continued)

---

[7]The term "deferred approval items" refers to systems or components that have yet to be approved by the building department having jurisdiction over the project. In the interest of time, governing entities will allow construction to commence without a complete description of the project, with the understanding that the items identified on the deferred approval items list will be submitted for approval prior to their installation, whenever that occurs. Roof trusses, fire suppression systems, fire alarm systems, and specialized windows and doors are examples. In the case of fire suppression systems, it makes little sense to design them without complete project drawings. Their performance is substantially affected by the number, location, and type of fittings as well as changes in pipe size or direction, so it is critical that the fire system engineer know where all the spaces and key elements are (room boundaries; structural, HVAC, and plumbing systems; and electrical lighting and equipment elements) prior to designing the system. Changes are not viewed with equanimity by the fire marshal.

Discipline	Title	Subject Matter	Comments
			In addition, the construction type is of interest; both the structures in the example are Type V (wood light frame with some glued-laminated beams and manufactured lumber and structural elements). Building codes have specific fire safety requirements in all classifications, and because of its structural composition (a Type V is the most combustible of the five types), the fire safety requirements are of particular interest. (A load-bearing wall in a wood light frame contains numerous cavities, unlike a cast-in-place reinforced concrete bearing wall, for example, so many of the measures the code requires are focused on isolating the oxygen within the walls from a fire. This is true of fires that start within the wall as well—the oxygen that might otherwise be available is limited.) Construction drawings for this type of building do not always describe all the requirements (see commentary on items that are often missing from construction drawings following this table). Worth mentioning as well is the location of the project. San Francisco exists in the most restrictive seismic zone in the U.S., and consequently the structural requirements are the most extensive found. In particular, shear transfer connections between foundations and walls, walls and floors, and walls to roofs are significant.
Civil Sheets	C1—Existing Conditions and Demolition (2)	Graphic and written description of the site as it currently exists and instructions as to the focus and extent of the required demolition. Existing municipal utilities in the adjacent streets are described as well.	What is of interest here is the space available for the secure storage of lumber and manufactured wood after the demolition is complete and site access during construction. Portions of the chain-link fence in this project as well as a large linear paved area are to be preserved; however, no construction access is shown. Another issue worthy of investigation is the work environment: What space for prefabricating portions of the project might there be? What will the site immediately surrounding the foundation be like? Will it be reasonably level, a surface one might work on? (Some framers will demand that grade around a slab be level.) Is it easy to get on and off the structure? These and similar issued should be clarified with the prime contractor.
	C2—Grading Plan (3)	Site plan showing the footprint of the new structures, new parking, and the location (but not elevations) of the new offsite improvements Spot elevations of the finished floor (a slab-on-grade) and adjacent parking are shown.	Not much of great interest to a framer here, except the orientation of the streets, which the vicinity map did not show. The two streets bounding this property are both one way, which affects how materials will be delivered to the site. Such a circumstance probably has a greater effect on a structural steel sub on a high-rise structure than a framer, for example, because of the size of the elements and the vehicles required to transport them.
	C3—Layout Plan (4)	Plat of property with property line setback requirements, location of public improvements such as surrounding streets and sidewalks.	This sheet adds nothing to previous sheets.
	C4—Details (5)	17 details with varying scales describing site elements and components such as the plan for the dumpster enclosure, the typical concrete curb, sign footings, and the like. It is worth mentioning that there are two details called "16," and they are not redundant.	Nothing in a framer's scope on this sheet.

Discipline	Title	Subject Matter	Comments
	C5—Details (6)	A partial plot plan showing underground utilities on the west side of the lot, details regarding the handicap parking space and signage, a typical section of chain-link fence, landscape headerboard detail, and two details of different backflow preventer valves.	Nothing in a framer's scope on this sheet.
	C5.1—Fence Details (7)	Plan of rolling chain-link gate, elevation of gate and CMU wall adjacent to it, and section of gate footing.	Nothing in a framer's scope on this sheet.
	CM-1—Site and Utility Plan (8)	Underground utility drawing showing sanitary sewer, storm drain, domestic water system, fire service loop, gas piping, fuel, a pneumatic system line and new offsite improvements (curb, sidewalk, and gutter).	Nothing in a framer's scope on this sheet, but the absence of electrical underground (other than a 110 KV line in the center of one of the streets adjoining the property) prompts an investigation of the source of electrical power on the site. Sheet E7—Site Plan Electrical reveals the extent of the underground electrical system. The number, type, and location of the power panels—main and subpanels—may affect a framer's work, since the panels may be integrated within walls. Even surface-mount panels might affect a framer's work, since the substrate for the panel would be within or on the face of the wall.
	CM 1.1—Site and Utility Plan (9)	Appears to repeat CM 1, with the exception that spot elevations for the offsite improvements have been added, following revisions required by the building department, which stipulated that the sewer service from the property should increase from 4" to 6", that the driveways should conform to the city standard, and that spot elevations at the top of curb (TC) and back of walk (BW) should be shown.	Nothing in a framer's scope on this sheet.
Abatement	AB-1—OSA Building #61, 1st Floor Asbestos Abatement Plan (10)	This drawing describes the asbestos abatement requirements for the project. It appears to be architectural floor plans of the (existing) structure to be demolished and instructions pertaining to the abatement of asbestos in the structure.	Nothing in a framer's scope on this sheet.
	AB-2—OSA Building #61, 2nd Floor Asbestos Abatement Plan (11)	This drawing includes what appears to be the second-floor plan of the existing structure and the original room finish schedule, for use in identifying the location of asbestos to be abated.	Nothing in a framer's scope on this sheet.

(Continued)

Discipline	Title	Subject Matter	Comments
Architectural	A2.1—Schedules (12)	Matrix of rooms on the left (vertically oriented) and elements of rooms across the top (floor, base, wall, and ceiling requirements with a comments column) oriented horizontally. A matrix of door types, consisting of 10 details of doors, is also offered.	The room finish schedule is always worth reviewing, since ceiling elevations and materials for every room are shown in these matrices. In this project the ceilings are listed as 8'-6" in thirty-one of the forty-five spaces; 10'-6" in eight; 11'-6" in one; and "varies" in five, which usually means the ceiling is sloped. The door schedule, being specific, may take precedence over floor plans showing doors, so it makes sense to review the door details for information such as net height and width of openings.
	A2.2—Door Frame Type Storefront/Window/Louver (13)	Eight details describing storefront, windows, and louver types with dimensions, the symbols representing each type, and reference details on a subsequent sheet. Details of four door frame types with some dimensions, identifying symbols, and reference symbols are also included.	Though they do not have much of an effect on a material take-off, noting the window and door opening requirements is essential in the determination of actual construction costs (among them the labor costs associated with the installation of the door and window openings). Structural engineers tend to add elements to an opening as it grows in size: for example, openings in a wood light frame building that are over six feet often require twice the number of trimmers under the header.
	A2.3—Main Building Floor Plan (14)	Architectural floor plan with grid lines, dimensions, major building section references (2); wall section reference details (16); partial plan designations (1); room numbers (45); interior elevations (13); window designators (24); door designators (60); a legend for three different wall types, and a note that nonbearing partitions are to be braced to the roof structure on eight-foot centers, and where cabinetry attaches to walls the spacing of the bracing is cut in half.	Architectural floor plans are useful for a variety of purposes, not the least of which is a general size and space layout. The subject project is a rectangle 123'-6" × 107'-6", with a contiguous rectangle (Public Lobby and Special Duty rooms) 23' × 44' connected to the main building in the lower right quadrant. The long axes of both rectangles are parallel. As noted earlier in this chapter, the floor plan identifies the length, width, and location of the walls of the structure as well as the type. In this project, the walls consist of 2 × 4 and 2 × 6 studs (and CMU for the Evidence Room, Armory, and the Enclosed Patio contiguous to the Employee Breakroom). Whenever wall widths differ, issues of alignment when a single wall uses two different wood members arise, as do the connections between masonry and wood. Nothing to be alarmed about, but worth tucking in the back of one's mind. Most of the critical information a framer needs is found in the structural drawings. When architects employ a design consultant, they normally defer to the specialist, except for aspects of the work they maintain control over (wall location, for example, because it may affect space layout). In a material quantity take-off, using the given designations for rooms and other spaces and the polar notation on the drawings is the most effective methodology for mapping the take-off, because anyone possessing a copy of the drawings can relate to the take-off without additional information. The floor plan provides this information, whereas structural drawings typically show only information relevant to the structure.
	A2.4—Office Building Roof Plan (15)	Roof plan showing the roof slope (combines a 4-in-12 partial hip with two low-slope roofs, one in the mechanical well and the other over the entry lobby), roof coverings, mechanical well, crickets, roof drain and overflow locations, ventilation (18 locations), mechanical equipment (6 units) with return air and supply duct penetration locations, roof access hatch location, 13 detail references, two different louvers required in the wall of the mechanical well, and the mechanical equipment walkway.	The roof plan is critical to understanding the configuration of the roof and it informs the study of the structure itself. Roof plans of commercial buildings are normally fairly simple.

Discipline	Title	Subject Matter	Comments
	A2.5—Shop Building Floor Plan and Roof Plan: Canopy Plan and Section (16)	Maintenance facility floor and roof plans, with a roof plan and section of the fuel island canopy.	The shop building is simpler, but ditto the preceding comments on floor and roof plans.
	A3.1—Office Building Ext. Elevations (17)	North, south, and west elevations of the office building, with a plan of a screen wall, and elevations for the building sign/screen wall and the portion of the perimeter wall hidden behind the wall enclosing the patio.	Elevations are the views that most closely approximate a photograph in that they show the surface features of the object as if the viewer were an observer standing at a distance, so they are very useful for general object description and information on the elevation of window heads and sills, and exterior finishes (a framing contract for a small office building may include exterior wood finishes such as board or plywood siding and trim).
	A3.2—Office Building Sections and Elevations (18)	East exterior elevation, transverse and longitudinal building sections, section through the screen wall shown on previous page.	As above regarding the elevations. These sections clearly show the basic shape of the structure (including the braces from the nonbearing walls to the roof structure first noted on the floor plan) and the exterior finishes of the walls of the mechanical well.
	A3.3—Shop Building Ext. Elevations and Bldg. Sections (19)	As in A3.1 and A3.2 above.	As in A3.1 and A3.2 above.
	A3.4—Wall Sections (20)	Three wall sections, one at the mechanical well wall, one at the main perimeter wall showing the pitched roof, and one of the perimeter wall at the low-slope roof over the lobby.	Architects that have delegated the structural design to a subconsultant focus on the things they remain responsible for, namely space and finishes, so they are very useful to understanding the configuration of the building and to framers for understanding any special requirements pertinent to the finishes; for example, nonstructural blocking within the wall to attach the suspended ceiling to and sometimes draft stopping. Though a framer will actually construct according to the instructions in the structural drawings, architectural sections show the space well. In the example structure, the perimeter walls are level walls about 14' high with a 5' tall parapet wall mounted on the outside face of the structural wall adding another 3' to the overall height of the wall. Questions as to how best to construct the wall begin to arise. Stick build (construct a wall a piece at a time with one worker at the base and another at the top of the wall)? Platform frame and erect with machinery (construct wall on the floor, and tilt into place on the foundation with power equipment instead of labor)? Prefabricate portions and hoist them into place (the parapet for example)? Scaffolding? Hydraulic personnel lifts?
	A3.5—Wall Sections (21)	Ditto above, but 3 other locations	Ditto above.
	A3.6—Wall Sections (22)	Ditto above, but 2 other locations	Ditto above.
	A3.7—Wall Sections (23)	Ditto above, but 3 other locations	Ditto above.
	A3.8—Wall Sections (24)	Ditto above, but 2 other locations	Ditto above.
	A3.9—Wall Sections (25)	Ditto above, but 3 other locations	Ditto above.
	A4.1—Enlarged Floor Plan (26)	This plan is a larger scale floor plan than in A2.3 ($\frac{1}{4}$" = 1'0" instead of $\frac{1}{8}$" = 1'0") of a portion of the building focusing on men's and women's restrooms and locker rooms, the Armory, the Evidence Room, and the Janitor and Storage Room, and it includes 3 wall section references, 14 references to interior elevations, and numerous dimensions.	These drawings are useful for identifying the location of backing requirements (elements within walls that support things like grab bars, partitions, cabinets, suspended ceilings, and the like) or requirements to construct voids (for paper dispensers, for example). Combined with the elevations themselves, the help locate these elements in three dimensions.

(Continued)

Discipline	Title	Subject Matter	Comments
	A5.1—Interior Elevations (27)	17 elevations of restroom and locker room walls primarily at ¼" = 1'0".	As above.
	A5.2—Interior Elevations (28)	11 elevations of Briefing Room, Security Evidence, Lobby Counter, Employee's Breakroom, and a pass-through window, primarily at ¼" = 1'0".	As above.
	A5.3—Interior Elevations (29)	11 elevations of yet more walls (e.g., Armory, EMT Storage and Issue Room), primarily at ¼"= 1'0".	As above.
	A6.1—Reflected Ceiling Plan (30)	Description of ceiling features including lighting, soffits, projection screen locations, the location of the ridge in rooms with sloped ceilings, supply and return air duct locations, and the like.	Depending on the ceiling type, a framer may have the responsibility to furnish and install the backing for elements such as supply and return air ducts and registers, lighting soffits or valances, suspended ceiling elements, or even the ceiling structure itself.
	A8.1—Exterior Details (31)	12 details pertaining primarily to the Exterior Insulation and Finish System (EIFS), drawn at 1½" = 1'-0" or 3" = 1'-0" scale.	Other than the fundamental elements of the substrate for exterior finishes, not much of significance to the framer is offered.
	A8.2—Exterior details (32)	11 details pertaining to door jambs, thresholds, and heads, and a couple other items.	Ditto above.
	A8.3—Exterior details (33)	11 details pertaining to roof vents, penetrations through the roof (plumbing drains), and connection details for the aluminum and glass roof over the Enclosed Patio.	About half the details shown affect the framer, in particular the manner in which the architect wants block-outs for roof penetrations constructed.
	A8.4—Interior Details (34)	10 sections through assemblies such as the shower seat, shower floor, cabinets, lavatory counter, acoustic ceiling.	As noted in eight foregoing drawings, these drawings show how the architect wants certain elements (the lavatory counter brackets, for example) supported. The framer normally integrates blocking or other support requirements, if they are wood, into the walls during the framing phase. Depending on how the subcontracts are written, the framer may install elements furnished by other subcontractors; for example, metal brackets supporting a countertop that are bolted through the stud, or anchorage hardware provided by the concrete subcontractor.
	A8.5—Interior Details (35)	One horizontal (not exactly a plan) and one section of the Security Evidence cabinetry, three sections through the Public Lobby cabinetry, one through a suspended metal gypboard ceiling structure, and one through the soffit in the toilet rooms.	Of these seven details, one (the section through the light soffit) may affect the framer. The soffit is a wood structure, however, because the connection to a supporting structure over head is not shown (break lines terminate the drawing a couple inches above the soffit), and the ledger connected to the stud wall is shown over a layer of gypsum wallboard. The question arises as to whether the wood structure would be included in a framer's subcontract.
	A8.6—Miscellaneous Details (36)	Eleven sections through window and louver sills, heads, and jambs, recessed light fixtures, and a fire extinguisher cabinet.	Of these details, two are relevant; the well for the recessed light fixture in a fire rated corridor comprising a wood frame, and the cavity required in walls for the fire extinguishers.

Discipline	Title	Subject Matter	Comments
Structural	S0.1—General Notes (37)	Generic notes pertaining to nails; the nailing schedule for typical connections; wood species and grade; minimum plywood requirements; cutting and notching guidelines; symbols for wood;[8] information on hardware; guidelines on masonry, structural steel, composite metal deck, concrete and reinforcing steel, rebar welding, and a series of notes pertaining to administrative issues (order of precedence clause,[9] submittal requirements, admonitions regarding safety, design of temporary structures, inspection requirements, footing design information, and the like). Notes as to the allowable load of the soil and the design loads the structural engineer used are listed as well. It's worth pointing out that this page alone contains some three thousand words.	For a specialty contractor involved with some part of the structure (primarily foundation and frame), it is in the structural sheets that the real work begins. All the information relevant to one's task must be carefully reviewed and studied. The first several pages of a structural drawing set often begin with generic details pertaining to the common materials used in buildings, namely concrete, masonry, steel, wood, combinations of these materials, and earth. Dubbed "S1 Sheets" by some (see "Graphics and Specifications" in Chapter 4), these specifications are sometimes a source of confusion because they may differ from the specifications book that forms one of the five fundamental contract components (refer back to "The Construction Contract: Five Components" in Chapter 2). The structural engineer's work commonly involves four specifications divisions: Division 3 (concrete and reinforcing steel); Division 4 (masonry), Division 5 (metals), and Division 6 (wood, plastics, and composites)[10] but it is the responsibility of the prime design professional to generate the specifications for the project. Consequently, discrepancies occur, and the prudent framer (and builder) will clarify how these circumstances are to be dealt with prior to submitting a price for the work. As with all other specifications, the builder should understand what the design professional requires. For example, it is not unusual for structural engineers to direct the contractor to predrill for larger nails and for lag screws and to limit the tolerance of bolt holes to $1/16$" larger than the shank of the bolt. If one prices the work with full compliance in mind, one may end up with a profitable project, but going into the job thinking that the requirement might not be insisted upon would be a mistake. Considerable variety in the solutions structural engineers arrive at exists. At the very least, such requirements should be clarified with the prime contractor and prime design professional.
	S0.2—Typical Details (38)	17 details describing common conditions in stepped foundations, interior nonbearing footings, slab-on grade-joints, pipe sleeves through foundations, reinforcing steel bends and splices, requirements pertaining to openings in masonry and stud walls, restrictions on drilling studs, masonry wall reinforcement, lapping of top plates, welding requirements for tube-steel-to-base-plate connections, and manufacturing specs for glued-laminated timbers. As with sheet C4 above, this page has two different details with the same label.	Of interest here are the details pertaining to requirements for openings in walls, restrictions on drilling studs, and the required lap splicing for top plates in a stud walls.

(Continued)

[8]Interestingly, the symbols for continuous wood, blocking, and "end of member" used by the structural engineer and referenced in these general notes are, respectively, finish wood, continuous wood, and blocking in the standard sets of symbols.

[9]The typical order of precedence clause establishes the priority of information in the event of conflicts in the drawings and specifications. Here, the clause is contained in the structural engineer's notes and most likely pertains just to the information on the structural drawings: Discrepancies between the specifications on the drawing (generated by the structural engineer) and those in the specifications book, which are generated by the architect, create a different legal entanglement to resolve. Whose instructions should the builder follow? The design professional in charge (the architect), or the design professional who is the specialist?

[10]Both the 16 division (prior to 2005) and 49 division (2004 and later) Construction Specifications Institute MasterFormat use the same numerical designation for these four divisions. Refer to Appendix F for the complete list.

Discipline	Title	Subject Matter	Comments
	S0.3—Typical Details (39)	18 details describing the proper size and number of elements in a stud wall, the proper terms for the elements, and the manner in which they are connected to each other and the foundation; framing requirements for openings in a diaphragm comprising manufactured wood I-joists, mid-span blocking requirements of those diaphragms, splicing requirements for ledgers, and blocking for nonbearing walls, among others.	All of these details address framing issues, and in this particular drawing set, some of the details represent a considerable amount of work. For example, the bracing for nonbearing walls, as noted on the architectural floor plan, is to be connected to the wall and the roof structure above. With a plate height starting at 14 feet or so at the perimeter (see A3.4 above), the first brace in a nonbearing wall parallel to the rafters of the roof would consist of two T-shaped struts perpendicular to the wall at 45° some 8'-6" long to blocks between rafters 17' above the slab (the nonbearing walls vary in height from 10'-9" to 11'-6"). The second and third sets of braces would require, respectively, 12' and 16' long pairs of struts tying into blocks at just over 19' and 22'. Since there is no working platform in this project except at the mechanical well some 32' from the perimeter on all sides, this means the carpenters would need to install the blocks in the rafters from rolling scaffold, scissor lifts inside the building, or from the roof structure, and with pairs of 2 × 4's ranging from 9' to 16' forming the struts, the task would require two carpenters. For nonbearing walls to which cabinets are connected, twice the bracing is required (see Figure 4.19). Another example is the truss blocking panel shown in Figure 4.20 (Detail 12 S0.3). Here the engineer is calling for a special block about 18" × 23", consisting of a plywood panel with a perimeter frame the same size and material as the chords of the truss, rabbeted and glued to the panel, inserted between the I-joist rafters. Each corner of the block has to be cut out to accommodate the top and bottom chords of both rafters. The project calls for 340 of these panels in two rows of mid-span blocking in the sloped portion of the roof. The elevation of the roof at the first row is approximately 17'-6" and 20'-8" above the finished floor at the second. The low-slope roof framing contains the same blocking requirement around the perimeter of the roof structure, adding some 20 blocks. Although one could have the truss manufacturer produce such a block for the standard bay, there are likely to be (owing to the vagaries of wood construction, minor imperfections in the work, and weather) instances where these blocks would have to be produced in the field or perhaps in the framer's shop.
	S2.1—Foundation Plan, Building A (40)	Foundation plan of slab-on-grade foundation, grid lines, dimensions, 28 wall section references, hold-down types and locations, extent of depressed slabs in locker rooms and restrooms.	This drawing is essential to identifying the location and nature of all the bearing components in the foundation. It describes the structural engineer's solution to transferring the loads acting on the structure and dissipating them into the earth. The foundation plan is the map that leads to the wall sections (drawings) that show the required elements in greater detail.Of particular interest to a framer is information on the hold-down hardware—the number, type, and location of all the elements anchoring the walls to the foundation. This and many other questions about critical connections must be asked and answered. Issues involving building constraints arise too. For instance, the structural engineer for this project calls for the walls to sit on a 6" curb for most of the perimeter and on several interior walls. Curbs pose problems for a framer intending to assemble the walls on the slab and tilt them into place (a common method on the typical slab-on-grade foundation; see Figure 7.15b). The bottom of the wall has to be far enough from the curb so that a hammer or pneumatic nailer can be used to connect the studs and sole plate, and the ends of the wall

Discipline	Title	Subject Matter	Comments
			being framed either sit on top of the curb, making the assembly of studs and blocks more difficult, or the sole plate and bottom end of the studs must be set and assembled on the curb. If the wall is framed on the flat portion of the slab, it has to be dragged back to the curb and erected—not always an easy task. A 14'-high wall of any length is heavy and very awkward to handle manually, and it comes apart easily unless it has been prepared for hoisting and movement. (Plate and stud connections depend on the friction between the nail shank and the wood to stay together, and with common wire nails particularly, the friction resistance is negligible.) Such a wall requires some sort of erection equipment for a safe and effective installation. When the work space is limited by interior walls with curbs, the framing and erection challenges are exacerbated.
	S2.2—Roof Framing Plan Building A (41)	Plan of roof structure, with grid lines, pertinent roof dimensions, 63 wall section or detail references, and 12 key notes.	The roof structure plan is the drawing that identifies the location and number of the key elements in the roof, namely joists ("horizontal" framing members) and rafters (sloped framing member) and any vertical elements such as columns that support girders and walls supporting portions of the roof. In the example project, the roof is simple: for the first 30'-9" on all four sides, it is a pitched roof, which is supported by walls on the periphery of a 45' × 60' rectilinear mechanical equipment well (low sloped roof) in the center of the structure. The public entrance and lobby sits under a low-sloped roof over the 23' × 44' wing attached to the southeast corner of the main building. The pitched roof portion consists of 18" deep wood I-joist rafters at a 4-in-12 slope (4" rise in 12" of run). The low-sloped structural members are 18" deep in the mechanical well, and 12" deep over the lobby. The plan shows the basic configuration of the roof, the main structural elements, and the location of the critical relevant details.
	S2.3—Foundation and Roof Framing Plans, Building B (42)	Foundation and roof framing plans for the second building on the site.	As noted above, except the roof is simpler yet: a roof sloping gradually from the front of the shop to the rear.
	S2.4—Shearwall Plans and Schedule (43)	Simplified plan views of buildings A and B, with shearwalls, correlated to a shearwall schedule and notes.	This drawing focuses just on the foundation and the critical shearwalls that are installed on it. The plan is accompanied by a shear wall schedule that identifies 7 different wall types, their symbols, the sheathing and edge nailing requirements, the sill bolting requirements, pertinent notes (9 of them), and the engineer's remarks.
	S4.1—Sections (44)	Four different wall sections describing the foundation, the wall itself, and roof structure. Two other details describe the wall-to-machine-well sheathing connection and one the connection between jack rafters and the four hips in the structure.	Herein lie the numerous wall elements and the manner in which they are all connected. Among many other things, wall sections shed light on the sequence within which work might be done. Regarding the quantity take-off that is the focus of this review, one would carefully study the wall section, determine the elements in it, list them, then determine (by returning to the roof or wall framing plan) the total lineal footage of wall with those characteristics. Having determined that, one can determine the total quantity of material required for that type of wall. If the general notes fail to, details may identify elements that must be completed after the structure has been loaded. In this project, for example, the bottom end of the metal cross bracing required in the low-slope roof joists must be fastened after the roof has been loaded.

(Continued)

Discipline	Title	Subject Matter	Comments
	S4.2—Sections (45)	Three more wall sections and three details, one describing the connection between machine well joists and walls below and above the joists, one describing a machine well wall that sits over the joists in one portion of the project, and one describing the connection between the low-slope roof joists, the low end of the pitched roof rafters over the lobby, and the bearing wall underneath them both.	As above.
	S4.3—Sections (46)	Three wall sections and four details showing still other critical connections.	As above.
	S4.4—Sections (47)	Three sections through the CMU walls in the project, two framing elevations (showing beam-wall-column connections), and six sections.	As above, however the sections through the CMU walls show how the wood wall should connect to the masonry wall.
	S4.5—Sections and Details (48)	Three more wall sections and four details.	Ditto S4.1 above.
	S4.6—Details (49)	Twelve details of critical connections, for example, the hips and king rafters at each of four corners of the mechanical well, numerous beam-to-beam connections.	Details often show critical connections that are first introduced in wall details in yet greater detail (¾" = 1'-0" and 1½" = 1'-0" scale). It is in these details that the intricacies of the connection and the sequence within which the work will be done are revealed. They are critical to the complete understanding of the engineer's requirements.
	S4.7—Details (50)	Seven details, one of column bases, three of floor diaphragm to shearwall connections, one tube steel column to glued laminated beam connection, one describing the connection between a column and the low-slope roof over the lobby, and the connection between a wall and beam separated by a beam running perpendicular to those components.	As above.
Mechanical	M1—Plumbing Plan (51)	Floor plan showing walls and the drain waste and vent plan, a room schedule, plumbing fixture pipe size schedule, and nine sheet notes (about 250 words).	Though plumbing work may affect a framer's work (widened or double walls behind plumbing fixtures lavatories, urinals, carriers, and the like, backing for those fixtures, framing of the wells for roof drains for example—see Figures 10.4i, 12.4c, 12.6a and 12.6b), in this project, there are no details on the subject, but it would be worth clarifying with the prime contractor whether such things are expected.
	M2—Roof Drain Plan (52)	Floor plan with numbers and a dashed line identifying the location of the mechanical well and roof drain location, as well as a room schedule and mechanical legend (abbreviations primarily, with some symbols).	The only detail relevant to this issue is an architectural detail shown on page A8.3 in the drawings. Again, it would be worth the effort to clarify this issue, since considerable work may be involved in framing the roof drains.

Discipline	Title	Subject Matter	Comments
	M3—HVAC Plan (53)	Floor plan with room numbers showing duct runs, damper locations, supply and return air register locations, a main building air balance schedule, wall exhaust fan schedule, HVAC unit schedule, and 10 key notes.	Since all but a few of the registers are located in suspended ceilings (which are not normally in a framer's work scope), little regarding the HVAC system that has not already been addressed in structural drawings is shown on this drawing.
	M4—Shop Building Mechanical Plans (54)	Roof plan showing the drains and overflows, with a floor plan showing the underfloor plumbing, the pneumatic equipment (compressor and air lines), emergency power generation equipment and fuel, and HVAC system.	As above.
	M5—Mechanical Details (55)	Nine details addressing through-wall penetrations by pipe in a fire-rated wall, gas regulator assembly, backflow preventer, water heater strapping detail, sprinkler valve riser assembly, and fire damper detail, and a detail on the balance damper assembly.	Of these details, the through-wall pipe penetration, fire damper in the wall, and water heater strapping protocol might fall into a framer's scope, for the supporting elements concealed within the wall.
	M6—Mechanical Details (56)	Eight details pertaining to pipe bracing, water heater platform, above-ground fuel tank details, the air compressor and the like, and a pipe bracing schedule.	One detail shows how the metal support for piping connects to the I-joists, but whether it would be part of the framer's scope of work is up to the prime contractor's management personnel. Subcontractors frequently cooperate on such things; one sub will do the layout, and the other will do the installation.
	M7—Underground Storage Tank Details (57)	Plan and sections of underground fuel storage and tank with anchorage.	Nothing for the framer here.
	M8—Underground Storage Tank Details (58)	Five details of the vapor recovery, fill connection, and similar details for the fuel tank.	As above.
Electrical	E1—Fixture Schedule Symbol List (59)	Fixture schedule, mounting detail, switching detail, general electrical notes, and list of symbols.	Since most of the fixtures in this project are located in a suspended ceiling system, there is little of interest for the framer, except at the service panel locations, which on this project are shown on the site electrical plan (p. E7).
	E2—Floor Plan Lighting (60)	Room schedule, reflected ceiling plan of location of light fixtures.	Nothing of interest or concern to the framer here.
	E3—Floor Plan Power (61)	Room schedule, floor plan and power outlet locations.	As above, except the panel locations are shown. Surface or flush-mounted panels may require backing.
	E4—Signal Plan (62)	Room schedule, floor plan with signal locations (telephone and data).	As in E2 above.
	E5—Shop Building Floor Plan (63)	This drawing shows a fixture and cabinet plan and a power plan side by side.	As above.
	E6—One-Line Diagram Panels (64)	Wiring instructions for the six panels and a one-line diagram of the fire alarm system.	As above.

Discipline	Title	Subject Matter	Comments
	E7—Site Plan Electrical (65)	Site plan showing existing utilities in the street (including power and telephone) and the locations of the main panel, telephone terminal, site lighting, and electronic gate.	As above.
	E8—Electrical Details (66)	Seven details showing the light fixture at the fuel island canopy, the treatment required for conduit penetrating a fire wall, a recessed fixture installation detail, and several other things.	Although the penetration detail shows studs, no special treatment is required.
Landscape	L1—Planting Plan (67) L2—Irrigation Plan (68) L3—Irrigation and Planting details (69)	Planting instructions. Irrigation layout. Details involving tree and shrub planting, a remote control irrigation valve, irrigation controller and footing, and various sprinkler head details.	No work pertinent to framing.
Summary	69 pages of notes and drawings measuring 24" × 36"	Thirty full plan views, 6 partial plans, 6 major (building) sections, 9 exterior elevations, 39 interior elevations, 32 wall sections, 221 details, and 18 schedules.	Hundreds of instructions in thousands of words and graphics, with the site work, concrete, rough frame, HVAC, and electrical, specialty contractors playing key roles. Most of the wall sections and details affect a framing contractor's work scope.

TAKE OFF

Project __C&E Facility__   Sheet __1__ of __9__
__02200 Earthwork__   Est. No. __80-7__
__Structural Earthwork__   By __RAN__

	No	ℓ (FT)	w (FT)	d (FT)	Footing Exc (CF)			10/S4	12/S4
					(ℓ×w×d)				
								54.5	36.4
Footing Excavation								15.2	5.3
2/S3	2	92.3	5.0	4.1	3784			12.1	12.1
		69.7	5.0	4.1	1429			12.1	6.1
10/S4		113.5	4.0	2.8	1271			19.6	
12/S4		59.9	4.0	2.1	503			113.5 LF	59.9 LF
Excavate from S.G.									
Exc. footing width +24"					6987 CF				
Assume vertical cut.					259 CY ✓				
Common earth: 30% swell									
10% shrinkage									

	No.	ℓ (FT)	w (FT)	d (FT)	Footing Backfill (CF)	
					(ℓ×w×d)	
Footing Backfill						
Excavation		(See above)			6987	
Less: Footing concrete		(See concrete)			(⊖ 1285)	
Less: Found walls to S.G.						
2,3/S3		243.7	.67	2.8	(⊖ 466)	
10/S4		109.5	.67	1.8	(⊖ 132)	
12/S4		55.9	.67	1.1	(⊖ 41)	
					5063 CF	
					188 CY ✓	

Surplus Excavation			Gravel Drain Rock – under interior SOG
Exc (see above)	= 259 BCY		
Backfill $\frac{188\ CCY}{.90}$	= 209 BCY		Inter SOG (see concrete) 17098 SF
Surplus exc	50 BCY ✓		× .5
			8549 CF
			317 CY ✓

CSUS F8

Figure D.1 This page shows one approach to performing a quantity take-off for structural earthwork.

TAKE OFF

Project __C & E Facility__   Sheet __2__ of __9__
__02200 Earthwork__   Est. No. __80-7__
__Site excavation__   By __RAN__

Refer to plotted cross-sections for area calculations

Station	Dist. (FT)	CUT Area (SF)	CUT Average Area (SF)	CUT Volume (CF)	FILL Area (SF)	FILL Average Area (SF)	FILL Volume (CF)
0+00		0					
	50		157.5	7875			
0+50		315					
	50		519.0	25950			
1+00		723					
	50		917.5	45875			
1+50		1112					
	50		1270.5	63525			
2+00		1429					
	50		1491.5	74575			
2+50		1554					
	13		1643.5	21366			
2+63		1733					
	37		1587.5	58738			
3+00		1442					
	50		1263.5	63175			
3+50		1085					
	50		873.0	43650			
4+00		661					
	50		474.5	23725			
4+50		288					
	33		144.0	4752			
4+83		0					
Common earth				433206 CF			
Swell 30%							
Shrinkage 10%				16045 CY ✓			

CSUS F8

Figure D.2 This sheet shows the traditional format for earth quantity take-offs performed by the average end-area method, manually done.

TAKE OFF

Project _C & E Facility_
03100/03300 CIP Concrete & Forms

Sheet _3_ of _9_
Est. No. _80-7_
By _RAN_

	No.	ℓ (FT)	h (FT)	t (FT)	Found Wall Forms (SFCA) (2×ℓ×h)	Found Wall Conc (CF) (3000psi) (ℓ×h×t)		3/S3	10/S4	12/S4
Foundation Walls										
2/S3	2	87.3	3.8	.67	1327	445		46.7	15.2	32.4
3/S3		74.1	3.8	.67	563	189		27.4	12.1	5.3
10/S4		109.5	2.8	.67	613	205			12.1	12.1
12/S4		55.9	2.1	.67	235	79		74.1 LF	19.6	6.1
					2738 SFCA	918 CF Ⓐ 34 CY ✓			50.5 109.5 LF	55.9 LF

Patch walls - unexposed 2738 SF ✓

[Check: 34 CY × 81 SFCA/CY = 2754 SFCA]

	No.	ℓ (FT)	h (FT)	w (FT)	Cont. Ftg Forms (SFCA) (2×ℓ×h)	Footing Conc (CF) (3000psi) (ℓ×h×w)		3/S3	10,12/S4
Continuous Footings								46.7	5.3
2/S3	2	90.3	1.25	3.0	452	677		25.0	12.1
3/S3		71.7	1.25	3.0	179	269			6.1
10,12/S4		169.4	1.0	2.0	339	339		71.7 LF	15.2 12.1 12.1 19.6 52.5
					970 SFCA	1285 CF Ⓐ 48 CY ✓			36.4 169.4 LF

	No	ℓ (FT)	h (FT)	t (FT)	High Wall Forms (SFCA) (2×ℓ×h)	High Wall Conc (CF) (3000psi) (ℓ×h×t)	Bucks (LF) 2(ℓ+h)	Patch Walls Exposed (SF) (2×ℓ×h)
High Building Walls								
N & S 2/S3	2	87.3	13.3	.67	4644	1556	—	4644
Internal 3/S3		46.7	15.7	.67	1466	491	—	1466
		27.4	14.5	.67	795	266	—	795
Less door openings	3	3.2	7.0	.67	—	(⊖ 45)	61	(⊖ 134)
	2	6.4	7.0	.67	—	(⊖ 60)	54	(⊖ 179)
No deduct for wall form @ door openings					6905 ✓ SFCA	2208 CF Ⓐ 82 CY ✓	115 ✓ LF	6592 ✓ SF

Check: 82 CY × 81 SFCA/CY = 6642 SFCA

CSUS F8

Figure D.3 Concrete foundations and walls are the subject of this take-off.

TAKE OFF

Project __C & E Facility__   Sheet __4__ of __9__
__03100/03355 CIP Concrete & Forms__   Est. No. __80-7__
                                                                      By __RAN__

	No.	ℓ	w	t	Finish		SOG	Edge Forms	
					Interior	Exterior	Concrete	4"&6"	12"
		(FT)	(FT)	(FT)	(SF)	(SF)	(CF)	(LF)	(LF)
					($\ell \times w$)	($\ell \times w$)	(2500si) ($\ell \times w \times t$)		
Slabs on Grade									
Interior – North Wing		83.2	41.6	.5	3461	–	1731	42	–
– Main Bldg		107.3	62.7	.5	6728	–	3364	170	–
		153.2	35.6	.5	5454	–	2727	–	–
– Entry		15.0	12.0	.5	180	–	90	27	–
– South Wing		63.1	20.2	.5	1275	–	637	20	–
Thickened slab		323	1.5	.5	–	–	242	–	–
Exterior – Walks		74.1	6.0	.33	–	445	147	148	–
		52.0	3.0	.33	–	156	51	104	–
– Stoops	2	10.0	4.0	.33	–	80	26	–	–
	3	6.0	4.0	.33	–	72	24	–	–
Thickened Edge		78	1.0	.67	–	–	52	–	78

Check:
6" slab 17098 sf ÷ 54"/cy = 317 cy
4" "   753 sf ÷ 81"/cy = 9
                              326 cy
plus thickened

					17098 SF ✓	753 SF ✓	9091 CF ✓	511 LF ✓	78 LF ✓
					Trowel	Broom	② 337 cy ✓		

Expansion Joints in Exterior Walks
½" x 3'   llll   4x3 = 12 LF
½" x 6'   llll l   6x6 = 36
                                48 LF

Hardener – Interior   (see above)   17098 SF
        Less VA tile   (see Sh 9)   (⊖ 1358 SF)
                                            15740 SF

Vapor Barrier   (see above)   17098 SF
             (")   753
                                            17851 SF

CSUS F8

Figure D.4 A different format for concrete slabs-on-grade is used, and in this sheet the expansion joints, hardener, and vapor retarder required as part of this concrete work are taken off as well.

TAKE OFF

Project ___C & E Facility___ Sheet __5__ of __9__
___03200 Reinf Steel___ Est. No. __80-7__
___Rebar for Tilt Up & Closures___ By ___RAN___

	No.	Size	L (FT)	#2 (LF)	#3 (LF)	#4 (LF)	#5 (LF)	#6 (LF)
Rebar - Tilt Up Screen Walls								
End panels (8) Horiz	8 x 13	#4	24.6			2558		
	8 x 4	#5	24.6				787	
Vert	8 x 24	#4	13.8			2650		
Typ panels (32) Horiz	32 x 13	#4	24.8			10317		
	32 x 4	#5	24.8				3174	
Vert	32 x 24	#4	13.8			10598		
Add @ openings Vert	4 x 4	#6	13.8					221
Horiz	4 x 4	#5	9.0				144	
Corners	4 x 4	#5	4.0				64	
Add @ pick points	80 x 2	#5	6.0				960	
						26123	5129	221
23132 LBS						x .668	x 1.043	x 1.502
Plus 5% 1157						17450 LBS	5350 LBS	332 LBS
24289 LBS ✓								
[÷ 234 CY = 104 LB/CY]						TOTAL 23132 LBS		

	No.	Size	L (FT)	Total Length (LF)	Factor LB/LF	Total Weight (LBS)		
Rebar - Closure Columns								
Type 1 (2x4) Verts	8 x 4	#6	14.0	448	1.502	673		
Stirrups	8 x 15	#2	5.8	696	.167	116		
Type 2 (5x4) Verts	20 x 2	#6	14.0	560	1.502	842		
Type 3 (4x4) Verts	16 x 4	#6	14.0	896	1.502	1346		
Stirrups	16 x 15	#2	4.7	1128	.167	188		
						3165 LBS		
3165 LBS								
Plus 5% 158								
3323 LBS ✓								
[÷ 32 CY = 104 LB/CY]								

CSUS F8

Figure D.5 One effective approach to determining reinforcing steel quantities is shown here.

TAKE OFF

Project _C & E Facility_
03400 Tilt Up Concrete

Sheet _6_ of _9_
Est. No. _80-7_
By _RAN_

	No.	ℓ (FT)	h (FT)	t (FT)	Finish (SF) (ℓxh)	Panel Concrete (CF) (ℓxhxt) (3000psi)	Edge Forms (LF)		
							Plain (ℓ)	Sloped Top (ℓ)	Keyed (2h)
Tilt Up Screen Walls									
(4-Identical walls									
x 10 panels each)									
End panels	4x2	22.8	14.0	.50	2554	1277	184	184	224
Typ panels	4x8	23.0	14.0	.50	10304	5152	736	736	896
Less openings	4x2	4.0	7.0	.50	(⊖224)	(⊖112)	* 176	—	—
					12634 SF ✓	6317 CF (A)234CY	920 * 176 1096 LF ✓	920 LF ✓	1120 LF ✓
Panels	40EA ✓								
Max.wt. 161CFx150=24150LBS							* opening form 2(ℓ+h)		
Double lift inserts 10x2	80EA ✓								
Brace inserts 10x4	160EA ✓								
Braces	80EA ✓					Check: 12634 SF ÷ 54 SF/CY = 234 CY			
Chamfer @ keyed edge 1120x2	2240								
@ openings 176x2	352					40 x 2(23+14) = 2960 LF			
	2592 LF ✓					or, 2(920)+1120 = 2960 LF			

	No	ℓ (FT)	h (FT)	w (FT)	Column Concrete (CF) (ℓxhxw) (3000psi)	Closure Forms		Chamfer (LF) (4h)	
						Contact Area (ℓxh) (SFCA)	Vert LF of form (2h) (LF)		
Closure Columns									
End Type 1 Concrete	4x2	1.67	14.0	1.50	281	—	—	—	
Form	4x2	5.86	14.0	—	—	654	224	448	
Intermed Type 2 C	4x5	1.00	14.0	.50	140	—	—	—	
F	4x5	2.00	14.0	—	—	560	560	1120	
Intermed Type 3 C	4x4	1.33	14.0	1.50	447	—	—	—	
F	4x4	4.66	14.0	—	—	1044	448	896	
					868 CF (B)32 CY ✓	2258 SFCA ✓	1232 Vert LF ✓	2464 LF ✓	

CSUS F8

Figure D.6 In this take-off the amount of concrete required for the walls of a tilt-up concrete project (site precast tilt construction) is determined. "Closure columns" (bottom portion of take-off) were used to connect wall panels. More common now are plates of steel connected to embedments in each panel edge.

TAKE OFF

Project __C&E Facility__
__06100 Rough Carpentry__

Sheet _7_ of _9_
Est. No. _80-7_
By _RAN_

	No	Size	L (FT)	Total L (LF)	FG Rwd (FBM)	Constr. D.F. Studs Plates Hdrs (FBM)	Block (FBM)		Interior Walls LF	Studs EA	Drive Pins EA
Frame Interior Walls (2x4@16")											
Bottom plates	1	2x4	200	200	133	—	—		55	65	27
Studs	211	2x4	8	1688	—	1125	—		45	50	20
Top plates	2	2x4	200	400	—	267	—	6@10	60	54	30
Blocking	1	2x4	200	200	—	—	133		12	14	6
Headers	10	4x12	4	40	①160			4@6	24	24	16
									4	4	3
									200 LF	211 EA	102 EA ✓
					133	1392 ✓	133 ✓				
					FBM	①160 ✓ FBM	FBM				
① 4X12 Constr. D.F.						1552 FBM			Pins @ 32"O.C. max		
					TOTAL 1813 FBM ✓						

	No	Size	L (FT)	Total L (LF)	FG Rwd (FBM)	Constr. D.F. Studs Plates Hdrs (FBM)	Block (FBM)		Exterior Walls LF	Studs EA	Anchor Bolts EA
Frame Exterior Walls (2x6@16")											
Bottom plates	1	3x6	162	162	243	—	—		26	31	8
Studs	190	2x6	8	1520	—	1520 ✓	—		55	63	15
Top plates	2	2x6	162	324	—	324 ✓	—		55	63	16
Blocking	1	2x6	162	162	—	—	162		26	33	8
Headers	19	6x12	4	76	②456 ✓		—		162 LF	190 EA	47 ✓ EA
					243	1844	162 ✓				
② 6X12 Constr DF					FBM	②456 FBM	FBM				
						2300 FBM			Bolts @ 48"OC max		

CSUS F8

Figure D.7 Lumber take-offs vary considerably from individual to individual, but this approach is an effective one.

TAKE OFF

Project __C & E Facility__
__06100 Rough Carpentry__

Sheet __8__ of __9__
Est. No. __80-7__
By __RAN__

	No.	Size	L	Total L	Select Struct DF	Constr.DF Block	Constr.DF Bridge				Blocking		LF
			(FT)	(LF)	(FBM)	(FBM)	(FBM)				3@55		165
Frame Roof (2X10@16")													
Joists	43	2X10	18	774	1290	—	—						45
	43	2X10	12	516	860	—	—						12
Blocking	1	2X10	222	222	—	370	—						222 LF
Bridging 83 pair✓	166	2X3	1.5	249	—	—	125				Bridging	LF	Pair
												25	19
					2150	370	125✓					18	14
					FBM ✓✓	FBM ✓✓	FBM ✓✓					55	42
												10	8
													83 Pair

1" = 1'-0"

	No.	Size	L	Total L	Constr. DF	Bolts			
			(FT)	(LF)	(FBM)	(EA)			
Bolted Ledgers									
3/S2	6	3X8	12	72	144	24			
5,6/S2	1	3X6	176	176	264	55			
5/S3	5	3X6	16	80	120	25			
					528 FBM ✓✓	104 EA ✓			

Roof Sheathing   Net area 54.83 × 25.75 = 1412 SF
Buy 4' × 8' sheets
 ℓ: 54.83 ÷ 8 = 6.85, or 7 sheets long
 w: 25.75 ÷ 4 = 6.44, or 7 sheets wide

Plywood ⅝" CDX   7×7= 49 sheets @ 4'×8' = 1568 SF ✓✓

CSUS F8

Figure D.8 The use of simple drawings and tallies for repetitive elements results in a clear description and fewer calculation errors.

TAKE OFF

Project __C & E Facility__
__Doors - Hardware - Floor Covering__

Sheet __9__ of __9__
Est. No. __80-7__
By __RAN__

08100 Steel Doors & Frames

	Openings	Hardware Type 1	2	3	4
HM doors - single X PM frames in wood studs	⦀⦀⦀⦀ ✓ (22)	⦀⦀⦀⦀ (12)	⦀⦀ (4)	⦀⦀⦀ (6)	
HM doors - single X PM frames in concrete	⦀⦀⦀ ✓ (8)	⦀⦀ (4)		⦀⦀ (4)	
HM doors - in pairs X PM frames in concrete	⦀⦀⦀ ✓ (4)				⦀⦀ (4)
	34				
6²x3⁰ PM borrow light frames in wood studs	⦀⦀ ✓ (6)	—	—	—	—
		16 Sets	4 Sets	10 Sets	4 Sets

08700 Hardware

	No. Openings	Closers	Kick Plates	Stops Floor Mtd.	Stops Wall Mtd.	Panic Devices
Hardware Type 1	16	16	16	16		
2	4	4			4	
3	10	10		10		10 ✓
4	4	8	8	8		8 ✓
	34	38 EA ✓	24 EA ✓	34 EA ✓	4 EA ✓	18 EA ✓

09650 Resilient Flooring

	L (FT)	W (FT)	Vinyl Asb. Tile ① (SF)	4" Base ② (LF)	Base Corners (EA)	
						① 12"x12"x.080"
Entry	16	10	160	46	4	② Door openings not deducted.
Corridor	45	6	270	96	2	
Rooms 1-4	45	10	450	170	16	
Rooms 5-6	18	10	180	98	10	
	6	4	24			
Rooms 7-8	25	10	250	102	10	
	6	4	24			
			1358 SF ✓	512 LF ✓	42 EA ✓	

CSUS F8

Figure D.9 Odds and ends are quantified in a variety of ways, as is evident here.

APPENDIX E
Welding in Construction

Background

Welding is common in manufacturing processes of all sorts. In the construction industry it is widely used for connecting elements in structural steel frames, composite structural systems, piping systems, ornamental iron, and light gauge metal framing among other things. It is also widely used in the maintenance and repair of construction equipment and in preparing some equipment for heavy use, for instance in machinery involved in excavation, hauling, and similar tasks. In a high-rise construction project comprising a structural steel frame in a seismically active zone, the sequence of the welding has to be carefully planned and executed—extensive welding causes the structure to shrink—and there may be enough welding required that it drives the schedule, sometimes for months at a time.

The American Welding Society's Master Chart of Welding and Allied Processes identifies some seventy welding processes (e.g., gas metal arc welding, stud arc welding, and friction welding), and thirty-three allied processes whose primary orientation is cutting (e.g., oxygen, arc, laser beam, and electron beam cutting). In the construction industry, gas cutting (oxygen-acetylene), arc, and electric resistance welding are common. Lesser-known processes are used, such as friction welding of reinforc-

ing steel bars to the mushroom-shaped heads that replace tensile stress-reducing rebar hooks.

Welding Instructions

Considering that welding is a process that connects metal elements, it is not surprising that the instructions on a drawing set and in specifications come from structural engineers hired by architects or employed by departments of transportation. Engineering students are subjected to rigorous professional educations involving a variety of subjects including structural steel design and analysis, detailing, strength of materials, materials testing, and chemistry before they graduate and are pronounced competent to begin practice. Those who work for structural engineering firms after graduation continue their educations under the supervision of licensed civil and structural engineers until such time as they are prepared to take the Fundamentals of Engineering exam, followed by the Professional Engineer exam. After considerable supervised practical experience, some sit for the Structural Engineer exam, a grueling two-day exam that allows those that pass the test to use the title Structural Engineer (SE). For example, in California, which has active seismic zones, the structural design

Construction Graphics: A Practical Guide to Interpreting Working Drawings, Third Edition. Keith Bisharat.
© 2025 John Wiley & Sons Inc. Published 2025 by John Wiley & Sons Inc.

for schools and hospitals is limited to holders of the SE license.

Participants in the modern construction industry face rising costs, increasingly complex projects, and compressed schedules, and those that are successful are attuned to the decisions that limit costs and increase productivity. In addition to having the required technical knowledge, the structural engineering firm with current practical knowledge—familiarity with indigenous construction practices and challenges, construction costs as they related to structural engineering, and the work of the skilled trades—distinguishes itself from the firms that are simply technically competent. Spacing reinforcing steel somewhat narrower than calculations indicate to coincide with the stride of the cement finisher, thoughtful selection of the elements that stabilize the bottom chord of wood I-joists in a diaphragm, access to the joints that have to be welded, allowing chord bars and panel connection plates in concrete tilt-ups with two reinforcing steel mats to be lapped rather than installed as a single assembly, specifying single bevel groove rather than double bevel, U- or J-groove welds, taking the time to select the generic details commonly used in an office so that their applicability is assured, and prescribing single-pass instead of multiple-pass fillet welds whenever possible are all examples of the kind of decision a structural engineer can make to enhance productivity in the field.

One way structural engineers can enhance productivity is by specifying what are known as "prequalified welded joints," which are widely accepted standardized specifications related to welded connections that have been developed by the American Welding Society, promulgated by the American Institute of Steel Construction in its design manual. These specifications allow both the structural engineer and the contractor to forego tests that would otherwise be required. Visual inspection and some simple field tests are permitted when the welds are selected from the Prequalified Welded Joint tables.

Weld Types

As noted earlier, there are numerous welding processes, and when combined with the orientation of the various pieces being connected, the shapes and thicknesses of the pieces, the different metals, and the strength requirements of the joint(s) the possibilities seem endless. Fortunately, there are a few popular types for construction. The basic configurations are butt joints, corner joints, and T joints. The elements in the joint to be connected may be prepared in a variety of ways including

butted end-to-end with or without a space between the pieces, and one or more may be beveled or made concave on one side or both. In the gas and electric arc family of welding processes, the welds include the fillet, the plug or slot weld, the back weld, and the groove weld. The groove family has five "children": the square groove, the V groove, the bevel groove, the U groove, and the J groove. The electric resistance welding family has four members: the spot weld, the seam weld, the flash or upset weld, and the projection weld.

Six Common Construction Industry Weld Types

The overwhelming favorite among gas and electric arc-welded joints in systems specified by structural engineers is the **fillet weld**. It is a multipurpose joint that is easy to produce and inspect visually, and welds as large as $5/16"$ can be created in a single pass. It is used in hundreds of applications, and is popular in other construction systems as well. The **complete joint penetration** and **partial joint penetration bevel groove weld** (known colloquially as full-pen and partial-pen welds) are widely used in, respectively, moment frame connections and connections where the full potential of the joint is not required. The **flare bevel groove** weld is useful when the curved shape of one or both members is joined to a flat plate or to another curved member (e.g., reinforcing steel rod-to-flat-plate, tube steel columns or angle stock welded to plates or other angles). The **square groove weld** is a simple weld that requires little preparation: the square ends of the pieces are butted and joined with the weld. It is useful as an all-purpose joint for thinner metal pieces. **Puddle welds** are commonly used to connect metal deck sheets to structural steel members. Popular for connecting headed studs to wide flange beams through a metal deck in a composite structural system is the electric resistance weld known as the **flash or upset weld**.

Example Structural Details[1]

Figures E.1 through E.8 are details from the structural sheets of projects designed and constructed in California. For a table of welding symbols, see Table E.1.

---

[1]Drawings courtesy of Hope Engineering Inc., San Diego, CA and Buehler and Buehler Structural Engineers, Sacramento, CA.

Detail 9/S5.5 — 1"=1'-0"

Figure E.1 This drawing shows four fillet welds, a ¼" weld 4" long on the top of the angle attached to the column, two welds the same size as the weld above joining a framing strap to the other leg of the angle, and a fourth weld the same size as the other three connecting the bottom of the angle to the column. Figure E.2 is a photograph of the actual condition. (Drawing courtesy of Buehler and Buehler Structural Engineers.)

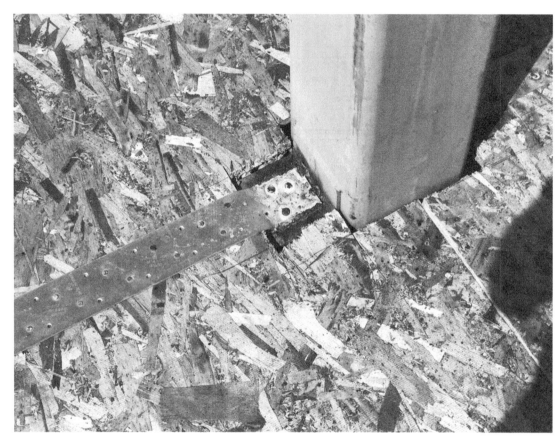

Figure E.2 The condition as it was constructed does not exactly match the drawing (Detail 955.5 in Figure E.1). The angle on the column has been rotated 180° from the orientation described in the detail. (Photo by the author.)

BRACED FRAME COLUMN
BASE PLATE DETAIL

SCALE: 1" = 1'-0"

Figure E.3 This detail shows a section through the base plate of a column, an elevation of the column and brace (lower half of drawing), and a section through the column above the base plate (upper half of the drawing). Starting from the upper left corner of the lower half of the drawing, the engineer calls for four fillet welds: two ⁵⁄₈" welds 26" long connecting the tube steel brace to the gusset plate on the arrow side of the drawing the side visible in the elevation), and two in exactly the same spot on the other side of the gusset plate. A review of the upper half of the drawing and the elevation makes it clear that the engineer requires the tube steel brace to be slotted to fit over the gusset plate. When a length of weld is specified, as in the first symbol, it generally means that more length is available, that is, the interface between the tube steel column and gusset plate is more than 26" long. The engineer has calculated that the weld must be that long minimally; if you look closely at the note right above the title of the drawing, you will see the engineer admonishing the user to verify that adequate length exists. The welding symbol to the right of the column calls for a total of six fillet welds ⁵⁄₈" in size; on both sides of the gusset plates where they attach to the column and on both sides of the column flange where it meets the base plate. The symbol immediately below this one calls for a single bevel groove weld with a 90° groove face angle between both column flanges and the base plate. This may be an error: the drawing shows a 45° angle, and one wonders how a 90° face groove is possible in this connection. (A double bevel composed of two 45° face groove bevels would constitute a 90° bevel groove weld). Perhaps what was intended is a 45° face groove angle. When the length of the weld is not specified, as in this case, it is assumed that the weld will be continuous for the length of the members being joined. The fourth and final weld symbol in this detail calls for a ⁵⁄₈" fillet weld between the base plate and gusset plate on both sides.

Figure E.4 This photograph shows the connection between a wide flange beam and a concrete tilt-up panel. The drawing establishing the parameters for the connection is shown in Figure E. 5. (Photo by the author.)

Figure E.5 On the left is a horizontal section through the concrete panel at its end. To the right is an "elevation" of the same connection looking toward the interior of the building (to the right of the steel beam and panel shown in the photograph in E.4). The concrete has been "removed" to expose the embedment to which the two 10'-long horizontal and one short reinforcing steel bar about 30" long are welded (left half of the elevation). Starting from the left, the engineer calls for two ³/₁₆" fillet welds, one on the arrow side and one on the opposite side of what looks like an L-shaped reinforcing steel bar (containing the jamb bars) tied to an 8" × 8" × ½" angle cast into the concrete. Since the tail is used and the abbreviation "typ" (typical) is shown, one may conclude that wherever this condition exists, the engineer wants a pair of fillet welds. There is the connection to which the arrow points, and another at the other end of the bar (catty corner to this one). To the right and slightly above this symbol is one calling for a pair of ⁵/₁₆" fillet welds connecting the flag plate that carries the W16 beam to the embedded 8 × 8 angle. Staggered fillet welds are shown—the welds are separated by 5". The dark triangle on top of the perpendicular line that looks like a flag is the field-weld symbol, meaning the welding should be done on site. The third symbol shows flare bevel groove welds on both sides of the two horizontal reinforcing steel bars as well as both sides of the 30" drag bar bent at 30° to the edge of the panel. (Drawing courtesy of Buehler and Buehler Structural Engineers.)

FRAME CONNECTION AT
METAL DECK WITHOUT CONCRETE TOPPING

Figure E.6 The elevation (upper half) and horizontal section (lower half) describe the post to beam connection for the top beams in a structural steel moment frame. (Moment is the tendency of a force to cause rotation around a point or an axis.[1] A moment frame employs rigid joints to control the effects of moments in a structure, and it is one way of bracing buildings, the other two being shear panels and braces, or some combination of the three methods). Starting with the upper left quadrant in the top drawing, the engineer is instructing the builder to "Fill Valley (Typ)" that is, fill the space between the "doubler" plate and the flanges of the column with a bevel groove weld. (The rectangular plate between the column flanges, the bottom of which is below the bottom flanges of the adjacent beams, is the doubler plate. The horizontal section below shows the two doubler plates, one on each side of the column web.) These welds will be done in the shop. To the right and above that symbol is one with two reference arrows. The first reference line (closest to the head of the arrow) calls for the installation of a permanent back-up bar (a thin steel plate that prevents molten metal from dripping out of a joint and enhances electrical conductivity between the pieces to be welded) and a bevel groove weld with a 30° groove face and a $^3/_8$" root opening (space between the two elements to be welded–see "Location of Elements of a Welding Symbol" in Table E1). The second operation, described directly above the first, calls for the contractor to attach the back-up bar to the column with a $^5/_{16}$" fillet weld on the side opposite the one the arrow is pointing to. These welds will be done in the field (the small black flag is the field weld symbol). To the right and just below that symbol is a shop weld instruction, directing the builder to connect the erection plate (the plate with the bolts in it) to the column flange with a ¼" bevel groove shop weld. The typical erection plate is $^3/_8$", so in effect this weld is a partial penetration weld. In the center of the beam to the right of the column are instructions to field-weld a complete penetration bevel groove weld between the beam web and the column, on the side opposite the arrow. In the lower right quadrant, the engineer calls for another two-operation field weld, starting with the installation of a temporary back-up bar on the side opposite the arrow (under the bottom flange), and a bevel groove weld with a 30° groove face and a $^3/_8$" root opening (gap between members to be welded) on the arrow side (connecting the bottom flange of the beam and the column flange). After that operation is complete, the contractor is instructed to remove the bar, clean the joint, have it inspected, and after approval, complete welding the joint with a $^5/_{16}$" fillet weld (overhead). Notice how the engineer has set up the situation where the welder can complete most of the welds while sitting astride the beam. Welding downward is known as "down-hand" welding; it is far more efficient than welding overhead, which is the other way the engineer could have required the work be done. The better access and clearer visibility of the joint tying the top flange to the column, which is the same weld type as that tying the bottom flange, may be the reason that greater care and an inspection is required in the lower weld. The last welding symbol shown, in the lower left quadrant, calls for a fillet weld on the bottom of the continuity plate that is ¾ of the thickness of the doubler plate, welded in the shop. (Drawing courtesy of Hope Engineering, Structural Engineers.)

[1] James Ambrose, *Simplified Engineering for Architects and Builders*, 9th edition, John Wiley & Sons, 2000.

1/2"∅ PUDDLE WELD

METAL DECK PER PLAN

CLOSURE PL.

3" MIN.
1 1/4" MIN.
1 1/4" MIN.
1 1/2" MIN.

BM PER PLAN

36" STEEL SHT., TYP.

NOTE:
1. SEE NOTES ON SHT. S1.1 FOR "AUTOMATIC END WELDED STUDS."

TYP. METAL DECK ATTACHMENT WITH SHEAR STUDS
5300—03UC

Figure E.7 This drawing shows the connection between headed studs and beams in a composite structural system, wherein the concrete deck becomes part of the structural system. Rather than showing welding symbols at each stud, the engineer directs the user to Sheet S1.1 for information on "Automatic End Welded Studs." These studs are flash welded using special equipment that passes an electric current through the studs into the beam, creating fusion through hea.. The deck metal itself is attached by puddle welding (applying the electrode in the leads of an arc welder to the metal pan, thereby welding it to the beam). The beams that support the deck between the girders (the I-shaped elements, top and lower left) have been eliminated for clarity. (Drawing courtesy of Hope Engineering, Structural Engineers.)

TYP. BEAM SPLICE
5000–05

Figure E.8 In this connection the engineer calls for three welds. The symbol on the top and to the right is a full penetration V groove weld connecting the top and bottom beam flanges to each other. The third weld is a square groove weld on both sides of the seam between the two webs. Notice how the access hole in the webs, just above the bottom flanges, provides the welder access with the electrode to both sides of the web. (Drawing courtesy of Hope Engineering, Structural Engineers.)

TYP. NON-FRAME COLUMN SPLICE
5000–07

Figure E.9 This column splice is a simple column splice. The welding is less intensive than that in E.10 and consequently less expensive. Two of the three weld symbols call for field-welded partial penetration bevel groove welds (lower left and upper right), and the third calls for ⁵/₁₆" fillet welds around three sides of the column splice plate, welded in the shop. The lower left symbol calls for a partial-penetration bevel weld between webs with a 45° face groove on the opposite side of the arrow, field welded. In the weld symbol in the upper right, the depth of the bevel groove weld is given as a proportion of "tf," or column flange thickness. (Drawing courtesy of Hope Engineering, Structural Engineers.)

TYP. FRAME COLUMN SPLICE

5100-01UC

Figure E.10 Here the engineer is calling for six welds, one $^{5}/_{16}$" fillet shop weld on each of three sides of the column splice plate, two full-penetration bevel groove field welds with back-up bars tying the column flanges together, and a bevel groove field weld joining column webs with a 45° groove face and a ¼" root opening (gap between elements to be joined). The rectangle above the reference line in the upper right symbol directs the welder to use back-up bars on the side opposite the arrow. (Drawing courtesy of Hope Engineering, Structural Engineers.)

Figure E.11 This detail shows the welding requirements for two intersecting pieces of tube steel (square pipe with radiused corners). The upper symbol denotes a ¼" fillet weld on the top and bottom edges of the intersecting tube, with flare bevel groove welds on the two sides. To get a better idea of why the two different welds are used, draw a horizontal section through the vertical member on the left. The elements are 6" × 6" × 5/16" (six inches square, 5/16" wall thickness). For purposes of this exercise, figure a ¼" radius on the outside corners of the tube steel. (See "Weld Types" above for the application of the various common welds). (Drawing courtesy of Hope Engineering, Structural Engineers.)

TABLE E.1 American Welding Society Welding Symbol Chart

The American Welding Society's welding symbols table is a complete list of welding symbols, orientation of components to be joined, and an explanation of them all. Table E.1 is reprinted with the permission of the American Welding Society.

APPENDIX F

MasterFormat™ 2004 Division Numbers and Titles

The outline on the following page is excerpted from CSI's 2004 edition of MasterFormat. It illustrates the numbering system CSI employs. Each of MasterFormat's divisions complies with this numbering methodology. This document is published by the Construction Specifications Institute (CSI) and is used with permission from CSI, 2008. For those interested in a more in-depth explanation of this document and its use in the construction industry, contact the Construction Specifications Institute, 99 Canal Center Plaza, Suite 300, Alexandria, VA 22314, 800-689-2900, 703-684-0300, CSINet URL: www.csinet.org.

Construction Graphics: A Practical Guide to Interpreting Working Drawings, Third Edition. Keith Bisharat.
© 2025 John Wiley & Sons Inc. Published 2025 by John Wiley & Sons Inc.

Division Numbers And Titles

The outline on the following page is excerpted from CSI's 2004 edition of MasterFormat. It illustrates the numbering system CSI employs. Each of MasterFormat's divisions complies with this numbering methodology. This document is published by the Construction Specifications Institute (CSI) and is used with permission from CSI, 2008. For those interested in a more in-depth explanation of this document and its use in the construction industry, contact the Construction Specifications Institute, 99 Canal Center Plaza, Suite 300, Alexandria, VA 22314, 800-689-2900, 703-684-0300, CSINet URL: www.csinet.org.

PROCUREMENT AND CONTRACTING REQUIREMENTS GROUP
Division 00   Procurement and Contracting Requirements

Division Numbers and Titles

PROCUREMENT AND CONTRACTING REQUIREMENTS GROUP
Division 00   Procurement and Contracting Requirements

SPECIFICATIONS GROUP

GENERAL REQUIREMENTS SUBGROUP
Division 01   General Requirements

FACILITY CONSTRUCTION SUBGROUP
Division 02   Existing Conditions
Division 03   Concrete
Division 04   Masonry
Division 05   Metals
Division 06   Wood, Plastics, and Composites
Division 07   Thermal and Moisture Protection
Division 08   Openings
Division 09   Finishes
Division 10   Specialties
Division 11   Equipment
Division 12   Furnishings
Division 13   Special Construction
Division 14   Conveying Equipment
Division 15   Reserved
Division 16   Reserved
Division 17   Reserved
Division 18   Reserved
Division 19   Reserved

FACILITY SERVICES SUBGROUP
Division 20   Reserved
Division 21   Fire Suppression
Division 22   Plumbing
Division 23   Heating, Ventilating, and Air Conditioning
Division 24   Reserved
Division 25   Integrated Automation
Division 26   Electrical
Division 27   Communications
Division 28   Electronic Safety and Security
Division 29   Reserved

SITE AND INFRASTRUCTURE SUBGROUP
Division 30   Reserved
Division 31   Earthwork
Division 32   Exterior Improvements
Division 33   Utilities
Division 34   Transportation
Division 35   Waterway and Marine Construction
Division 36   Reserved
Division 37   Reserved
Division 38   Reserved
Division 39   Reserved

PROCESS EQUIPMENT SUBGROUP
Division 40   Process Integration
Division 41   Material Processing and Handling Equipment
Division 42   Process Heating, Cooling, and Drying Equipment
Division 43   Process Gas and Liquid Handling, Purification, and Storage Equipment
Division 44   Pollution Control Equipment
Division 45   Industry-Specific Manufacturing Equipment
Division 46   Reserved
Division 47   Reserved
Division 48   Electrical Power Generation
Division 49   Reserved

Procurement Requirements

Solicitation – consisting of the Advertisement to Bid (public works projects) and the Invitation to Bid (private works), both documents contain enough relevant project information that the bidding community can determine whether to bid a project. In California, the Advertisement to Bid, published for at least two weeks in a periodical of general circulation, is required by the Public Contract Code. The Invitation to Bid pertains to private works and is typically directed toward a select list of bidders.

Procurement Instructions

Identify the requirements for a responsive bid; designed to standardize responses to bids. Include the list of bid documents, requirement to examine the bid documents, how questions during bidding should be pursued and answered, guidance on substitutions, type of bid required, instructions on how to prepare bids, bid payment and performance bond requirements, sub listing requirements, bid submission requirements, information on modification of bids prior to opening and bid withdrawal terms, disqualification, applicable laws, prebid meeting requirements, liquidated damages, bid opening and evaluation of bids, and the requirements for executing the contract.

Available information

Includes geotechnical reports, hazardous materials information, resource drawings, existing site conditions, and property survey information. Does not relieve the bidder from verifying the appropriateness of the information.

Procurement Forms and Supplements

The bid form, while not as detailed as the instructions, is the form that standardizes responses for easy comparison of bids by the owner. Essential information includes Project ID, name and contact information of person to whom the bid is directed, bidding entity's name, acknowledgements of specific owner requirements, time that bids will be held open, acknowledgement of addenda, price for the work, alternates (if used), allowances, unit prices, schedule for completion, liquidated damages, (if any), bid supplements (e.g., bid security, bidders' qualification statement), and signature line with dates and contact information.

Agreement

The agreement sets forth the legal relationship of the parties and includes: Parties' names; the agreement type (e.g., lump sum, negotiated); brief work description; consideration; payment plan including retention; liquidated damages; incentives and penalties; documents making up the contract; start and substantial completion dates; miscellaneous provisions, (references to termination or suspension clauses in the general conditions), and parties' signatures.

Bonds

Required under the Miller Act (federal) and the Little Miller Acts (states) on public works projects of any consequence. Performance bonds assure the owner that the contractor will complete the work as described, or the surety will. Payment bonds are security for the owner and those who improve the owner's property, including subcontractors, suppliers, and workers. Claims by subcontractors, suppliers, or workers for lack of payment by the prime contractor may be filed against the payment bond. Those claims, when accompanied by a bond (at 125% of the value of the claim), are referred to as bonded stop notices, and they require the holder of the funds to withhold undisbursed proceeds of the project funds. Stop notices function the same way on private and public works, and protect the same people that lien laws do.

Certificates

Refers to certificates of compliance with applicable laws and regulations and certificates of insurance. Certificates of insurance are evidence that a policy of insurance, with policy type and coverage amounts, has been written.

Conditions

Consisting of general (boiler-plate documents) and supplementary or special conditions (modifications to the general conditions that tailor a project to the specific needs of an owner), conditions are guidelines to the administration of a project. They describe the rights and responsibilities of the parties and include basic definitions, interpretation of the documents, limits of use of the documents, parties' roles, contract administration guidelines, changes in work, time and notice issues, payment process, protection of people and the project, insurance and bonds, and correcting defective work, among other things.

Specifications

Consist of administrative guidelines and guidelines to the quality of work and materials required in a project. Division 1 consists of General Requirements (e.g., construction schedule, measurement and payment, temporary facilities, project close-out) and procedures that apply to all spec sections. CSI's SectionFormat prescribes the format for each of the 49 spec sections, and consists of three parts: **Part 1-General**, describes information pertinent to the particular spec section, including administrative requirements and reference to other spec sections. **Part 2-Products**, describes the required products, and **Part 3-Execution**, describes the desired preparation and installation requirements for a component or assembly.

Specification types

Prescriptive, aka descriptive specs describe the preparation and installation requirements for a component or assembly. The burden for performance is borne by the specifier. Performance specs describe the performance the owner desires, placing the burden of a favorable outcome on the respondent. Open specs are specifications that are open to fulfillment of the spec by any respondent. Closed specs, or sole-source specs, identify a single provider of some aspect of a project, and, due to their anti-competitive nature, are illegal on public works projects except in specific circumstances.

Restricted specifications make the most of the situation by identifying several acceptable suppliers and adding "or equal" language to the spec.

---

[1] Most of the information in these columns and the graphics are derived from the 5th edition of CSI's Manual of Practice.

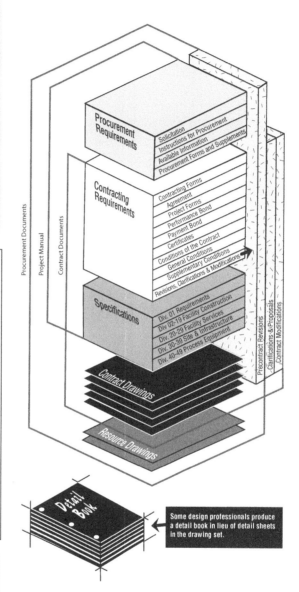

The Construction Document Family[1]

Some design professionals produce a detail book in lieu of detail sheets in the drawing set.

Contract Drawings
Drawings are the graphic depiction of the work, and are particularly effective for size and shape description, the correlation of elements, and the manner in which the elements are connected.

Pre-contract Revisions
Most commonly referred to as addenda, these instruments may modify any aspect of the project before the bids are opened.

Clarification and Proposals
Documents initiating changes or clarifications that have not been incorporated into the contract by formal contract modifications.

Contract Modifications
Refers to change orders, construction change directives, architect's supplemental instructions, or written amendments, and similarly-named adjustments to the contract terms.

Figure F.1

APPENDIX G
Executing a Site Section

Purpose of Site Sections

Site sections are vertical cuts through the surface of the earth to some point below it. Their purpose is to describe the characteristics of a site in profile such that a design professional can identify whether improvements (in the form of earthwork) are required, and if they are, the sort that should be made. For the builder, site sections are critical to determining the quantity of earth that must be transported to or from a project or moved around on it, and there are different ways of calculating those quantities. When doing earth quantity take-offs manually, one superimposes a design profile over an existing profile in the same site section and measures the difference in area between existing and planned grades in that section. In the adjacent section the same task is performed; the areas of cut and fill in the two stations are separately averaged and multiplied by the distance between the sections to determine a volume of earth to be filled or cut. There are other methods of determining earth quantities but this one, known as the average end-area method, is popular (see Chapter 6 for a discussion on average end-area method).

Executing a Site Section

Executing a site section is relatively simple as long as one bears in mind that *what gets plotted on the profile is the elevation of the ground **at the section line,*** which

(for convenience and accuracy) is usually drawn directly below the contour map. Every contour line that crosses the section line establishes the ground level at that point. The other tidbit to keep in mind is that the lines from the section down to the profile must be perpendicular to the section line. The number, frequency, and location of sections is 1) a function of the level of accuracy required (the more sections, the more accurate the take-off), 2) the rate at which the terrain changes elevation (one might take sections every 25 linear feet when calculating the quantities of earth for a highway in the mountains, and every hundred or more feet across a desert, for instance), and 3) the time one has to do the task.

Once one has determined where to take sections (they are the easiest to produce when taken more or less perpendicular to the contour lines of a topographic map), the parameters of the drawing have to be established. Normally, one draws a section only through the areas where change is planned. After all, there's no sense in quantifying the amount of earth that does not have to be changed. So the horizontal extent of the work has to be delineated, then the vertical extent. At the very least, all of the elevations (existing and planned) should be identified on the resulting profile—the idea is to identify all the work required at that section. It is worth pointing out that overexcavation—removing earth well below the foundation of a planned structure—is common. (For buildings with foundations comprising continuous footings or slabs-on-grade, this could mean removing the earth for five, ten, or more feet under and

Construction Graphics: A Practical Guide to Interpreting Working Drawings, Third Edition. Keith Bisharat.
© 2025 John Wiley & Sons Inc. Published 2025 by John Wiley & Sons Inc.

around the whole building depending on the weight of the structure). Earth is a natural material; that is, it is subject to some performance uncertainly, and consequently the only way to manage the challenge with any predictability is to remove it and replace it under carefully prescribed conditions. The geotechnical engineer's testing will show the condition and nature of the in-situ soil, and the subsequent analysis and recommendations (prescription) will be described in the geotechnical report that follows. A prudent contractor will read the report carefully and identify the extent of the work accurately.

The scale chosen for the horizontal expression of distance usually matches the scale used for the topographic map (e.g. 1" = 50' or 1" = 100'); in fact, sections are often taken with the profile sheet laid over the plan. A good general rule for the vertical scale is l0 percent of the horizontal scale, or 1" = 5' or 1" = 10' using the above scales. Figures G.1–G.4 show the steps involved in taking a site section.

A survey grid, with elevations 133.8, 132.3, 136.7, and 140.5 establishes known elevations at 4 points. By using graphic or mathematical interpolation techniques (refer back to Chapter 6 to read about interpolation), the whole-numbered contour elevations can be determined and a topographic map drawn. It helps to imagine what the site looks like in reality: here, a creek is shown on the left, and the slope from left to right is gradual and uniform as it approaches the lower right corner, where it begins to get steeper. If the grid is a hundred-foot grid (no scale is given), that makes the distance between elevations 132.3 and 140.5 (or, for that matter, between 133.8' and 136.7'), about 141 feet. The difference in elevation is 8.2 feet (140.5' − 132.3' = 8.2'), or about 6 percent (8.2' ÷ 141'). Wheelchair ramps are limited to 6 percent, so if you imagine such a ramp, you'll have a sense of the slope of this site. Between the other two points, the slope is 2 percent.

Figure G.1

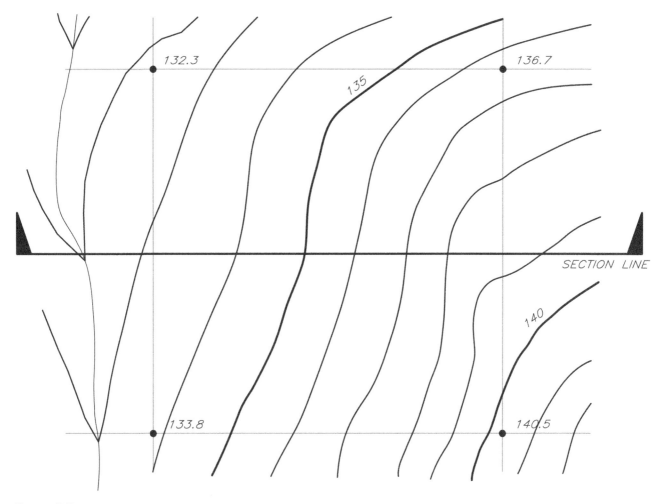

132.3

135

136.7

SECTION LINE

140

133.8

140.5

Figure G.2

The section line was selected at random, and no design contours are shown on this map, so the section will ultimately just expose the existing terrain. If there were design contours, say of a pad the same size and location as the grid, one could probably take just three sections (due to the uniformity of the slope and the grade: one at each of the two pad edges, and one in the center of the pad. The beginning and ending of the work has to be determined however, because the pad, regardless of the elevation, will have some sort of slopes created for it (unless a retaining wall is constructed). Given a pad with an elevation of 135 and two to one slopes, for example (run to rise), the work would begin about 15' away from the survey gridline with elevation points 133.8 and 140.5 on it (in other words, toward the bottom of the map). The new slopes from the pad cut into the existing terrain to the right and toward the bottom of the page from elevation point 140.5 take up at least 11 feet horizontally ($140.5' - 135' = 5.5' \times 2 = 11'$). Fifteen feet was selected as a likely distance from the grid intersection because the ground rises to the right of the intersection, which means the embankment

continues to take two horizontal feet for every single foot in elevation until such time as the daylight line (the point at which the planned contours and the existing contours are the same) is found. Table G.1 shows how the sections for this site (by station number) would be formatted for this example pad. The section at Station 0+00, which is where we anticipate the work will begin, has no area of cut or fill. Station 0+15 is mostly cut, with a little fill from contour 135 left. Station point 0+65, which is about where the section line in Figure G.2 is drawn, intersects with four contours higher than 135 (136, 137, 138, and 139) and 3 lower (134, 133, 132) and consequently has slightly more cut than fill. The section at Station 1+15 (115' from the starting station), which occurs at the other edge of the pad on the survey grid line, consists largely of fill. There is just one contour higher than 135, and there are four lower points (134, 133, 132, and the creek bed which is about 131.1). Station 1+21 (six feet toward the top of the map from section 1+15, or 121' from the starting station) is where we believe the work ceases, and like the starting station, it has no cut and fill area.

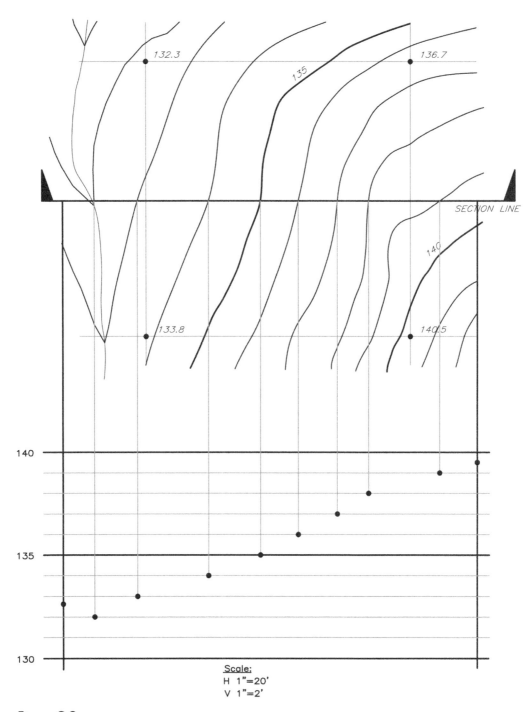

Figure G.3

Figures G.3 and G.4 show how to complete site sections once the locations have been decided upon. One must develop the elevations on the profile (a series of horizontal lines with elevations correlated to them). Notice that the elevation lines on the profile are sufficient in number to encompass the entire vertical dimension of the profile. The scales for a profile are selected for practical reasons—we sometimes represent large areas on a single sheet, but very large sheets are awkward to work with—but a reasonable level of accuracy is required. The horizontal scale of the profile usually corresponds to the scale of the topographic map, and the vertical scale is whatever the drafter chooses. One caveat: the potential for errors in a section with small scales (for example, 1" = 100') is significant. One square inch of a section using a 1" = 100' horizontal scale and a 1" = 10' scale represents 1,000 square feet. A square inch on a section using 1" = 10' and 1" = 1' scales represents 10 square feet. Consequently an error in measurement of even 1/16" in both dimensions would result

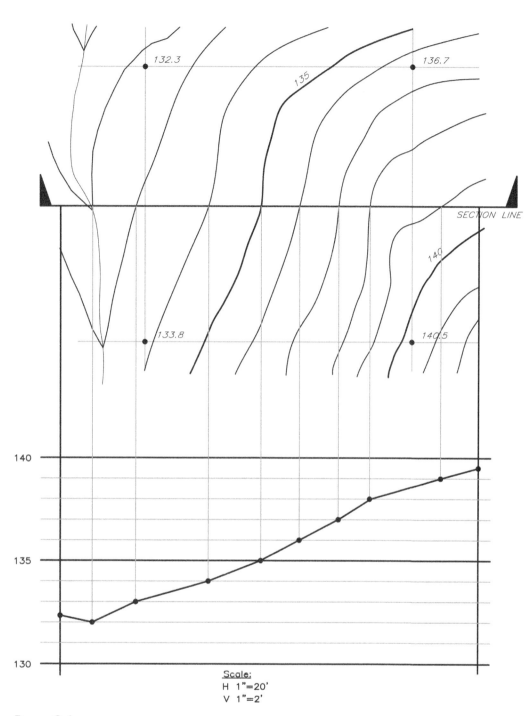

132.3   135   136.7

SECTION LINE

140

133.8   140.5

140

135

130

Scale:
H 1"=20'
V 1"=2'

Figure G.4

in an overall miscalculation of 120 square feet on the first section and 12 square feet in the second, if the square is reduced to $^{15}/_{16}$" on a side. With a distance of just 50 feet between sections, that represents an error in the volume of earth of over 222 cubic yards, as opposed to 22 cubic yards. Once the scales have been selected, the drafter draws perpendicular lines from the two sides of the section, making sure to incorporate the areas where work will be done. The elevation lines in the profile may

be drawn next, also perpendicular to the section line, from the intersection of the section and contour lines. Once the intersection points have been traced down to the profile, where dots are drawn to represent the intersection, the profile itself can be drawn by connecting the dots.

To determine the volume of earth to be moved, one would compare existing and planned profiles and determine the difference in area of the two. This can be

Table G.1 Standardized Format for Earth Quantity Take-offs Using the Average End-Area Method

Station	Distance (ft)	Cut			Fill		
		Area (sf)	Ave. area (sf)	Vol. (cf)	Area (sf)	Ave. area (sf)	Vol. (cf)
1	2	3	4	5	6	7	8
0 + 00		0			0		
	15		$\dfrac{0 + A^1}{2}$	Product of Col. 2&4		$\dfrac{0 + A^1}{2}$	Product of Col. 2&7
0 + 15		A^1			A^1		
	50		$\dfrac{A^1 + A^2}{2}$	Product of Col. 2&4		$\dfrac{A^1 + A^2}{2}$	Product of Col. 2&7
0 + 65		A^2			A^2		
	50		$\dfrac{A^2 + A^3}{2}$	Product of Col. 2&4		$\dfrac{A^2 + A^3}{2}$	Product of Col. 2&7
1 + 15		A^3			A^3		
	6		$\dfrac{A^3 + 0}{2}$	Product of Col. 2&4		$\dfrac{A^3 + 0}{2}$	Product of Col. 2&7
1 + 21		0			0		
				Sum of above, total volume of cut in cf			Sum of above, total volume of fill in cf

achieved by using a planimeter to determine the area, or simply by breaking the section into geometric shapes and calculating the area of them. The areas of cut and fill are listed for each section and the areas of adjacent sections are averaged and multiplied by the distance between sections (or station points). Table G.1 shows the proper format for earth quantity take-offs using the average end area method.

APPENDIX H

Trends in Design and Construction

Construction Graphics: A Practical Guide to Interpreting Working Drawings, Third Edition. Keith Bisharat.
© 2025 John Wiley & Sons Inc. Published 2025 by John Wiley & Sons Inc.

Key Terms
Integrated Project Delivery
Modular Construction
Modular Design
Prefabricated buildings, systems, sub-systems, and components

Key Concepts
- The methods by which the design professional presents instructions to the builder vary widely, from sketches on construction materials or scratch paper to construction drawings to sophisticated models using advanced digital modeling technology, but one thing remains constant; the builder must translate those instructions into discreet processes and sequences of work to which a value in labor, equipment, materials, money, and time can be attributed.
- Even a perfect set of drawings and specifications or an accurate and complete digital model does not assure a successful construction project. And a flawed model or estimate, in the hands of a competent manager, can result in a successful project. Although the design professional's instructions may clearly set forth project requirements, the transposition of drawings and models into a building, bridge, or highway requires a builder who not only fully understands their content, but can organize the work logically and methodically and, probably most importantly, communicate the plan for construction clearly and completely to all the participants.

- Design professional offices commonly use a varied combination of drawings and models to describe their desired outcome. So far, the use of models alone, at least in construction, is not common, except in complicated projects with substantial budgets such as hospitals.

Objectives

- Supplement the understanding of the value of drawings, specifications, and models in the construction process.
- Emphasize the value of the imagination both in design and construction.
- Recognize the importance of accepting that construction is a dynamic process that presents numerous opportunities for improvement.

■ Introduction

Chapter 15 focuses on the interpretation of the instructions for the construction of an elevator pit given by the architect and engineers for a two-story educational building, and how those instructions might be translated into a construction plan with associated costs and a schedule. The general analysis offered shows how the plan might be developed, first through a careful review of the drawings and specifications and the site investigation report, then by segregating the work into operations[1] to which a value in labor, equipment, and materials can be attributed, and finally by organizing those operations into a reasonable sequence.

Both the design and construction professions require a remarkable imagination. For design professionals, this means translating an owner's wants or needs into a design program and translating that program into the graphic expression of the desired outcome, all of which occurs in the design professional's imagination—wholly in the imagination. Contractors, on the other hand, as Figure H.1 makes clear, must also exercise their imaginations, but in a very different way; they must understand what the design professionals are instructing them to do, which they do by analyzing the project as in Chapter 15, then imagining the project under construction and correlating a reasonable time and cost to the project. Consider all the things in this photograph that are not shown on the drawings: the trenches, the stakes, the foam around the ground plumbing, the temporary supports, the dobies, caps on the reinforcing steel, the snap ties, and the formwork. It is quite a list.

---

[1] Operations in construction—form walls, place and consolidate concrete, install reinforcing steel, strip forms, for example—because they involve discernible steps which are critical to understanding the nature of the work, provide enough detail for the estimator to safely correlate a cost, yet not so much that determining construction costs becomes unwieldy. A contractor's historical data is based on a database of actual operations costs over a period of time; the operation is the unit used in construction cost databases as well.

Figure H.1 (photograph: Elevator pit forms + rebar).

Digital Modeling

Digital modeling is now in much wider use than when earlier editions of this book were published, but it remains a technology that is used in different ways and to varying extents. For very complicated projects such as hospitals, modeling is used extensively by particularly MEP contractors (Mechanical, Electrical, Plumbing). MEP contractors frequently design systems on behalf of architects, who provide performance standards, or they may use the model to manufacture components designed by the mechanical engineer. Drone technology is now commonly used to develop topographic maps from an existing site, as well as for determining monthly progress on a project. LIDAR (light directing and ranging) is remarkably useful in mapping existing structures, to the extent that the distance between existing floors can be determined within a millimeter (1/25")—even images, graffiti in one project viewed by the author—can be recorded. It is an indispensable tool for the creation of resource drawings.

Building information modeling, a collection of object databases, is a departure from older technology such as CAD (computer aided design), which is oriented toward lines and line development. For designers, BIM offers the opportunity to evaluate the development of a design as it is being created, through renderings and "fly-throughs," and is especially useful when changes are made. Prior to BIM, design professionals were required to evaluate the changes they made and how those changes telegraphed through the project, then follow through on making all the changes indicated. With BIM, the software "knows" where the changes need to be made and will make them automatically.

All models manifest the design professional's intentions; and in all their forms—scale models in wood, paper, cardboard, plastic, or other materials—strive to help all those involved with a project understand their responsibilities. Mockups are normally full-scale models of parts of a project, a window and cladding system, for instance, and by virtue of their full size, and of course the testing they are often subjected to, are tremendous aids in conceiving how a finished project will look and behave upon completion. Additionally, builders can use various kinds of digital technology to explain, if in a rudimentary way, how they plan to organize and build a project. Modeling is being used to account for costs as well—imagine the convenience of delivering a load of materials with a bar or QR code to the site; the costs are accurately and promptly posted to the cost accounts for the project. Construction accounting has always been challenging, due to the percent-complete approach to recognizing costs.

The task of managing all those work processes in the field, however, remains out of the reach of the technology. While modeling can help understand the outcome, it is up to the field personnel to identify, populate, and maintain a site with the necessary people and accommodations (e.g., office trailers, storage bins, chemical toilets and wash-ups.), procure the materials for the site, and select the means of transporting them, (railroad, trucks, on-site storage and hoisting and conveying equipment, to name a few), install and remove myriad temporary structures, manage the work, and monitor progress. Daily tasks include organizing the work, monitoring whether and at what rate it is taking place, determining the extent of completed work, verifying that materials that have been ordered will arrive at the site or storage facility in time to be properly integrated into the project, and many other tasks that are best accomplished by people.

While modeling has been helpful to design and construction, of perhaps greater value is the increasingly common adoption of delivery systems that account for the shortcomings of both the design and construction professions, notably the integrated project delivery and design-build delivery systems.

Integrated Project Delivery (IPD)

IPD involves putting owners, design professionals, general contractors, and key specialty contractors under a single, collaborative contract to design and build a project. It is distinguished from other delivery systems in that signatories have "skin in the game," resulting in each having the interests of the others in mind throughout the process of designing a project and bringing it to fruition. It is an involved process, and it is popular with complex projects; hospitals being one very good example. The contents of the contract reflect the participants' respective roles, obligations, and compensation. Samples of various contract forms used by the American Institute of Architects are available at https://learn.aiacontracts.com, and the Lean Construction Institute's web page at https://leanconstruction.org/lean-topics/integrated-project-delivery-ipd/.

While not difficult conceptually, design and construction are complex undertakings, with significant potential for ambiguity and error. Having a team of experts—architects, landscape architects, engineers, design consultants, general and specialty contractors—collaborate on a project from inception to commissioning, reduces misunderstandings and resolves conflicts before they become truly detrimental.

Design-Build Project Delivery

The design-build delivery system has its benefits, in that the design professional and constructor form an entity and sign a legal arrangement with an owner to design and construct a project. This approach gives the owner a single point of contact for the project and puts both the design professional and builder on comparable footing when it comes to liability to the owner. The tendency to assign blame for problems during construction, a common feature of Design-Bid-Build projects dissolves, since both the design professional and builder are liable to the owner for contract performance. As with IPD, design-build encourages collaboration between the key participants in design and construction, leading to innovative solutions; many common problems are mitigated before they manifest in the form of errors, defective work, and schedule delays. A peek at Figure F.1 reveals just how complex contracts for design and construction may be, but a review will reveal opportunities for streamlining the document delivery process. For instance, the builder and architect might agree on the number and type of drawings required for a satisfactory outcome, a decision normally made by the architect alone, and the design-build entity might choose to modify other aspects of the contract as well.

Modular Design and Construction

Modular design and construction consists of designing and manufacturing modular units that serve a particular purpose, small office or retail spaces for example, that are, once fabricated in a factory, assembled in the field. This approach to infrastructure growth offers many advantages. Any aspect of construction that can be created in a controlled environment obviously avoids the problems associated with the exact same construction when it is performed in the field. Increasingly extreme and unpredictable weather clearly affects field work dramatically. Worker fatigue, of particular concern in extreme climates, is less of an issue when work is performed in a factory. The work itself takes place in a conditioned environment and is often placed at waist height, for instance, reducing the stress imposed on workers' bodies in a field situation, and materials can be conveyed to the point of assembly. Construction materials for multiple projects would be shipped to a factory rather than sites scattered around a particular region, and they are stored in a protected environment until integrated into a building module. The number of elements, components, and systems normally assembled on a site is vastly reduced with modular construction, and when they arrive, they are at least partially if not fully protected from the elements. On-site construction requires the use of materials such as mold-resisting underlayment, which is less of an issue in a factory. Critically, the schedule can be compressed, since site work including grading, underground utilities, paving, and the foundation work, can occur concurrently with the manufacturing of the modules.

Even complex projects such as medical office and health care buildings are being designed and constructed using some form of modular construction. Prefabricated components such as movable wall partitions produced in a shop or factory are incorporated into the project in the field; other projects might use major and minor components selected from a standardized list of options archived in a premanufactured building library. The east viaduct of the San Francisco Bay Bridge consisted of segmented box girder sections, cast and stored months ahead in a casting yard in Stockton, California, then shipped by barge to the San Francisco Bay, where they were hoisted into place and post-tensioned. The whole effort of modular construction is aimed at making costs and the schedule more predictable.

One of the advantages of a project designed for and constructed on a particular site is the near-total freedom offered the design professional in laying out and organizing the work or living space. In modular design, regulations imposed by state highway agencies that limit the size of truck loads, for example, constrain the design of the modules, particularly as to width and height. And structural engineers must evaluate and mitigate the stresses imposed on the module during hoisting at the factory and during transportation to the site. In contrast, contractors building site-precast (tilt-up) projects take responsibility for designing and locating the reinforcing steel at pick point and bracing connections, as well as the lifting steel or bracing required to resist bending during erection. Improvements to modular design and construction will proliferate the more it is employed in infrastructure development. How will it affect construction work? The challenge of understanding what is required, formulating a plan to accomplish the work, implementing and communicating the plan, and monitoring progress in the field will remain the essential characteristics of construction work.

Glossary

Addendum As used in the construction industry, an addendum is any change to a bid package, including adjustments to its bidding requirements, agreement, general and supplementary conditions, drawings, or specifications that occur prior to a contract being signed by the parties. Addenda (the plural of addendum) can add to, remove, qualify, change, or clarify any aspect bidding or construction document requirements.

Ambient air The air surrounding or encompassing an object.

Angle of repose The maximum angle at which a material will remain at rest without sliding. This term applies most frequently to excavations and stockpiled materials such as soil and aggregate.

Area separation assemblies Any UL-approved wall or floor system of varying fire-resistance capability (typically, one to four hours) that segregates a building into zones. The idea is that a fire starting anywhere in a structure will be contained in that area for the rated period.

Backer rod Also called *foam rope*, backer rod is used to support the installation of wet sealants (gunnable sealants) in joints in exterior walls. A piece of foam rope that is larger than the gap being filled is inserted between the two sides of the joint, and the sealant is installed and pressed into the joint and smoothed out with a special tool.

Backing A generic term describing any material used as a substrate for elements or components that are connected to a system, for example, sheet metal or wood blocking installed in a wall to which grab bars or toilet partitions are attached.

Ballast (1) Any heavy material that serves to protect a roof membrane from damage by light and the corrosive effect of water falling on a dusty surface. It also prevents the membrane from being lifted up by wind. Stones, rock, and interlocking concrete blocks are examples of ballast. (2) In railway construction, ballast refers to the large coarse aggregate that sits under and around the railroad ties. (3) Any material installed in a container to stabilize it during transportation or placement. (4) The device in a fluorescent lamp that limits the current flowing through the luminescent gas.

Base In interior construction, base is the general term for the material used to cover the joint between a wall and the floor. Various woods, ceramic tile, and synthetic rubber are commonly used as base in commercial and residential construction.

Batter boards A temporary structure consisting of stakes and horizontal boards placed in proximity to the corners of buildings during foundation construction, used to locate building lines for foundation work.

Best management practices (BMPs) Techniques to control erosion, manage sediment, and contain chemicals on construction project sites that might otherwise make their way to "receiving bodies" of water in proximity to a site (stormwater ditches, streams, rivers, ponds, lakes, sloughs, bays, the ocean).

Blocking Refers primarily to wood blocks installed between studs, joists, or rafters at midpoints in a span. Blocks join structural members, preventing them from distorting under loading and causing them to act together and provide backing for nailing, among other things.

Construction Graphics: A Practical Guide to Interpreting Working Drawings, Third Edition. Keith Bisharat.
© 2025 John Wiley & Sons Inc. Published 2025 by John Wiley & Sons Inc.

Branch circuitry   The electrical wiring and protective conduit that connects the circuit protection devices in panels (fuses and circuit breakers) to light fixtures and receptacles.

Break metal   Vernacular for aluminum sheet that is custom-fabricated to a specific project.

Building code   Widely accepted minimum standards for health and safety in building construction.

Bulkhead   A dam placed between wall forms in the ends of formed walls that separates one concreting effort from another.

Bus   A metal bar, usually copper or aluminum, that conducts electricity and serves as a common connector for two or more circuits. Bus bars usually carry substantial electric current.

Cable termination   The device or devices connecting the ends of electrical wiring to electrical equipment.

Cable tray   A ladderlike framework of steel, aluminum, fiberglass, stainless steel, or PVC-coated metal that provides rigid open support for power or communications cable. Cable tray is most commonly used in horizontal applications to carry cables in the interstitial space of buildings.

Cant strip   Generally describes something at an angle; a cant strip is a mineral fiber material, triangular in cross section, installed between a wall and the roof substrate to ease the transition from horizontal to vertical in a built-up or single-layer roof system; for example, when a roof covering terminates at a parapet wall. Cant strip also describes the horizontal wood installed to support the first course of a concrete shingle roof; subsequent courses rest on the top edge of the shingles in the course below.

Cantilever   A vertical or horizontal structural member supported at one end only.

Casework   A term describing cabinetry and milled or manufactured trim.

Catch basin   A cistern, receptacle, or reservoir that collects surface runoff from gutters, swales, or drain systems, frequently constructed of precast concrete or concrete masonry units with a framed grate where the runoff enters.

Centroid   The center of mass of an object.

Chair rail   An element, frequently milled lumber, attached to a wall at the height of the backs of chairs to protect the wall surface from damage.

Chase   A fully enclosed space within a structure that frequently houses pipes, ducts, conduit, wiring, and the like.

Chord bars   The horizontal structural element installed in buildings at or near the roof deck that resists bending in the wall caused by lateral forces acting on the roof diaphragm of a building. In a concrete tilt-up building, the chord bars consist of two or three reinforcing steel rods (aligned in the same plane as the wall panel, that is, one on top of the other) spanning the entire wall panel. The chord bars are installed midway between the faces of the panel, and are welded to plates of steel at each panel edge. After the wall panels have been erected, a steel plate is welded to each of the embedded chord bar plates across the panel joint to assure the continuity of chord bar from corner to corner. The top plates in a wood light frame act as the chord bar.

Cleanout   A fitting in piping systems that, when the threaded plug is removed from the fitting, provides access to the drain system to clear the line of obstructions. The term applies to similar devices in ducts and fireplaces.

Clear and grub   The first step in an earthwork operation in which land is cleared and trees, stumps, roots, and shrubs are removed.

Closure strip   The periphery of a slab-on-grade in tilt-up construction ranging from 3' to 10' in width that is cast after the wall panels have been erected.

Conductor   Metal bar or wire with low resistance to the flow of electrical current. Copper and aluminum are commonly used as conductors.

Conduit   A channel or pipe for conveying fluids; a tube or duct for enclosing electric wiring.

Contour interval   The vertical distance between contour lines.

Contour line   A line connecting points of equal elevation, commonly used on maps to describe the topography of a site.

Control joint   A joint created to control where cracking due to shrinkage occurs in concrete. It can be created by inserting plastic strips in a slab, by using a deep joint trowel, or by saw-cutting while the concrete is still green (uncured). Control joints create a weakened plane, making it easier for cracking to occur where the joint has been created.

Convection   (1) The physical phenomenon through which heat is exchanged by fluid motion between regions of unequal density that is itself caused by nonuniform heating. (2) Fluid motion caused by an unequal force such as gravity.

Couplers   Cylindrical sections of steel with clamping mechanisms which are used to join sections of scaffolding, also called *coupling*.

Coupling   Cylindrical sections of steel, copper, plastic, or other piping material that connect sections of pipe together. Couplings may be threaded, glued, soldered, brazed, or mechanically fastened.

Crib wall   A retaining structure comprising open boxes constructed of heavy timbers filled with rock or pervious fill; a wall constructed of commercially produced precast concrete interlocking elements that create cavities that are filled with rock or pervious fill.

Crickets The term used to describe the structures or materials that divert water or snow around obstructions on a roof, or that guide the water to a particular point, for example, the plywood roofs constructed on the high side of a mechanical equipment pad.

Curtain wall A cladding system supported by the structural frame it covers.

Daylight line The point at which the design contours and existing contours on a site plan are the same.

Deconstruct To break down into constituent parts; dissect.

Deferred approval items Any component or system to be installed in a construction project that has yet to be designed, fully described, and approved by the prime design professional and the government entities having jurisdiction over the project. Roof trusses, special doors, fire suppression, and fire alarm systems are examples that are commonly identified as "deferred approval items."

Delivery systems The legal package that identifies the contractual responsibilities of the key participants involved in designing and constructing projects. Owner, design professional, and constructor responsibilities are, or should be, spelled out in the delivery system documentation.

Device In electrical specialty work, a device is a component of an electrical system that does not consume electricity; it merely carries or controls it. Wall receptacles and switches are examples.

Direct costs Costs that can be reasonably and accurately correlated to discrete operations in a construction project. The wages of craft personnel, materials used in the construction of the project, equipment, and subcontract costs are examples of direct costs.

Door schedule *See* schedule.

Dowels Cylindrical lengths of material (commonly wood or steel) inserted into holes drilled in separate elements or components, which the dowel aligns or connects. In cast-in-place concrete, dowels are used to span separate concrete pours occurring at different times, for example, between two adjacent slabs or between a footing and the stem wall connected to it.

Draft stop Any material that contains oxygen that might otherwise be available to a fire; for example, a row of blocks in a stud bay acts as a draft stop. The blocks, the studs themselves, and the wall surface act to prevent the air in the bay from being drawn to another location. Although in this example it is the blocks that are thought of as the "stop," if any one of these three elements is missing, the air can be drawn to a fire, so it is actually the assembly that is the draft stop.

Drop inlet A catch basin top that sits lower than the surrounding pavement.

Duct coil A heat exchanger within a duct run.

Embedment Derived from the word *embed,* which means to envelop, enclose, or surround tightly, this term most frequently refers to objects that are installed in concrete, although it may pertain to other components in construction. Also spelled *imbedment.*

Equipment-paced operation An operation is a series of actions that results in an end, or a course, or procedure of productive or industrial activity. An equipment operation is a series of actions performed by a machine that accomplishes work of some kind. Equipment operations are characterized by continuous activity, few tasks, and repetition.

Evaporator The component within a refrigeration system where the refrigerant is vaporized, thereby drawing heat from the surrounding medium, usually air or water.

Fan coil An assembly consisting of a heat exchanger and a fan that transfers heat or cold from a liquid passing through the heat exchanger (a fin-tubed device).

Feeder A term applied to the electrical conductors that provide the power source to electrical panelboards.

Fenestration The design and placement of windows in a building.

Field nailing Field nailing refers to the fasteners installed through plywood or wallboard to the framing members between the edges of the board; seam nailing refers to the nailing required around the perimeter of plywood or wallboard.

Fire rating Also known as *fire endurance rating,* fire rating is the capability of a material or assembly to continue to perform as designed for a period of time while subjected to a "standard" fire. A one-hour wall, for example, is designed to resist a standard fire for a minimum of an hour before it ceases to act effectively as a barrier.

Fixture whip Electrical conductors encased in flexible metal conduit that run from a junction box to a light fixture.

Flange A rim or collar protruding from (usually) a pipe that is used to connect similarly fitted pipes, pumps, or motors. Also, the top and bottom edges of a steel beam.

Flashing A loosely used term for all materials that are installed in cladding or roof systems to repel water; commonly, asphalt-impregnated papers and sheet metal. In walls, the flashing normally acts as a second line of defense, the first being the exterior finish (plaster, wood siding, masonry, etc.). In roof systems, the flashing is often the first line of defense, and is used to seal larger openings; sealants are often used with the flashing to complete the barrier to water and air.

Gabion Compartmentalized containers of galvanized wire mesh, cylindrical or rectangular in shape, containing rocks or cobbles that are used to retain earth, prevent erosion, or dissipate the energy of water.

Galvanic action Accelerated corrosion occurring between two dissimilar metals in the presence of an electrolyte such as water.

Geocomposites Any of a combination of geosynthetics used in a particular soil application.

Geogrid Synthetic products (usually inert plastics) used to stabilize or reinforce soil, usually installed in layers in a backfilling operation

Geomembrane Synthetic products (usually high-density polyethylene (HDPE) or polyvinyl chloride (PVC) plastics) that are used to separate contiguous soil areas, for example as liners under a solid waste disposal site. The joints in geomembranes are either heat or chemically fused to form a continuous unbroken membrane.

Geonet Synthetic products (usually plastics) that are designed to extract and transport liquid from soils to collection points.

Geosynthetics The generic term for mostly inert plastic products used in soils to improve or otherwise control drainage and sedimentation, to separate or isolate soils, or to reinforce them.

Geotechnical engineer Registered professional engineer who investigates project sites and produces comprehensive evaluations of their soil conditions, which are recorded in a geotechnical report.

Geotextiles Woven, needle-punched, or heat-bonded synthetic products (usually plastics) that are used to separate water from soil, for example, behind retaining walls.

Girder A primary structural beam supported by columns that carries other beams.

Hachure A symbol used in topographic maps to distinguish depressions from mounds, consisting of numerous short lines drawn perpendicular to a contour line. The side on which the hachure appears slopes downward.

Handing The identification of door hinge location and the direction that a door swings. The definition of a right- or left-hand door varies; the prudent constructor will identify how a particular supplier refers to the handing of doors. For example, one definition of a "left-hand door" is a door that opens to the left, away from the viewer, with the knob on the right side of the door; and "left-hand reverse" is a door that opens to the left, toward the viewer, with the knob on the right. Conversely, a "right-hand door" is a door that opens to the right, away from the viewer, with the knob on the left; and "right-hand reverse" is a door that opens toward the viewer, with the knob on the left. It is

worth mentioning that the right-hand and left-hand reverse doors swing the same; the difference is in the orientation of the exterior to the frame.

Hard costs Costs that are payable in cash within periods ranging from a few days to several months.[1] Also used by developers to describe building costs (the contractor's direct, indirect, general and administrative costs, and profit).

Heat exchanger A device that exchanges heat between two fluids that are physically separated.

Hinge point In embankments, the point at which a level or slightly sloped section turns downward to form a slope.

Hydronic system Heating and air conditioning systems that transport hot or chilled water, or both, through pipes to heat exchangers in duct runs.

Hydrostatic head The pressure at any point in a liquid at rest.

Imbedment Derived from the word *imbed,* which means to envelop, enclose, or surround tightly, this term most frequently refers to objects that are installed in concrete, although it may pertain to other components in construction. Also spelled *embedment.*

Index contour Contour lines labeled with elevations; usually every fifth contour line.

Indirect costs Costs associated with a project that are not easily correlated to a specific operation. Project personnel, temporary power, site office space, telephone, and similar costs are examples of indirect costs.

In situ earth Earth in its natural state.

Interceptor A device in a soil piping system that separates grease, oil, or sand from waste water.

Intermediate contour Contour lines between index contours.

Interstitial space Frequently refers to the space between a suspended ceiling and the floor above it in a commercial building; also applies to wall cavities.

Integrated Services Delivery Term describing the delivery of design and construction services on behalf of an owner by a single entity; synonymous with design-build.

Isotropic Identical in all directions; invariant with respect to direction.

King rafter The three common rafters that form a hip roof. All three king rafters "kiss" the two hips adjacent to them. Jack rafters, or hip jacks, fill the space between king rafters and hips.

King stud The stud attached to the header (beam) supporting a load over an opening.

[1] From *Construction Cost Estimating and Bidding: A Managerial Approach*, an unpublished manuscript of Richard A. Nickles.

Labor-paced operation An operation is a series of actions that results in an end, or a course or procedure of productive or industrial activity. A labor-paced operation is work that is performed by people, characterized by discontinuous activity and little repetition.

Lifts Layers of earth varying from several inches to several feet in depth that are placed in a backfilling operation; the concrete placed between horizontal construction joints; the grout placed in concrete masonry unit walls in a typical grouting operation; a hoist.

Mean radiant temperature The collective thermal effect of the radiant surfaces within a room, as experienced by its occupants.

Mullion The vertical member supporting glazing in a window or metal and glass cladding system.

Muntin The slender member separating panes (lights) of glass in a (usually small) window.

On center (o.c.) An expression referring to the spacing between centers of any element, for example, the spacing between studs in load-bearing partitions is commonly 16" on center (occasionally referred to as c.c., or center to center).

Operation A process or series of acts performed with a specific purpose or result.

Overexcavation The practice of excavating well beneath a structure and backfilling under controlled conditions to create suitable fill for slab-on-grade, continuous, or spread footing foundations, or for other purposes.

Panelboard Metal boxes in which electrical circuit switching is contained. In residential applications, the panelboard is referred to as the *loadcenter*.

Panic hardware Door hardware consisting usually of a bar that spans the width of the door at waist height. In a panic situation such as a fire, people function less thoughtfully so panic hardware enables the door to open just by a body falling against it; in other words, it is not necessary to turn a knob and open the door.

Parapet In the context of architecture and construction, this term usually refers to a low wall at the edge of a roof. It also refers to a low wall on a dam, balcony, or terrace.

Plenum A chamber into which conditioned air is forced or drawn out of, from which ducts or registers emanate.

Preconstruction services Services provided by a constructor on behalf of an owner during the predesign or design development phases of a construction project. Estimating, scheduling constructability analysis, and value engineering are examples of preconstruction services.

Prime design professional A licensed architect, landscape architect, structural, mechanical, electrical or other engineer who signs an agreement for design services with an owner.

Project manual The collection of documents prepared by the design professional for a construction project, consisting of bidding requirements, general and supplementary conditions, specifications, and addenda, if they exist.

Purlin A secondary horizontal framing element that carries tertiary framing members or that supports primary framing members at midspan.

Reducer A connector that connects conduit (piping or ductwork) of different sizes. Reducers may be eccentric or concentric.

Refrigerant Any fluid used to transfer heat in a refrigeration system. Refrigeration involves converting a refrigerant from one state to another and deriving the benefit of the physical change; a refrigerant in its gaseous state is compressed, converting it to a high-temperature liquid, which is circulated in a condenser, which draws heat from the liquid. The warm, pressurized liquid is propelled to a thermal expansion valve, which relieves the pressure on the liquid. Allowing the liquid to expand results in considerable heat being extracted from the surrounding medium (air or water). Once in its gaseous state again, the refrigerant is pumped through the condenser again, where the cycle repeats.

Register An opening into a room through which conditioned air passes, often fitted with a grille and integrated damper for the regulation of air direction and volume.

Reglet A narrow groove cut or cast into a concrete, masonry, or stone wall that receives metal flashing; the flashing that fits into a narrow groove in such a wall; a narrow, flat molding; a channel or groove that holds or guides a window sash or panel.

Relative humidity The measure of the amount of water vapor in the air as compared with the maximum that the air could contain at a given temperature. Fifty percent relative humidity means that the air contains 50 percent of its maximum potential water vapor; 100 percent relative humidity means the air has achieved its saturation point, also known as the *dew point*.

Reveals A term generally describing the decorative surface treatment (frequently grooves) that are cast into the faces of concrete elements; rustication.

Rhombus An equilateral parallelogram.

Roll A quantity of material wound around a cylinder; in piping, the term refers to hangers with horizontal rods that are curved to accept the pipe, thereby increasing the area of support of the pipe.

Roof jack A common term describing the two-part sheet-metal flashing devised to keep water from intruding into the roof system around penetrations in the roof caused by plumbing, furnace flues, electrical conduit, and the like. The jack itself covers the hole in the roof substrate and connects to the penetrating element; the counterflashing (also called the *storm collar*) seals the connection between the jack and the penetrating element.

Runner The metal channel into which metal studs are installed and to which they are fastened. Also called *track*.

Safing A highly fire-resistant mineral batt material used to block the passage of fire from one area to another. Cladding systems commonly exist in their own design zone outside the floor line, leaving voids between the vertical members of the cladding system and the floors. Safing is installed in these voids.

Schedule (1) The proposed sequence and duration of the principal tasks involved in the construction itself, usually presented in graphic form. (2) A door and hardware, equipment, or room finish schedule is a matrix that identifies types, numbers, and locations of these products in the project. (3) A schedule of values refers to the line-item list of costs established at the beginning of the project that the owner and contractor agree will be used to allocate monthly progress payments. The summation of all the line items results in the total contract value (other than unforeseen changes that occur during the course of the work).

Seam nailing Seam nailing refers to the nailing required around the perimeter of plywood or wallboard; field nailing refers to the fasteners installed through plywood or wallboard to the framing members between the edges of the board.

Shake Unit roofing material comprising hand-split wood (commonly of Western Red Cedar) installed over spaced sheathing; a type of multilayer roof covering that makes use of pressure-equalized construction.

Shims Any material used to fill in a gap in a connection; used to adjust the vertical or horizontal fit of a system quickly and dependably. Shims may be temporary or permanent.

Shop drawings Contractor-produced drawings that manifest how a portion of the work in a construction project will be executed.

Single-line drawings Drawings, usually schematic in nature, that depict the relationship between components in a system using a single, usually medium-width line. Plumbing drain waste and vent plans, riser diagrams, and isometric drawings of the system are single-line drawings. Double-line drawings, which are more realistic, are used in details of pipe, valve, and pump configurations.

Snap ties Hardware used in cast-in-place concrete formwork that holds the two sides of wood forms together during the concreting process. The ties are cast into the structure. After the forms have been removed, the part of the ties that protrudes is snapped off at the face of the concrete on both sides of the wall.

Soffit A term widely applied in construction to describe any horizontal surface (except a ceiling) extending outward from a vertical surface such as a wall. The underside of an arch, boxed eaves, and the finished underside of a cantilevered second floor are examples of soffits.

Spandrel beam A beam on the perimeter of a structure that supports floor and roof loads between columns.

Spandrel panel A panel in a curtain wall that covers elements (spandrel beam and nonstructural framing) between the head of a window on one story to the sill of the window above it.

Spoils Soil or rock generated in grading and excavation operations that is cast off (in a trenching operation), temporarily stored, or hauled away.

Storefront A term loosely used to describe large window systems like those used in retail shopping centers behind which merchandise is displayed. Storefront systems fit within and attach to structural elements, as compared with curtain walls, which attach to the structural frame but exist in a zone outside of it.

Stringers In formwork, a stringer is a horizontal support for joists that support deck sheathing (plywood). Refers also to the sloped beams between floors or landings in a staircase that support the stair treads.

Surface metal raceway Ducting for electrical power cable that mounts on the surface of walls, floors, and ceilings, used to wire a space when installing wiring in the normal manner is too costly. Also known as *wire mold*.

Switchboard A modular assembly of metal cabinets used to house electrical switches, transformers, motor control centers, and wiring in buildings whose electrical load is too substantial for a single panelboard.

Takeoff The process that determines the quantity of materials required in construction projects.

Tenant improvements Known generically as TIs, tenant improvements consist of improvements installed at the behest of a tenant (lessee or renter of property), who takes responsibility for the design and financing of the improvements. The construction services are provided by companies that perform this sort of work for a fee.

Terrazo A flooring material made of marble or other stone chips and Portland cement, polished to a glossy finish when cured.

Thrust block A crudely cast mass of concrete used in pressurized underground water systems to resist thrust where piping branches or changes direction.

Tie-back A rod or cable that prevents a vertical element, such as form, retaining wall, or the wall of an excavation, from lateral movement. When used as restraints for an excavation, tie-backs are commonly drilled into the earth that the wall is retaining and grouted. The other end of the cable or rod is often connected to a horizontal member (wale) that spans the wall of the excavation.

Tie-down A tie-back that is installed vertically, to resist the uplift of a floor caused by the hydrostatic pressure exerted by groundwater at a level higher than the floor.

Trammel A shackle used to teach a horse to amble; anything that restricts free movement; a device for drawing circles.

Trap Piping in proximity to plumbing fixtures which, when connected, forms a U-shape, thus trapping a quantity of water in the pipe, which acts as a seal to the sewer gasses that could otherwise make their way into a building.

Trench duct An enclosed steel trough set into a concrete floor, with a top cover set to the elevation of the finished floor, used to house cables, particularly in labs and computer rooms. The duct is accessible through the lid, which provides the building user with wiring flexibility that does not exist for other types of conduit.

Underfloor duct A two-way metal ducting system cast into concrete floors in office and similar spaces that provides access to the wiring on 2' centers through access holes. The system provides maximum flexibility in wiring for specific rooms.

Value engineering The search for alternatives to costly materials, components, and assemblies without compromising the structural or esthetic integrity of a construction project.

Vapor barrier Materials that impede the passage of water vapor across a plane. Unpunctured aluminum foil and unvented galvanized iron decking are two materials that qualify as vapor barriers.

Vapor diffusion The transportation of water vapor across planes due to vapor pressure differentials.

Vapor retarder Materials of varying permeance that retard the passage of water vapor across a plane. Polyvinyl chloride (PVC) and polyethylene sheets with thicknesses ranging from 4 to 10 mil (thousandths of an inch), building paper, and some paints have the capacity to retard the flow of vapor. Vapor retarders are commonly used under concrete slabs-on-grade, inside walls, and in roof structures.

Variable air volume (VAV) box A device in the supply air ducts in an HVAC system that allows the user of a space to control air temperature by determining the volume of heated or chilled air coming through the duct.

Wainscot Wood paneling or other material on a portion of a wall installed to protect the wall from damage or to add a certain architectural style to a room. Restrooms in commercial buildings frequently have tile wainscots in areas with a high potential for damage.

Wale, waler A horizontal member made of wood, aluminum, or steel, used to stiffen a form wall, sheet piling, or soldier beams and lagging. In addition to making a wall stiffer, wales help make the vertical elements act together to resist the horizontal pressures caused by soil and water behind the wall, or by wet concrete during the casting process.

Wattle A large rope of straw contained in fine plastic mesh that retards the progress of water down a slope and filters sediment from runoff, used to control erosion and contain sediment.

Wedge anchor The assembly that grips post-tensioning tendons at the stressing end of the tendon. The two half-cone clamps have ribbed semi-cylindrical shafts at their centers that fit over the high-tension cable, and the resulting cone-shaped device fits into a conical void in the anchor body. As the cable shortens, the clamps are forced into the void and clamp the tendon, preventing it from slipping.

Wire duct A duct with slotted sides and cover, used to organize wiring in enclosed spaces such as panels. The slots allow wiring to be turned out of the duct for a connection, and the lid is removable, which makes access easy.

Wire mold Ducting for electrical power cable that mounts on the surface of walls, floors, and ceilings, used to wire a space when installing wiring in the normal manner is too costly. Also known as *surface metal raceway*.

Wireway Sheet metal troughs with hinged or removable covers that house and protect electrical cables.

Selected References

Chapter 1

American Institute of Architects. 2001. *The Architect's Handbook of Professional Practice,* 13th ed. New York: John Wiley & Sons, Inc./American Institute of Architects.

This publication, nearly a thousand pages long, includes all of the boilerplate documents produced by the American Institute of Architects (on the CD that accompanies the book) as well as samples of forms (in writing) that are commonly used, and commentary on every facet of architectural practice.

Liebing, Ralph W. 1999. *Architectural Working Drawings,* 4th ed. New York: John Wiley & Sons, Inc.

This book is a comprehensive summary of the design process, with critical background information and a frank discussion of the challenges facing the architectural profession. The comparison of effective and not-so-effective drawing, and the narration that accompanies many of the drawings, are informative and helpful.

National Institute of Building Sciences/American Institute of Architects/Construction Specifications Institute/Tri-Service and the U.S. Coast Guard. 2001. *National CAD Standard,* Version 2.0. Washington, DC: National Institute of Building Sciences.

The National Institute of Building Sciences is a nongovernmental, nonprofit organization authorized by Congress to improve efficiencies in building and design, and the National CAD Standard is one of its works in process. Any interested party may become involved in the development of the NCS by joining the National CAD Standards Project Committee (see Internet Resources below).

Ramsey, Charles George, and Sleeper, Harold Reeve. 2000. *Architectural Graphic Standards,* 10th ed. New York: John Wiley & Sons, Inc./American Institute of Architects.

AGS has evolved into a wide-ranging compendium of design data that is useful to designers and contractors alike.

Internet Resources

Information relevant to the National CAD Standard or the Uniform Drawing System (UDS):

www.nationalcadstandard.org

Chapter 2

Associated General Contractors of America (AGC), 2004. *Project Delivery Systems for Construction.*

A useful book that simplifies the process of identifying and selecting project delivery systems and discusses the merits of the basic types.

Construction Specifications Institute (CSI). 1996. *Manual of Practice.* Construction Alexandria, VA: Specifications Institute.

The *Manual of Practice* is the industry standard for the development and production of specifications for construction projects. It goes into considerable detail on construction documents, delivery systems, bidding requirements, and specifications writing.

Dorsey, Robert W. 2000. *Understanding Architects: A Constructor's Guide to Architectural Practice.* Cincinnati, OH: Frank Messer & Sons Construction Co.

For people who have not spent much time conversing with architects, this is a good, short work that describes one architect's view of the history of architectural design and the manifestations of architectural culture.

———. 1997. *Project Delivery Systems for Building Construction.* Washington, DC: Associated General Contractors of America.

A useful book that goes into some detail regarding contractual relationships in construction. It explains the typical roles of the various participants in the construction process.

Construction Graphics: A Practical Guide to Interpreting Working Drawings, Third Edition. Keith Bisharat.
© 2025 John Wiley & Sons Inc. Published 2025 by John Wiley & Sons Inc.

Internet Resources

Accreditation of two- and four-year construction management and construction technology programs:

American Council of Construction Education (ACCE)

www.acce-hq.org

Technology Accreditation Commission/Accreditation Board of Engineering and Technology (TAC/ABET)

www.abet.org

Information on professional licensing for architects:

National Council of Architectural Registration Boards (NCARB)

www.ncarb.org

Information pertaining to the accreditation of architectural programs:

National Architectural Accrediting Board (NAAB);

www.naab.org

Information on professional licensing for engineers:

National Council of Examiners for Engineering and Surveying

www.ncees.org

Information on constructor certification:

American Institute of Constructors

www.constructioncertification.org

Specifications for Construction projects:

Construction Specifications Institute

www.csinet.org

Chapter 3

National Institute of Building Sciences/American Institute of Architects/Construction Specifications Institute/Tri-Service and the U.S. Coast Guard. 2001. *National CAD Standard,* Version 2.0. Washington, DC: National Institute of Building Sciences.

The National Institute of Building Sciences is a nongovernmental, nonprofit organization authorized by Congress to improve efficiencies in building and design, and the National CAD Standard is one of its works in process. Any interested party may become involved in the development of the NCS by joining the National CAD Standards Project Committee.

Simmons, Leslie H., and Olin, Harold B. 2001. *Construction Principles, Materials, and Methods,* 7th ed. New York: John Wiley & Sons, Inc.

This comprehensive publication does a credible job of describing many of the materials and process of construction. Among its contributions is background information on the development of the metric system in the United States.

Internet Resources

Conversion to the metric system, or International System of Units (SI);

Metrication Council of the National Institute of Building Sciences:

www.nibs.org.

Chapter 4

Allen, Edward. 1993. *Architectural Detailing: Function, Constructability, Aesthetics.* New York: John Wiley & Sons, Inc.

As the title suggests, this book devotes considerable energy to the rationale behind the development of various details. It is noteworthy for its consideration of the builder's experience.

Giesecke, Frederick E.; Mitchell, Alva; Spencer, Henry Cecil; Hill, Ivan Leroy; Dygdon, John Thomas; and Novak, James E. *Technical Drawing,* 11th ed. Upper Saddle River, NJ: Prentice Hall.

Although little in the way of construction graphics is discussed in this textbook, it is nevertheless a fine resource for technical drawing in general.

Madsen, David A.; Folkestad, James; Schertz, Karen A.; Shumaker, Terence M.; Stark, Catherine; and Turpin J. Lee. 2002. *Engineering Drawing and Design,* 3rd ed., Albany, NY: Delmar.

Although the focus in this book is mechanical drawing, there are several chapters that address disciplines pertinent to construction, namely process piping, structural drafting, HVAC, and electrical drawing.

Muller, Edward J.; Fausett, James G.; and Grau, Philip A. 2002. *Architectural Drawing and Light Construction,* 6th ed. Upper Saddle River, NJ: Prentice Hall.

This and previous editions are focused entirely on presentation and working drawings for small building construction projects. There are a number of good sketches that run the gamut of projection type and style.

Wakita, Osamu A., and Linde, Richard M. 2003. *The Professional Practice of Architectural Working Drawings,* 3rd ed. Hoboken, NJ: John Wiley & Sons, Inc.

This book, directed toward students of architecture, contains a fine collection of drawings, particularly freehand sketches, of building components, supplemented by an orderly explanation of the evolution of a construction drawing set.

———. 1999. *The Professional Practice of Architectural Detailing,* 3rd ed. New York: John Wiley & Sons, Inc.

This comprehensive publication sets forth detailing information using numerous drawings of details as well as pertinent photographs.

Chapter 5

Ching, Francis D.K. 1998. *Design Drawing.* New York: John Wiley & Sons, Inc.

Francis Ching has an uncommon skill with pencil and paper. This book is no doubt one of the finest collections of sketching and presentation drawings available.

Ching, Francis D.K., and Adams, Cassandra. 2001. *Building Construction Illustrated,* 3rd ed. New York: John Wiley & Sons, Inc.

Noteworthy primarily for its subject matter and the remarkable quality of its numerous drawings, this book explores a variety of basic building systems according to the Construction Specifications Institute's MasterFormat indexing system.

Cooper, Douglas. 2001. *Drawing and Perceiving,* 3rd ed. New York: John Wiley & Sons, Inc.

This soft-bound book, full of student drawing work, is an excellent resource for the individual who has an interest in developing good freehand drawing skill.

Giesecke, Frederick E.; Mitchell, Alva; Spencer, Henry Cecil; Hill, Ivan Leroy; Dygdon, John Thomas; and Novak, James E. *Technical Drawing,* 11th ed. Upper Saddle River, NJ: Prentice Hall.

This comprehensive technical drawing text has an especially fine section on sketching.

Edwards, Betty. 1989. *Drawing on the Right Side of the Brain.*

Aimed primarily at portrait sketching, this is an interesting read for drawing enthusiasts.

Nicolaides, Kimon. 1969. *The Natural Way to Draw.* Boston: Houghton Mifflin Company.

For those with an earnest desire to draw well freehand, this is an engaging treatise that treats the journey as the seminal experience. The importance of critical observation—key to the success of managers of the construction process—is made quite clear in this book.

Wang, Thomas C. 2002. *Pencil Sketching,* 2nd ed. Hoboken, NJ: John Wiley & Sons, Inc.

This marvelous small paperback addresses the essential elements of sketching, largely through graphic expression.

▪ Chapter 6

Bartholomew, Stuart. 2000. *Estimating and Bidding for Heavy Construction.* Upper Saddle River, NJ: Prentice Hall.

This is a straightforward, easy-to-read treatise on heavy construction practices, written by a seasoned heavy construction contractor.

Nunnally, S.W. 2004. *Construction Methods and Management,* 6th ed. Upper Saddle River, NJ: Pearson Prentice Hall.

This and previous editions of Nunnally's work explore site and heavy construction in some detail. It is a useful reference that focuses primarily on site construction, although information on several basic building frame systems is offered.

Parker, Harry; Macguire, John W.; and Ambrose, James. 1991. *Simplified Site Engineering,* 2nd ed. New York: John Wiley & Sons, Inc.

This book is short but nevertheless a very handy reference for surveying and site design.

Peurifoy, Robert L., and Schexnayder, Clifford J. 2002. *Construction Planning, Equipment, and Methods,* 6th ed. Boston: McGraw Hill.

This textbook is a comprehensive work focusing on the equipment used in engineering construction.

Walker, Frank R. Company, 1995. *Walker's Building Estimator's Reference Book,* 25th ed. Lisle, IL: Frank R. Walker Company.

Originally published in 1915, Walker's is an excellent reference on how estimating is performed for a variety of projects. Considerable historical information on industry practices is available in this book.

▪ Chapter 7

Das, Braja M. 2004. *Principles of Foundation Engineering,* 5th ed. Pacific Grove, CA: Brooks/Cole.

———. 2002. *Principles of Geotechnical Engineering,* 5th ed. Pacific Grove, CA: Brooks/Cole.

These two popular textbooks are widely used in civil engineering programs in the United States and abroad for their relevance and practicality.

Hatem, David J. 1998. *Subsurface Conditions—Risk Management for Design and Construction Management Professionals.* New York: John Wiley & Sons, Inc.

This book is an excellent reference on geotechnical engineering practice.

Liu, Cheng, and Evett, Jack B. 1998. *Soils and Foundations,* 4th ed. Upper Saddle River, NJ: Prentice Hall.

An easy-to-read text on the formation of soil deposits, engineering properties of soils, and foundation and retaining wall systems.

▪ Chapter 8

Allen, Edward.1999. *Fundamentals of Building Construction— Materials and Methods,* 3rd ed. New York: John Wiley & Sons, Inc.

Well written and beautifully illustrated, this text is one for the personal libraries of architecture and engineering students and construction enthusiasts.

Allen, Edward, and Thallon, Rob. 2002. *Fundamentals of Residential Construction.* Hoboken, NJ: John Wiley & Sons, Inc.

These two authors have a genuine love of architecture and construction that manifests itself in this and their other texts.

Hurd, M. K. 1995. *Formwork for Concrete,* 6th ed. Detroit, MI: American Concrete Institute.

M.K. Hurd's book has become the formwork bible for many a contractor and engineer.

Nunnally, S.W. 2004. *Construction Methods and Management,* 6th ed. Upper Saddle River, NJ: Pearson Prentice Hall.

This text contains useful chapters on concrete construction; formwork design, structural steel, masonry construction, and wood construction (both wood light and heavy timber framing).

Simmons, H. L., and Olin, Harold B. 2001. *Construction— Principles, Materials, and Methods.* New York: John Wiley & Sons, Inc.

Originally developed for the financial services industry as a reference, this lengthy text focuses largely on design principles and construction materials and their properties.

Spence, William P. 1998. *Construction Methods, Materials, and Techniques.* Albany, NY: Delmar Press.

A comprehensive treatise of over a thousand pages, this book runs the gamut of building systems and construction methods.

Thallon, Rob. 2000. *Graphic Guide to Frame Construction,* 2nd ed. Newtown, CT: The Taunton Press, Inc.

This beautifully illustrated collection of wood light framing details would be particularly useful to people whose experience with wood light framing is limited.

Zalewski, Waclaw, and Allen, Edward. 1998. *Shaping Structures: Statistics.* New York: John Wiley & Sons, Inc.

This book is a well written and beautifully illustrated text that offers highly unusual buildings, bridges, and other structures as subjects of statics problems, which are then solved graphically.

■ **Chapter 9**

Allen, Edward.1999. *Fundamentals of Building Construction— Materials and Methods,* 3rd ed. New York: John Wiley & Sons, Inc.

This author has produced five excellent chapters devoted to cladding and glazing systems and exterior doors, including critical background information on cladding design principles.

Harris, Samuel Y. 2001. *Building Pathology—Deterioration, Diagnostics, and Intervention.* New York: John Wiley & Sons, Inc.

This very well-written book is a thorough and detailed analysis of the destructive forces that act on buildings during the course of their lives. It should be required reading for both designers and managers of the construction process.

■ **Chapter 10**

Allen, Edward. 1999. *Fundamentals of Building Construction— Materials and Methods,* 3rd ed. New York: John Wiley & Sons.

The chapter on roofing in this book is very well done, replete with drawings and photographs that explain critical principles and practices in the roofing industry.

Patterson, Stephen, and Mehta, Madan. 2001. *Roofing Design and Practice.* Upper Saddle River, NJ: Prentice Hall.

This hard-cover publication thoroughly examines roofing design and construction and is simply but effectively illustrated and easy to read.

Simmons, H. L., and Olin, Harold B. 2001. *Construction— Principles, Materials, and Methods.* New York: John Wiley & Sons, Inc.

This book has an extensive chapter on roofing that includes numerous photographs and drawings.

Spence, William P. 1998. *Construction Methods, Materials, and Techniques.* Albany, NY: Delmar Press.

The chapter on roofing in this book is well done.

■ **Chapter 11**

Allen, Edward. 1999. *Fundamentals of Building Construction— Materials and Methods,* 3rd ed. New York: John Wiley & Sons.

Three well-written and illustrated chapters address the selection of interior finishes, the sequence of installation for interior finishes, and the construction of walls and ceilings.

Spence, William P. 1998. *Construction Methods, Materials, and Techniques.* Albany, NY: Delmar Press.

This book devotes several well-illustrated chapters to interior construction and finishes.

Walker, Frank R. Company, 1995. *Walker's Building Estimator's Reference Book,* 25th ed. Lisle, IL: Frank R. Walker Company.

Walker's devotes nearly 200 pages to interior construction and finishes.

■ **Chapter 12**

Means, R.S. Co. 1992. *Mechanical Estimating—Standards and Procedures.* R. S. Kingston, MA: Means Co., Inc.

This book offers valuable insight into estimating for plumbing and HVAC systems, and includes estimating methodology and project costing techniques for mechanical systems.

Simmons, H. L., and Olin, Harold B. 2001. *Construction— Principles, Materials, and Methods.* New York: John Wiley & Sons, Inc.

The chapters devoted to HVAC and plumbing systems tend to focus on the elements of the systems rather than the construction of them, but the information is nonetheless useful.

Spence, William P. 1998. *Construction Methods, Materials, and Techniques.* Albany, NY: Delmar Press.

The author of this book goes into considerable detail regarding plumbing, fire protection, and HVAC systems in several helpful chapters.

Walker, Frank R. Company, 1995. *Walker's Building Estimator's Reference Book,* 25th ed. Lisle, IL: Frank R. Walker Company.

Useful background information on mechanical systems and estimating methodology for these systems is recorded on some 45 pages of this book.

Wentz, Tim. 1997. *Plumbing Systems—Analysis, Design, and Construction.* Upper Saddle River, NJ: Prentice Hall.

This publication, just over 200 pages, does a credible job of describing plumbing systems and design criteria for conformance with building regulations.

■ **Chapter 13**

Means, R.S. Co. 1992. *Electrical Estimating Methods,* 2nd ed. Kingston, MA: R. S. Means Co. Inc.

This book offers valuable insight into estimating for electrical systems, and includes estimating methodology and project costing techniques for electrical systems.

Peurifoy, Robert L., and Oberlender, Garold D. 2002. *Estimating Construction Costs,* 5th ed. Boston: McGraw Hill.

Although the chapter on electrical estimating is brief, it is a good overview of electrical estimating methods and costs.

■ **Chapter 14**

Sweet, Justin. 2000. *Legal Aspects of Architecture, Engineering, and the Construction Process,* 6th ed. Pacific Grove, CA: Brooks/Cole.

This lengthy legal treatise discusses submittals including shop drawings, and liability for them, citing a number of interesting legal cases that address the issue.

Smith, Currie & Hancock, LLP. 2001. *Common-Sense Construction Law,* 2nd ed. New York: John Wiley & Sons, Inc.

Although the section on shop drawings is brief, there are a number of cases cited in this book that are worth reading.

Index

A

A-201, 16
Accessibility, 12
Agreement, 13, 16, 17, 19, 22, 379, 382, 481
AIA, *see* American Institute of Architects
American Institute of Architects, 4, 9, 16, 54, 489, 490
American Institute of Constructors, 8, 490
Approved set, *see* Drawing(s)
Architect, xi,
   education of the, 6, 7, 8
   engineer, 1, 22
   interior, 30
   landscape, 5, 7, 76
   responsibilities of, 19, 24, 27, 53, 54, 76, 83, 112, 121
Associated General Contractors, 13, 16, 489
Axonometric, *see* Projections

B

Balance a site, 86, 90
Balance of work, 359
Ballast, 196, 218, 384, 481
Best management practices, 79, 481
Bidding requirements, 24, 77, 382, 481, 485, 489
BMP, *see* Best management practices
Borrow pit, 86
Brace points, 154, 160
   hardware for, 154
Branch circuitry, 308, 309, 333, 334, 385, 482
Building codes, *see* Codes
Building module(s), 28
Buy quantity, 22, 60, 116, 425, 426

C

Cable trays, 267, 308, 311, 322
CADD, *see* Computer Aided Design and Drafting
Change orders, 22, 382
Chasing information, *see* Drawing(s)
Chord bars, 156, 159, 450, 482
Circuit breakers, 308, 334, 482
Circuit protective devices, 308
Civil engineer, *see* Engineer(s)
Civil engineering drawings, *see* Drawing(s)
Civil works, 26
Cladding, 33, 53, 58, 168–180, 185, 192, 193, 215, 220,
     270, 272, 337, 379, 380, 390, 483, 485, 486, 492
   physical phenomena affecting, 168
Clean Water Act, 76, 79, 383
Closure strip, 153, 159, 160, 482
Codes, 12, 13, 20
   building, 12, 13, 22, 255, 275, 308, 418, 421,
     423, 427, 428
   cost, 20
   fire, 27, 275
   zoning, 33, 390
Cognitive construction, 19, 40, 59
CAD, CADD (*see* Computer Aided Design and Drafting)
Computer Aided Design and Drafting, 4, 19
Concrete,
   architectural precast, 110, 190, 192, 337
   cast-in-place (CIP) post-tensioned, 120
   cast-in-place (CIP) reinforced, 115, 120, 129–139
   glass fiber reinforced, 168
   placing, consolidating, curing, 110

Construction Graphics: A Practical Guide to Interpreting Working Drawings,
Third Edition. Keith Bisharat.
© 2025 John Wiley & Sons Inc. Published 2025 by John Wiley & Sons Inc.